Y0-CDN-590

MARINE WEATHER

Nathaniel Bowditch

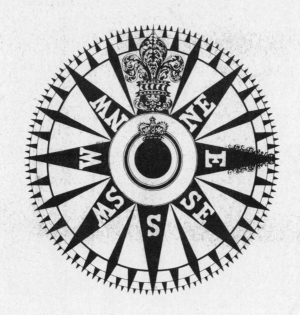

ARCO PUBLISHING, INC.
219 Park Avenue South, New York, N.Y. 10003

Reprinted from AMERICAN PRACTICAL NAVIGATOR, An Epitome of Navigation.
Originally by Nathaniel Bowditch, 1977 Edition published by Defense Mapping
Agency Hydrographic Center.

Published by Arco Publishing, Inc.
219 Park Avenue South, New York, N.Y. 10003

Copyright © 1979 by Arco Publishing, Inc.

All rights reserved.

Library of Congress Cataloging in Publication Data

Bowditch, Nathaniel, 1773–1838.
  Marine weather.

  "The edition reprints eleven pertinent chapters and
selected appendices from the 1977 edition of the
American practical navigator."
  1. Meteorology, Maritime.   2. Weather.   3. Oceanography.   I. Title.   II. Title: Marine weather.

QC994.B655   1979   623.89′2   79–799
ISBN 0–668–04455–1

Printed in the United States of America

CONTENTS

ABOUT THIS BOOK

MARINE WEATHER is a compendium of weather observation and elements of sky, wind, and wave material understandable to the yachtsman. *Meteorology* as the science that deals with weather and weather forecasting can be used effectively to improve the art of navigation and safety at sea only if a well-informed person of mature judgement and experience is on hand to interpret information as it becomes available. Thus, this book provides the basic facts of weather observation, tropical cyclones, ocean conditions, tides and tidal currents, ocean currents, ocean waves, breakers and surf and other related weather elements necessary to the mariner.

The material reprinted in this book is taken from the latest edition, 1977, AMERICAN PRACTICAL NAVIGATOR: AN EPITOME OF NAVIGATION (U.S. Defense Mapping Agency Hydrographic Center, Washington, D.C., 20390); "Bowditch," as it is called, has been a primary resource for yachtsmen since its first publication in 1802 by the U.S. Navy. Because of its voluminousness many boat owners are reluctant to keep it aboard for lack of space. *MARINE WEATHER* is designed to be taken aboard, includes essential information about weather, and can provide interesting reading on rainy days.

PREFACE

Before pages of technical information on meteorological phenomena, a few practical words to mariners are necessary. The yachtsman today has the ability, from close observation of the barometer and the general appearance of the sky, to forecast the weather with a certain degree of accuracy. The aneroid barometer is peculiarly sensitive to all atmospheric changes, and can indicate coming weather.

A rapid rise indicates unsettled weather.

A gradual rise indicates settled weather.

A rise with dry air and cold increasing in summer indicates wind from the northward, and if rain has fallen better weather may be expected.

A rise with moist air and a low temperature indicates a continuance of fine weather.

A rapid fall indicates stormy weather.

A rapid fall with westerly wind indicates stormy weather from northward.

A fall with northerly wind indicates storm with rain and hail in summer and snow in winter.

A fall with increased moisture in the air and increasing heat indicates southerly wind and rain.

A fall after very calm and warm weather indicates rain and squalls.

The barometer rises for a northerly wind, including from northwest by north to the eastward, for dry or less wet weather, for less wind, or for more than one of these changes, except on a few occasions when rain, hail or snow comes from the northward with strong wind.

The barometer falls for a southerly wind, including from southeast by south to the westward, for wet weather, for stronger wind, or for more than one of these changes, except on a few occasions, when moderate wind, with rain or snow, comes from the northward.

A fall, with a south wind, precedes rain.

A sudden and considerable fall, with the wind due west, presages a violent storm from the north or northwest, during which the glass will rise to its former height.

A steady and considerable fall of the barometer during an east wind indicates a shift of wind to the southward, unless a heavy fall of snow or rain immediately follows.

A falling barometer, with the wind at north, brings bad weather; in summer rain and gales; in spring snows and frosts.

If, after a storm of wind and rain, the barometer remains steady at the point to which it had fallen, severe weather may follow without a change in the wind. But on the rising of the barometer a change of wind may be looked for.

The following rhymes are familiar to most sailors:

> When the glass falls low,
> Look out for a blow.

> First rise after low,
> Portends a stronger blow.

> When the glass is high,
> Let all your kites fly.

> Long foretold—long last;
> Short notice—soon past.

The following notes may be relied on for forecasting the weather:

Red sky at sunset, fine weather.
Red sky in the morning, wind or rain, and often
both.
Gray sky in the morning, fine weather.
Hard, oily looking clouds, strong wind.
Yellowish green clouds, wind and rain.
Bright yellow sky at sunset, wind.
Pale yellow sky at sunset, rain.
Very clear atmosphere near the horizon is a
sign of more wind and often rain.

Here follow some old sailors' jingles which I heard when a boy in the forecastle:

When rain comes before the wind,
Sheets and halyards you must mind;
When wind comes before the rain,
Hoist your topsails up again.

Evening red and morning gray,
Are sure signs of a fine day;
But evening gray and morning red,
Makes a sailor shake his head.

Aneroid barometers of excellent quality, and of about the size of an ordinary watch, are offered for sale at a reasonable price, and a cruise should not be undertaken without one.

A phosphorescent sea is a certain sign of continuance of fine weather.

When porpoises come into shallow water and ascend the river stormy weather is near.

Sea birds fly far out to sea in fine weather, but if they fly inland bad weather may be expected.

A halo round the moon, especially if it appears distant and yet very distinct, indicates a gale of wind and probably rain.

When the wind changes it usually shifts with the sun from left to right. Thus an East wind shifts to West by way of Southeast, South and Southwest, and a West wind shifts to East by way of Northwest, North and Northeast. If the wind shifts the opposite way it is said to "back," but this it rarely does except in unsettled weather.

For more about the marine weather read on in this book.

JAMES MORRISON

CHAPTER I

WEATHER OBSERVATIONS

1.1. Introduction.—Weather forecasts are generally based upon information acquired by observations made at a large number of stations. Ashore, these stations are located so as to provide adequate coverage of the area of interest. Most observations at sea are made by mariners, wherever they happen to be. Since the number of observations at sea is small compared to the number ashore, marine observations are of importance in areas where little or no information is available from other sources. Results of these observations are recorded in the deck log (art. 1.19), or other appropriate form. Data recorded by designated vessels are sent by radio to centers ashore, where they are plotted, along with other observations, to provide data for drawing synoptic charts (art. 2.26). These charts are used to make forecasts. Complete weather information gathered at sea is mailed to the appropriate meteorological services for use in the preparation of weather atlases and in marine climatological studies.

A special effort should be made to provide routine synoptic reports when transiting those areas where few ships are available to make and report weather observations. Such effort is particularly important when vessels are transiting the tropical regions. A vessel's synoptic weather report may be the first indication of a developing tropical cyclone.

In many instances, the analysis of the surface weather map and subsequent forecasts can be no better than the weather reports received.

A knowledge of *weather elements* and the instruments used to measure them is therefore of importance to the mariner who hopes to benefit from weather *broadcasts*.

1.2. Atmospheric pressure measurement.—The sea of air surrounding the earth exerts a pressure of about 14.7 pounds per square inch on the surface of the earth. This **atmospheric pressure,** sometimes called **barometric pressure,** varies from place to place, and at the same place it varies with time.

Atmospheric pressure is one of the basic elements of a meteorological observation. When the pressure at each station is plotted on a synoptic chart, lines of equal atmospheric pressure, called **isobars,** are drawn to indicate the areas of high and low pressure and their centers. These are useful in making weather predictions, because certain types of weather are characteristic of each type area, and the wind patterns over large areas are deduced from the isobars.

Atmospheric pressure is measured by means of a **barometer. A mercurial barometer** does this by balancing the weight of a column of air against that of a column of mercury. The **aneroid barometer** has a partly evacuated, thin-metal cell which is compressed by atmospheric pressure, the amount of the compression being related to the pressure.

Early mercurial barometers were calibrated to indicate the height, usually in inches or millimeters, of the column of mercury needed to balance the column of air above the point of measurement. While the units **inches of mercury** and **millimeters of mercury** are still widely used, many modern barometers are calibrated to indicate the centimeter-gram-second unit of pressure, the **millibar,** which is equal to 1,000 dynes per square centimeter. A **dyne** is the force required to accelerate a mass of one gram at the rate of one centimeter per second per second. A reading in any of the three units of measure-

1

ment can be converted to the equivalent reading in either of the other units by means of table C, or the conversion factors given in appendix A. However, the pressure reading should always be reported in millibars.

1.3. The mercurial barometer was invented by Evangelista Torricelli in 1643. In its simplest form it consists of a glass tube a little more than 30 inches in length and of uniform internal diameter; one end being closed, the tube is filled with mercury, and inverted into a cup of mercury. The mercury in the tube falls until the column is just supported by the pressure of the atmosphere on the open cup, leaving a vacuum at the upper end of the tube. The height of the column indicates atmospheric pressure, greater pressures supporting higher columns of mercury.

The mercurial barometer is subject to rapid variations in height, called **pumping,** due to pitch and roll of the vessel and temporary changes in atmospheric pressure in the vicinity of the barometer. Because of this, the care required in the reading of the instrument, its bulkiness, and its vulnerability to physical damage, the mercurial barometer has been replaced at sea by the aneroid barometer.

1.4. The aneroid barometer (fig. 1.4) measures atmospheric pressure by means of the force exerted by the pressure on a partly evacuated, thin-metal element called a **sylphon cell (aneroid capsule).** A small spring is used, either internally or externally, to partly counteract the tendency of the atmospheric pressure to crush the cell. Atmospheric pressure is indicated directly by a scale and a pointer connected to the cell by a combination of levers. The linkage provides considerable magnification of the slight motion of the cell, to permit readings to higher precision than could be obtained without it.

An aneroid barometer should be mounted permanently. Prior to installation, the barometer should be carefully set to station pressure (art. 1.6). An adjustment screw is provided for this purpose. The error in the reading of the instrument is determined by comparison with a mercurial barometer or a standard precision aneroid barometer. If a qualified meteorologist is not available to make this adjustment, it is good practice to remove only one-half the apparent error. The case should then be tapped gently to assist the linkage to adjust itself, and the process repeated. If the remaining error is not more than half a millibar (0.015 inch), no attempt should be made to remove it by further adjustment. Instead, a correction should be applied to the readings. The accuracy of this correction should be checked from time to time.

1.5. The barograph (fig. 1.5) is a recording barometer. Basically, it is the same as a nonrecording aneroid barometer except that the pointer carries a pen at its outer end, and the scale is replaced by a slowly rotating cylinder around which a prepared chart is wrapped. A clock mechanism inside the cylinder rotates the cylinder so that a continuous line is traced on the chart to indicate the pressure at any time.

A **marine microbarograph** is a precision barograph with greater magnification of deformations due to pressure changes, and a correspondingly expanded chart. It is designed to maintain its precision through the varied and exacting conditions encountered in shipboard use. Two sylphon cells are used, one being mounted over the other in tandem. Minor fluctuations due to shocks or vibrations are eliminated by damping. Since oil-filled dashpots are used for this purpose, the instrument should not be inverted.

Ship motions are compensated by damping and spring loading which make it possible for the microbarograph to be tilted up to 22° without varying more than 0.3 millibars from true reading.

The barograph is usually mounted on a shelf or desk in a room open to the atmosphere, and in a location which minimizes the effect of the ship's vibration. Shock-

FIGURE 1.4.—An aneroid barometer.

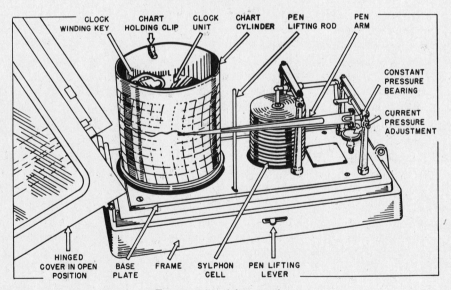

FIGURE 1.5.—A barograph.

absorbing material such as sponge rubber is placed under the instrument to minimize the transmission of shocks.

The pen should be checked and the inkwell filled each time the chart is changed, every week in the case of the barograph, and each 4 days in the case of the micro-barograph. The dashpots of the microbarograph should be kept filled with dashpot oil to within three-eighths inch of the top.

Both instruments require checking from time to time to insure correct indication of pressure. The position of the pen is adjusted by a small knob provided for this purpose. The adjustment should be made in stages, eliminating half the apparent error, tapping the case to insure linkage adjustment to the new setting, and then repeating the process.

1.6. Adjustment of barometer readings.—Atmospheric pressure as indicated by a barometer or barograph may be subject to several errors, as follows:

Instrument error. Any inaccuracy due to imperfection or incorrect adjustment of the instrument can be determined by comparison with a precision instrument. The National Weather Service provides a comparison service. In certain U. S. ports a Port Meteorological Officer carries a portable precision aneroid barometer for barometer comparisons on board ships which participate in the cooperative observation program of the National Weather Service. The portable barometer is compared with station barometers before and after each ship visit. If a barometer is taken to a National Weather Service shore station, the comparison can be made there. The correct sea-level pressure can be obtained by telephone. The shipboard barometer should be corrected for height, as explained below, before comparison with this telephoned value. If there is reason to believe that the barometer is in error, it should be compared with a standard, and if an error is found, the barometer should be adjusted to the correct reading, or a correction applied to all readings.

Height error. The atmospheric pressure reading at the height of the barometer is called the **station pressure** and is subject to a height correction in order to make it a sea level pressure reading. Isobars adequately reflect wind conditions and geographic distribution of pressure only when they are drawn for pressure at constant height (or the varying height at which a constant pressure exists). On synoptic charts it is customary to show the equivalent pressure at sea level, called **sea level pressure.** This is found by applying a correction to station pressure. The correction, given in table 2, depends upon the height of the barometer and the average temperature of the air between this height and the surface. The outside air temperature taken aboard ship is sufficiently accurate for this purpose. *This is an important correction which should be applied to all readings of any type barometer.*

Gravity error. Mercurial barometers are calibrated for standard sea-level gravity at latitude 45°32′40″. If the gravity differs from this amount, an error is introduced. The correction to be applied to readings at various latitudes is given in table 3. *This correction does not apply to readings of an aneroid barometer or microbarograph.* Gravity also changes with height above sea level, but the effect is negligible for the first few hundred feet, and so is not needed for readings taken aboard ship.

Temperature error. Barometers are calibrated at a standard temperature of 32°F. The liquid of a mercurial barometer expands as the temperature of the mercury rises, and contracts as it decreases. The correction to adjust the reading of the instrument to the true value is given in table 4. *This correction is to be applied to readings of mercurial barometers only.* Modern aneroid barometers are compensated for temperature changes by the use of different metals having unequal coefficients of linear expansion.

1.7. Wind measurement consists of determination of the direction *from* which the wind is blowing, and the speed of the wind. Wind direction is measured by a **wind vane,** and wind speed by an **anemometer.**

A wind vane consists of a device pivoted on a vertical shaft, with more surface area on one side of the pivot than on the other, so that the wind exerts more force on one side, causing the smaller end to point into the wind.

In its simplest form, an anemometer consists of a number of cups mounted on short horizontal arms attached to a longer vertical shaft which rotates as the wind blows against the cups. The speed at which the shaft rotates is directly proportional to the wind speed.

Several types of wind speed and direction recorders are available.

If no anemometer is available, wind speed can be estimated by its effect upon the sea and objects in its path, as explained in article 1.9.

1.8. True and apparent wind.—An observer aboard a vessel proceeding through still air experiences an **apparent wind** which is from dead ahead and has an apparent speed equal to the speed of the vessel. Thus, if the actual or **true wind** is zero and the speed of the vessel is 10 knots, the apparent wind is from dead ahead at 10 knots. If the true wind is from dead ahead at 15 knots, and the speed of the vessel is 10 knots, the apparent wind is 15+10=25 knots from dead ahead. If the vessel makes a 180° turn, the apparent wind is 15−10=5 knots from dead astern.

In any case, the apparent wind is the vector sum of the true wind and the *reciprocal* of the vessel's course and speed vector. Since wind vanes and anemometers measure *apparent* wind, the usual problem aboard a vessel equipped with an anemometer is to convert this to true wind. There are several ways of doing this. Perhaps the simplest is by the graphical solution illustrated in the following example:

Example 1.—A ship is proceeding on course 150° at a speed of 17 knots. The apparent wind is from 40° off the starboard bow, speed 15 knots.

Required.—The relative direction, true direction, and speed of the true wind.

Solution (figure 1.8a).—Starting at the center of a maneuvering board or other suitable form, draw a line in the relative direction *from* which the apparent wind is blowing. Locate point 1 on this line, at a distance from the center equal to the speed of the apparent wind (2:1 scale is used in figure 1.8a). From point 1, draw a line vertically *downward*. Locate point 2 on this line at a distance from point 1 equal to the speed of the vessel in knots, to the same scale as the first line. The relative direction of the true wind is *from* point 2 (120°) toward the center, and the speed of the true wind is the distance of point 2 from the center, to the same scale used previously (11 kn.). The true direction of the wind is the relative direction plus the true heading, or 120° + 150° = 270°.

Answers.—True wind from 120° relative, is 270° true, at 11 knots.

A quick solution can be made without an actual plot, in the following manner: On a maneuvering board, label the circles 5, 10, 15, 20, etc., from the center, and draw vertical lines tangent to these circles. Cut out the 5:1 scale and discard that part having graduations greater than the maximum speed of the vessel. Keep this equipment for all solutions. (For durability, the two parts can be mounted on cardboard or other suitable material.) To find true wind, spot in point 1 by eye. Place the zero of the 5:1 scale on this point and align the scale (inverted) by means of the vertical lines. Locate point 2 at the speed of the vessel as indicated on the 5:1 scale. It is always vertically *below* point 1. Read the relative direction and the speed of the true wind using eye interpolation if needed. The National Weather Service distributes a wind vector computer called a *Shipboard Wind Plotter* (fig. 1.8b). Solution by means of this plotter is illustrated in the following example:

Example 2.—A ship is proceeding on course 270° at a speed of 14.5 knots. The apparent wind is from 40° off the starboard bow, speed 20 knots.

Required.—The relative direction, true direction, and speed of the true wind by Shipboard Wind Plotter.

Solution (fig. 1.8b).—The true direction of the apparent wind is determined by adding the apparent wind direction to the ship's heading if the wind is from off the starboard bow and subtracting the apparent wind direction if the wind is from off the port bow. In this example, the true direction of the apparent wind is 310°. In this

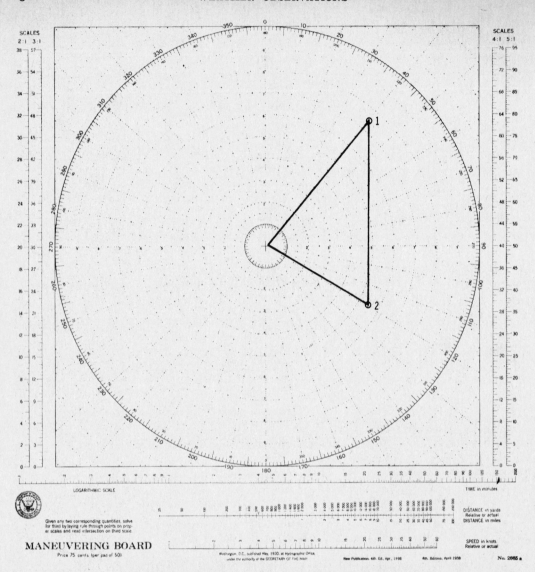

FIGURE 1.8a.—Finding true wind by maneuvering board.

solution the red arrowhead is considered the top of the plotter. Set ship's course, 270°, to the top of the plotter by rotating the protractor disk to set 270° at the red arrow. Using a convenient linear scale, measure vertically downward from the center peg of the plotting board a distance equivalent to 14.5 knots. Mark this point "S" for ship. Rotate the protractor disk of the plotting board until 310° is at the red arrowhead at the top of the plotting board. Using the same linear scale as for ship's speed, plot vertically downward from the center peg of the plotting board a distance equivalent to 20 knots. Mark this point "W." Rotate the protractor disk until the "S" is vertically above the "W," using the vertical lines on the plotting board to line up the two points. Read the true wind direction at the top of the plotting board. The distance between points "S" and "W" is the true wind speed, using the same scale as in plotting points "S" and "W."

Answers.—True wind direction is 357°, true wind speed is 13 knots.

Such problems can be solved by the use of true directions and a regular vector solution, but the use of relative directions simplifies the plot because that component

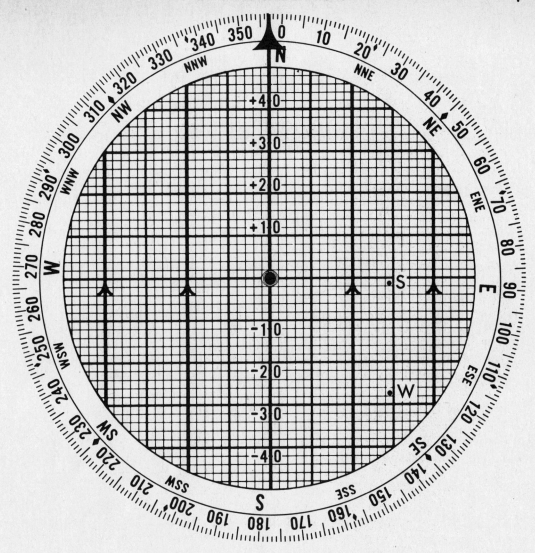

FIGURE 1.8b.—Finding true wind by Shipboard Wind Plotter.

of the apparent wind due to the vessel's motion is always parallel (but reversed) to the vessel's motion, and the apparent wind is always *forward* of the true wind.

A tabular solution based upon the same principle can be made by means of table 5. The entering values for this table are the apparent wind speed *in units of ship's speed*, and the difference between the heading and the apparent wind direction. The values taken from the table are the relative direction (right or left) of the true wind, and the speed of the true wind *in units of ship's speed*. If a vessel is proceeding at 12 knots, 6 knots constitutes one-half (0.5) unit, 12 knots one unit, 18 knots 1.5 units, 24 knots two units, etc.

Example 3.—A ship is proceeding on course 270° at a speed of 10 knots. The apparent wind is from 10° off the port bow, speed 30 knots.

Required.—The relative direction, true direction, and speed of the true wind by table 5.

Solution.—The apparent wind speed is $\frac{30}{10}$=3.0 ship's speed units. Enter table

5 with 3.0 and 10° and find the relative direction of the true wind to be 15° off the port bow (345° relative), and the speed to be 2.02 times the ship's speed, or $2.02 \times 10 = 20$ knots, approximately. The true direction is $345° + 270° = 255°$.

Answers.—True wind from 345° relative, 255° true, at 20 knots.

By variations of this problem, one can find the apparent wind from the true wind, the course or speed required to produce an apparent wind from a given direction or speed, or the course and speed to produce an apparent wind of a given speed from a given direction. Such problems arise in aircraft carrier operations.

Wind speed determined by appearance of the sea (art. 1.9) is the speed of the true wind. The sea also provides an indication of the direction of the true wind, because waves move in the same direction as the generating wind, not being deflected by earth rotation (art. 7.2). If the wind vane is used, the direction of the apparent wind thus determined can be used with the speed of the true wind to determine the direction of the true wind by vector diagram. If a maneuvering board is used, draw a circle about the center equal to the speed of the true wind. From the center, plot the ship's vector (true course and speed). From the end of this vector draw a line in the direction in which the apparent wind is blowing (reciprocal of the direction from which it is blowing) until it intersects the speed circle. This line is the apparent wind vector, its length denotes the speed. A line from the center of the board to the end of the apparent wind vector is the true wind vector. The reciprocal of this vector is the direction from which the true wind is blowing. If the true wind speed is less than the speed of the vessel, two solutions are possible. If solution is by table 5, the true speed, in units of ship's speed, is found in the column for the direction of the apparent wind. The number to the left is the relative direction of the true wind. The number on the same line in the side columns is the speed of the apparent wind in units of ship's speed. Again, two solutions are possible if true wind speed is less than ship's speed.

1.9. Wind and the sea.—The action of the wind in creating ocean currents and waves is discussed in chapters VI and VII respectively. There is a relationship between the speed of the wind and the state of the sea in the immediate vicinity of the wind. This is useful in predicting the sea conditions to be anticipated when future wind speed forecasts are available. It can also be used to estimate the speed of the wind, which may be desirable when an anemometer is not available.

Wind speeds are usually grouped in accordance with the **Beaufort scale** named after Admiral Sir Francis Beaufort, who devised it in 1806. As adopted in 1838, Beaufort numbers ranged from 0, calm, to 12, hurricane. The Beaufort scale, with certain other pertinent information, is given in appendix A. The appearance of the sea at different Beaufort scale numbers from 0 through 10 is shown in appendix B.

1.10. Temperature is the intensity or degree of heat. It is measured in degrees. Several different temperature scales are in use.

On the **Fahrenheit (F)** scale pure water freezes at 32° and boils at 212°.

On the **Celsius (C)** scale commonly used with the metric system, the freezing point of pure water is 0° and the boiling point is 100°. This scale, has been known by various names in different countries. In the United States it was formerly called the **centigrade** scale. The Ninth General Conference of Weights and Measures, held in France in 1948, adopted the name Celsius to be consistent with the naming of other temperature scales after their inventors, and to avoid the use of different names in different countries. On the original Celsius scale, invented in 1742 by a Swedish astronomer named Anders Celsius, the numbering was the reverse of the modern scale, 0° representing the boiling point of water, and 100° its freezing point.

Absolute zero is considered to be the lowest possible temperature, at which there is no molecular motion and a body has no heat. For some purposes, it is convenient to express temperature by a scale at which 0° is absolute zero. This is called **absolute temperature**. If Fahrenheit degrees are used, it may be called **Rankine (R)** temperature; and if Celsius, **Kelvin (K)** temperature. The Kelvin scale is more widely used than the Rankine. Absolute zero is at $(-)$ 459°69F or $(-)$ 273°16C.

Temperature by one scale can be converted to that at another by means of the relationship that exists between the scales. Thus,

$$C = \frac{5}{9}(F - 32),$$

$$F = \frac{9}{5}C + 32,$$

and

$$K = C + 273°16C.$$

A temperature of $(-)$ 40° is the same by either the Celsius or Fahrenheit scale. Similar formulas can be made for conversion of other temperature scale readings. Table 6 gives the equivalent values of Fahrenheit, Celsius, and Kelvin temperatures.

The intensity or degree of heat (temperature) should not be confused with the *amount* of heat. If the temperature of air or some other substance is to be increased (the substance made hotter) by a given number of degrees, the amount of heat that must be added is dependent upon the amount of the substance to be heated. Also, equal amounts of different substances require the addition of unequal amounts of heat to effect equal increase in temperature because of their difference of specific heat (art. 5.12). Units used for measurement of amount of heat are the **British thermal unit (BTU)**, the amount of heat needed to raise the temperature of 1 pound of water 1° Fahrenheit; and the **calorie**, the amount of heat needed to raise the temperature of 1 gram of water 1° Celsius.

1.11. Temperature measurement is made by means of a thermometer. Most thermometers are based upon the principle that materials expand with increase of temperature, and contract as temperature decreases. In its most usual form a thermometer consists of a bulb filled with mercury and connected to a tube of very small cross-sectional area. The mercury only partly fills the tube. In the remainder is a vacuum created during construction of the instrument. The air is driven out by boiling the mercury, and the top of the tube is then sealed by a flame. As the mercury expands or contracts with changing temperature, the length of the mercury column in the tube changes. Temperature is indicated by the position of the top of the column of mercury with respect to a scale etched on the glass tube or placed on the thermometer support.

Temperature measuring equipment should be placed in a shelter which protects it from mechanical damage and direct rays of the sun. The shelter should have louvered sides to permit free access of air. Aboard ship, the shelter should be placed in an exposed position as far as practicable from metal bulkheads. On vessels where shelters are not available, the temperature measurement should be made in shade at an exposed position on the windward side.

Sea surface temperature observations are used in the forecasting of fog and furnish important information about the development and movement of tropical cyclones. Commercial fishermen are interested in the sea surface temperature as an aid in locating certain species of fish. There are several methods of determining seawater temperature. These include engine room intake readings, condenser intake readings, thermister probes attached to the hull, and readings from buckets recovered from over the side. Although the condenser intake method is not a true measure of surface water temperature, the error is generally small. Measurement should be made near the entrance of the intake.

If the temperature of the water at the surface is desired, a sample should be obtained by bucket, preferably a canvas bucket, from a forward position well clear of any discharge lines. The sample should be taken immediately to a place where it is sheltered from wind and sun. The water should then be stirred with the thermometer, keeping the bulb submerged, until an essentially constant reading is obtained.

A considerable variation in sea surface temperature can be experienced in a relatively short distance of travel. This is especially true when crossing major ocean currents such as the Gulf Stream and the Kuroshio. Significant variations also occur where large quantities of freshwater are discharged from rivers.

1.12. Humidity is the condition of the atmosphere with reference to its water vapor content. **Relative humidity** is the ratio (stated as a percentage) of the pressure of water vapor present in the atmosphere to the saturation vapor pressure at the same temperature.

As air temperature decreases, the relative humidity increases. At some point, saturation takes place, and any further cooling results in condensation of some of the moisture. The temperature at which this occurs is called the **dew point,** and the moisture deposited upon natural objects is called **dew** if it forms in the liquid state, or **frost** if it forms in the frozen state.

The same process causes moisture to form on the outside of a container of cold liquid, the liquid cooling the air in the immediate vicinity of the container until it reaches the dew point. When moisture is deposited on man-made objects, it is usually called **sweat.** It occurs whenever the temperature of a surface is lower than the dew point of air in contact with it. It is of particular concern to the mariner because of its effect upon his instruments, and possible damage to his ship or its cargo. Lenses of optical instruments may sweat, usually with such small droplets that the surface has a "frosted" appearance. When this occurs, the instrument is said to "fog" or "fog up," and is useless until the moisture is removed. Damage is often caused by corrosion or direct water damage when pipes sweat and drip, or when the inside of the shell plates of a vessel sweat. Cargo may sweat if it is cooler than the dew point of the air. One of the principal problems of preserving ships of the reserve fleet is the protection against moisture. An important step is the draining of all water, sealing of compartments, and drying of the air.

Clouds and fog form by "sweating" of minute particles of dust, salt, etc., in the air. Each particle forms a nucleus around which a droplet of water forms. If air is completely free from solid particles on which water vapor may condense, the extra moisture remains in the vapor state, and the air is said to be **supersaturated.**

Relative humidity and dew point are measured by means of a **hygrometer.** The most common type, called a **psychrometer,** consists of two thermometers mounted together on a single strip of material, as shown in figure 1.12a. One of the thermometers is mounted a little lower than the other, and has its bulb covered with muslin. When

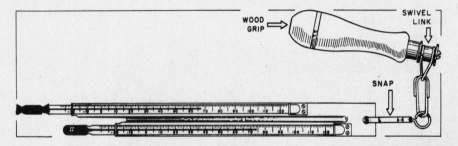

FIGURE 1.12a.—A sling psychrometer.

the muslin covering is thoroughly moistened and the thermometer well ventilated, evaporation cools the bulb of the thermometer, causing it to indicate a lower reading than the other. A sling psychrometer, illustrated in figure 1.12a, is ventilated by whirling the thermometers. **Dry-bulb temperature** is indicated by the uncovered **dry-bulb thermometer,** and **wet-bulb temperature** is indicated by the muslin-covered **wet-bulb thermometer.** The difference between these two temperatures, and the dry-bulb temperature, are used to enter **psychrometric tables** to find the relative humidity (tab. 7) and dew point (tab. 8). If the wet-bulb temperature is above freezing, reasonably accurate results can be obtained by a psychrometer consisting of wet- and dry-bulb thermometers mounted so that air can circulate freely around them without special ventilation. This type of installation is common aboard ship.

Example.—The dry-bulb temperature is 65°F and the wet-bulb temperature is 61°F.

Required.—(1) Relative humidity, (2) dew point.

Solution.— The difference between readings is 4°. Entering table 7 with this value and a dry-bulb temperature of 65°, the relative humidity is found to be 80 percent. From table 8 the dew point is found to be 58°.

Answers.—(1) Relative humidity 80 percent, (2) dew point 58°.

Also in use aboard many ships is the electric psychrometer (fig. 1.12b). This is a handheld, battery operated instrument with two mercury thermometers for obtaining dry- and wet-bulb temperature readings. It consists of a plastic housing that holds the thermometers, batteries, motor, and fan. Although the electric psychrometer is constructed primarily of non-corrodible materials, prolonged exposure to the marine environment will shorten the life of the instrument.

1.13. Clouds are visible assemblages of numerous tiny droplets of water, or ice crystals, formed by condensation of water vapor in the air, with the bases of the assemblages above the surface of the earth. **Fog** is a similar assemblage in contact with the surface of the earth.

The shape, size, height, thickness, and nature of a cloud depend upon the conditions under which it is formed. Therefore, clouds are indicators of various processes occurring in the atmosphere. The ability to recognize different types and a knowledge of the conditions associated with them are useful in predicting future weather.

Although the variety of clouds is virtually endless, they may be classified according to general type. Clouds are grouped generally into three "families" according to some common characteristic. **High clouds** are those having a mean lower level above 20,000 feet. They are composed principally of ice crystals. **Middle clouds** have a mean level between 6,500 and 20,000 feet. They are composed largely of water droplets, although the higher ones have a tendency toward ice particles. **Low clouds** have a mean lower level of less than 6,500 feet. These clouds are composed entirely of water droplets.

Within these 3 families are 10 principal cloud types. The names of these are composed of various combinations and forms of the following basic words, all from Latin:

Cirrus, meaning "curl, lock, or tuft of hair."

Cumulus, meaning "heap, a pile, an accumulation."

Stratus, meaning "spread out, flatten, cover with a layer."

Alto, meaning "high, upper air."

Nimbus, meaning "rainy cloud."

Individual cloud types recognize certain characteristics, variations, or combinations of these. The 10 principal cloud types are:

High clouds. **Cirrus (Ci)** are detached high clouds of delicate and fibrous appearance, without shading, generally white in color, and often of a silky appearance (figs. 1.13a and 1.13d). Their fibrous and feathery appearance is due to the fact that they

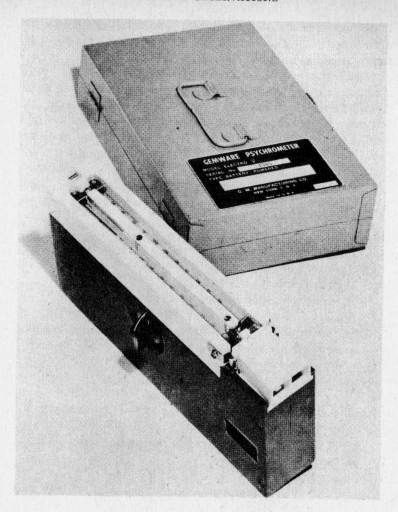

FIGURE 12b.—Electric psychrometer.

are composed entirely of ice crystals. Cirrus appear in varied forms such as isolated tufts; long, thin lines across the sky; branching, feather-like plumes; curved wisps which may end in tufts, etc. These clouds may be arranged in parallel bands which cross the sky in great circles and appear to converge toward a point on the horizon. This may indicate, in a general way, the direction of a low pressure area. Cirrus may be brilliantly colored at sunrise and sunset. Because of their height, they become illuminated before other clouds in the morning, and remain lighted after others at sunset. Cirrus are generally associated with fair weather, but if they are followed by lower and thicker clouds, they are often the forerunner of rain or snow.

Cirrocumulus (Cc) are high clouds composed of small white flakes or scales, or of very small globular masses, usually without shadows and arranged in groups or lines, or more often in ripples resembling those of sand on the seashore (fig. 1.13b). One form of cirrocumulus is popularly known as "mackerel sky" because the pattern resembles the scales on the back of a mackerel. Like cirrus, cirrocumulus are composed of ice crystals and are generally associated with fair weather, but may precede a storm if they thicken and lower. They may turn gray and appear hard before thickening.

Cirrostratus (Cs) are thin, whitish, high clouds (fig. 1.13c) sometimes covering the sky completely and giving it a milky appearance and at other times presenting, more or less distinctly, a formation like a tangled web. The thin veil is not sufficiently dense to blur the outline of sun or moon. However, the ice crystals of which the cloud is composed refract the light passing through in such a way that halos (art. 2.18) may form with the sun or moon at the center. Figure 1.13d shows cirrus thickening and changing into cirrostratus. In this form it is popularly known as "mares' tails." If it continues to thicken and lower, the ice crystals melting to form water droplets, the cloud formation is known as altostratus. When this occurs, rain may normally be expected within 24 hours. The more brushlike the cirrus when the sky appears as in figure 1.13d, the stronger the wind at the level of the level of the cloud.

FIGURE 1.13a.—Cirrus.

FIGURE 1.13b.—Cirrocumulus.

FIGURE 1.13c.—Cirrostratus.

FIGURE 1.13d.—Cirrus and cirrostratus.

Middle clouds. **Altocumulus (Ac)** are middle clouds consisting of a layer of large, ball-like masses that tend to merge together. The balls or patches may vary in thickness and color from dazzling white to dark gray, but they are more or less regularly arranged. They may appear as distinct patches (fig. 1.13e) similar to cirrocumulus (fig. 1.13b) but can be distinguished by the fact that individual patches are generally larger and show distinct shadows in some places. They are often mistaken for stratocumulus (fig. 1.13i). If this form thickens and lowers, it may produce thundery weather and showers, but it does not bring prolonged bad weather. Sometimes the patches merge to form a series of big rolls that resemble ocean waves, but with streaks of blue sky (fig. 1.13f). Because of perspective the rolls appear to run together near the horizon. These regular parallel bands differ from cirrocumulus in that they occur in larger masses with shadows. These clouds move in the direction of the short dimension of the rolls, as do ocean waves. Sometimes altocumulus appear briefly in the form shown in figure 1.13g, usually before a thunderstorm. They are generally arranged in a line with a flat horizontal base, giving the impression of turrets on a castle. The turreted tops may look like miniature cumulus and possess considerable depth and great length. These clouds usually indicate a change to chaotic, thundery skies.

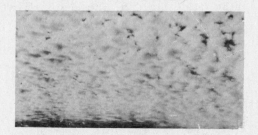

FIGURE 1.13e.—Altocumulus in patches.

FIGURE 1.13f.—Altocumulus in bands.

FIGURE 1.13g.—Turreted altocumulus.

FIGURE 1.13h.—Altostratus.

Altostratus (**As**) are middle clouds having the appearance of a grayish or bluish, fibrous veil or sheet (fig. 1.13h). The sun or moon, when seen through these clouds, appears as if it were shining through ground glass, with a corona (art. 2.19) around it. Halos are not formed. If these clouds thicken and lower, or if low, ragged "scud" or rain clouds (nimbostratus) form below them, continuous rain or snow may be expected within a few hours.

Low clouds. **Stratocumulus** (**Sc**) are low clouds composed of soft, gray, roll-shaped masses (fig. 1.13i). They may be shaped in long, parallel rolls similar to altocumulus (fig. 1.13f), moving forward with the wind. The motion is in the direction of their short dimension, like ocean waves. These clouds, which vary greatly in altitude, are the final product of the characteristic daily change that takes place in cumulus clouds. They are usually followed by clear skies during the night.

Stratus (**St**) is a low cloud in a uniform layer (fig. 1.13j) resembling fog. Often the base is not more than 1,000 feet high. A veil of thin stratus gives the sky a hazy appearance. Stratus is often quite thick, permitting so little sunlight to penetrate that it appears dark to an observer below it. From above, it looks white. Light mist may descend from stratus. Strong wind sometimes breaks stratus into shreds called "fractostratus."

Nimbostratus (**Ns**) is a low, dark, shapeless cloud layer, usually nearly uniform, but sometimes with ragged, wet-looking bases. Nimbostratus is the typical rain cloud. The precipitation which falls from this cloud is steady or intermittent, but not showery.

Cumulus (**Cu**) are dense clouds with *vertical development* (clouds formed by rising air which is cooled as it reaches greater heights). They have a horizontal base and dome-shaped upper surface, with protuberances extending above the dome. Cumulus appear in small patches, and never cover the entire sky. When the vertical development is not great, the clouds appear in patches resembling tufts of cotton or wool, being popularly called "woolpack" clouds (fig. 1.13k). The horizontal bases of such clouds may not be noticeable. These are called "fair weather" cumulus because they always accompany good weather. However, they may merge with altocumulus, or may grow to cumulonimbus before a thunderstorm. Since cumulus are formed by

FIGURE 1.13i.—Stratocumulus.

FIGURE 1.13j.—Stratus.

FIGURE 1.13k.—Cumulus.

FIGURE 1.13l.—Cumulonimbus.

updrafts, they are accompanied by turbulence, causing "bumpiness" in the air. The extent of turbulence is proportional to the vertical extent of the clouds. Cumulus are marked by strong contrasts of light and dark.

Cumulonimbus (Cb) is a massive cloud with great vertical development, rising in **mountainous towers to great heights (fig. 1.13l). The upper part consists of ice crys**tals, and often spreads out in the shape of an anvil which may be seen at such distances that the base may be below the horizon. Cumulonimbus often produces showers of rain, snow, or hail, frequently accompanied by thunder. Because of this, the cloud is often popularly called a "thundercloud" or "thunderhead." The base is horizontal, but as showers occur it lowers and becomes ragged.

1.14 Cloud height measurement.—At sea, cloud heights are often determined by estimate. This is a difficult task, particularly at night.

The height of the base of clouds formed by vertical development (any form of cumulus), if formed in air that has risen from the surface of the earth, can be determined by psychrometer, because the height to which the air must rise before condensation takes place is proportional to the difference between surface air temperature and the dew point. At sea, this difference multiplied by 236 gives the height in feet. That is, for every degree difference between surface air temperature and the dew point, the air must rise 236 feet before condensation will take place. Thus, if the dry-bulb temperature is 80° F, and the wet-bulb temperature is 77° F, the dew point (from tab. 8) is

76°F, or 4° lower than the surface air temperature. The height of the cloud base is $4 \times 236 = 944$ feet.

1.15. Visibility measurement.—**Visibility** is the extreme horizontal distance at which prominent objects can be seen and identified by the unaided eye. It is usually measured directly by the human eye. Ashore, the distances of various buildings, trees, lights, and other objects are measured and used as a guide in estimating the visibility. At sea, however, such an estimate is difficult to make with accuracy. Other ships and the horizon may be of some assistance.

Ashore, visibility is sometimes measured by a **transmissometer,** a device which measures the transparency of the atmosphere by passing a beam of light over a known short distance, and comparing it with a reference light.

1.16. Upper air observations.—Upper air information provides the third dimension to the weather map. Unfortunately, the equipment necessary to obtain such information is quite expensive, and the observations are time consuming. Consequently, the network of observing stations is quite sparse compared to that for surface observations, particularly over the oceans and in isolated land areas. Where facilities exist, upper air observations are made by means of unmanned balloons in conjuction with theodolites, radiosondes, radar, and radio direction finders.

1.17. Storm detection radar.—During World War II, it was found that certain radar equipment gave an indication of weather fronts (art. 2.11) and precipitation areas. It was of particular value near hurricanes and typhoons. Since the close of that war a great amount of work has been done in perfecting radar equipment for use in weather observation. It has proved of immense value in detecting, tracking, and interpreting weather activity out to a distance of as much as 400 miles from the observing station.

1.18. Automated weather stations and buoy systems provide regular transmissions of meteorological and oceanographic information by radio. They are generally used at isolated and relatively inaccessible locations from which weather and ocean data are of great importance. Depending on the type of system used, the elements usually measured include wind direction and speed, atmospheric pressure, air and sea surface temperature, spectral wave data, and a temperature profile from the sea surface to a predetermined depth.

1.19. Recording observations.—Instructions for recording weather observations aboard vessels of the U. S. Navy are given in OPNAV Instruction 3140.37C, *Manual for Ship's Surface Weather Observations*. Instructions for recording observations aboard merchant vessels are given in Weather Service Observing Handbook No. 1, *Marine Surface Observations*.

Problems

1.8a. A ship is proceeding on course 180° at a speed of 22 knots. The apparent wind is from 70° off the port bow, speed 20 knots.

Required.—The relative direction, true direction, and speed of the true wind by maneuvering board or the National Weather Service plotter.

Answers.—True wind from 231° relative, 051° true, at 24.3 knots.

1.8b. A ship is proceeding on course 050° at a speed of 13.5 knots. The apparent wind is from broad on the starboard bow, speed 20 knots.

Required.—The relative direction, true direction, and speed of the true wind by table 10.

Answers.—True wind from 086° relative, 136° true, at 14.3 knots.

1.8c. A ship is proceeding on course 020° at a speed of 16 knots. The true wind is estimated to be from 110° on the port bow, speed 10 knots.

Required.—The relative direction, true direction, and speed of the apparent wind by maneuvering board or the National Weather Service plotter.

Answers.—Apparent wind from 323° relative, 343° true, at 15.6 knots.

1.8d. A ship is proceeding on course 190° at a speed of 14 knots. The true wind is estimated to be from broad on the starboard quarter, speed 20 knots.

Required.—The relative direction, true direction, and speed of the apparent wind by table 5.

Answers.—Apparent wind from 090° relative, 280° true, at 14.0 knots.

1.8e. The true wind has been determined to be from 210°, speed 12 knots. The captain of an aircraft carrier desires an apparent wind of 30 knots from 10° on the port bow for launching aircraft.

Required.—The course and speed of the aircraft carrier.

Answers.—C 235°, S 18.6 kn. (The required apparent wind could also be produced by C 005°, S 40.5 kn.)

1.8f. A ship is proceeding on course 255° at a speed of 15 knots. The wind vane indicates the apparent wind is broad on the starboard beam. From the appearance of the sea the navigator estimates the speed of the true wind as Beaufort 5 (19 knots).

Required.—(1) Relative and true directions of the true wind, (2) speed of the apparent wind. Use the maneuvering board.

Answers.—(1) True wind from 142° relative, 037° true; (2) apparent wind speed 11.6 knots.

1.8g. A ship is proceeding on course 135° at a speed of 18 knots. The wind vane indicates the apparent wind is 40° on the starboard bow. From the appearance of the sea the navigator estimates the speed of the true wind as Beaufort 6 (24.5 knots).

Required.—(1) Relative and true directions of the true wind, (2) speed of the apparent wind. Use table 5.

Answers.—(1) True wind from 069° relative, 204° true; (2) apparent wind speed 36 knots.

1.8h. A ship is proceeding on course 330° at a speed of 20 knots. The wind vane indicates the apparent wind is 30° on the port bow. From the appearance of the sea the navigator estimates the speed of the true wind as Beaufort 4 (13.5 knots).

Required.—(1) Relative and true directions of the true wind, (2) speed of the apparent wind. Solve first by maneuvering board and then by table 10.

Answers.—Graphical solution: (1) true wind from 199° relative, 169° true or from 282° relative, 252° true; (2) apparent wind speed 8.5 knots or 26.3 knots. Table 10 solution: (1) true wind from 197° relative, 167° true or from 283° relative, 253° true; (2) apparent wind speed 8.0 knots or 26.0 knots.

1.12. The dry-bulb temperature is 41° F and the wet-bulb temperature is 35° F.

Required.—(1) Relative humidity, (2) dew point.

Answers.—(1) Relative humidity 53 percent, (2) dew point 26°.

1.14a. The dry-bulb temperature is 72° F and the wet-bulb temperature is 58° F.

Required.—Height of the base of cumulonimbus clouds formed in air which has risen from the surface of the sea.

Answer.—Height 5,900 feet.

References

Kotsch, W. J. *Weather for the Mariner*. Annapolis, United States Naval Institute, 1970.

National Weather Service. *Marine Surface Observations*, Weather Service Observing Handbook No. 1. Washington, U. S. Govt. Print. Off., 1974.

Royal Meteorology Office. *Meteorology for Mariners*. London, Her Majesty's Stationery Office, 1967.

CHAPTER II

WEATHER ELEMENTS

2.1. Introduction.—Weather is the state of the earth's atmosphere with respect to temperature, humidity, precipitation, visibility, cloudiness, etc. In contrast, the term **climate** refers to the prevalent or characteristic meteorological conditions of a place or region.

All weather may be traced ultimately to the effect of the sun on the earth, including the lower portions of the atmosphere. Most changes in weather involve large-scale, approximately horizontal, motion of air. Air in such motion is called **wind.** This motion is produced by differences of atmospheric pressure, which are attributable both to differences of temperature and the nature of the motion itself.

The weather is of considerable interest to the mariner. The wind and state of the sea affect dead reckoning. Reduced horizontal visibility limits piloting. The state of the atmosphere affects electronic navigation and radio communication. If the skies are overcast, visual celestial observations are not available; and under certain conditions refraction and dip are disturbed. When wind was the primary motive power, knowledge of the areas of favorable winds was of great importance. This consideration led Matthew Fontaine Maury, more than a century ago, to seek information from ships' logs to establish speed and direction of prevailing winds over the various trade routes of the world. The information thus gathered was shown on pilot charts. By means of these charts, the mariner could select a suitable route for a favorable passage. Even power vessels are affected considerably by wind and sea. Less fuel consumption and a more comfortable passage are to be expected if wind and sea are moderate and favorable. Pilot charts are useful in selecting suitable routes.

Optimizing ship speed and safety by taking advantage of favorable wind and sea conditions has become practicable with the advent of extended range forecasting techniques.

2.2. The atmosphere is a relatively thin shell of air, water vapor, dust, smoke, etc., surrounding the earth. The air is a mixture of transparent gases and, like any gas, is elastic and highly compressible. Although extremely light, it has a definite weight which can be measured. A cubic foot of air at standard sea-level temperature and pressure weighs 1.22 ounces, or about 1/817th part of the weight of an equal volume of water. Because of this weight, the atmosphere exerts a pressure upon the surface of the earth, amounting to about 15 pounds per square inch.

As altitude increases, less atmosphere extends upward, and pressure decreases. With less pressure, the density decreases. More than three-fourths of the air is concentrated within a layer averaging about 7 statute miles thick, called the **troposphere.** This is the region of most "weather," as the term is commonly understood.

The top of the troposphere is marked by a thin transition zone called the **tropopause,** immediately above which is the **stratosphere.** Beyond this lie several other layers having distinctive characteristics. The average height of the tropopause ranges from about 5 miles or less at high latitudes to about 10 miles at low latitudes.

The **standard atmosphere** is a conventional vertical structure of the atmosphere characterized by a standard sea-level pressure of 29.92 inches of mercury (1013.25

millibars) and a sea-level temperature of 59°F (15°C), the rate of temperature decreases with height (i.e., **standard lapse rate**) being a uniform 3°.6F (2°C) per thousand feet to 11 kilometers (36,089 feet) and thereafter a constant temperature of (−)69°.7F (−56°.5C).

With the aid of *weather satellite observations*, meteorologists are continually learning more of the atmospheric processes in the troposphere and stratosphere as they affect weather at sea. In recent years research has indicated that the **jet stream** is an important entity in relation to the sequence of weather. The jet stream refers to relatively strong (≥ 60 knots) quasi-horizontal winds usually concentrated within a restricted layer of the atmosphere. Although jet stream winds can occur at any level and geographic location and from any direction, the term is most often associated with mid-latitude winds with maximum speeds from 270° (±45°). Such winds, called **polar jet stream winds,** average 90 knots maximum, but speeds up to 200 knots may occur in the winter season.

2.3. General circulation of the atmosphere.—The heat required for warming the air is supplied originally by the sun. As radiant energy from the sun arrives at the earth, about 29 percent is reflected back into space by the earth and its atmosphere, 19 percent is absorbed by the atmosphere, and the remaining 52 percent is absorbed by the surface of the earth, much of which is reradiated back into space. This earth radiation is in comparatively long waves relative to the short-wave radiation from the sun, since it emanates from a cooler body. Long-wave radiation, being readily absorbed by the water vapor in the air, is primarily responsible for the warmth of the atmosphere near the earth's surface. Thus, the atmosphere acts much like the glass on the roof of a greenhouse. It allows part of the incoming solar radiation to reach the surface of the earth, but is heated by the terrestrial radiation passing outward. Over the entire earth and for long periods of time, the total outgoing energy must be equivalent to the incoming energy (minus any converted to another form and retained), or the temperature of the earth, including its atmosphere, would steadily increase or decrease. In local areas, or over relatively short periods of time, such a balance is not required, and in fact does not exist, resulting in changes such as those occurring from one year to another in different seasons and in different parts of the day.

The more nearly perpendicular the rays of the sun strike the surface of the earth, the more heat energy per unit area is received at that place. Physical measurements show that in the tropics more heat per unit area is received than is radiated away, and that in polar regions the opposite is true. Unless there were some process to transfer heat from the Tropics to polar regions, the Tropics would be much warmer than they are, and the polar regions would be much colder. Atmospheric motions bring about the required transfer of heat. The oceans also participate in the process, but to a lesser degree.

If the earth had a uniform surface and did not rotate on its axis, with the sun following its normal path across the sky (solar heating increasing with decreasing latitude), a simple circulation would result, as shown in figure 2.3a. However, the surface of the earth is far from uniform, being covered with an irregular distribution of land of various heights, and water; the earth rotates about its axis once in approximately 24 hours, so that the portion heated by the sun continually changes; and the axis of rotation is tilted so that as the earth moves along its orbit about the sun, seasonal changes occur in the exposure of specific areas to the sun's rays, resulting in variations in the heat balance of these areas. These factors, coupled with others, result in constantly changing large-scale movements of air. For example, the rotation of the earth exerts an apparent force, known as **Coriolis force,** which diverts the air from a direct path between high and low pressure areas. The diversion of the air is toward the right in the Northern

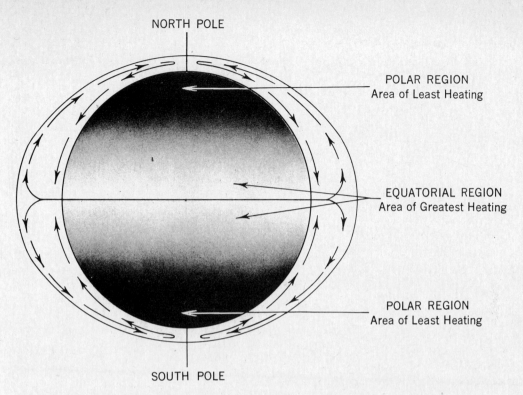

NORTH POLE

POLAR REGION
Area of Least Heating

EQUATORIAL REGION
Area of Greatest Heating

POLAR REGION
Area of Least Heating

SOUTH POLE

FIGURE 2.3a.—Ideal atmospheric circulation for a uniform and nonrotating earth.

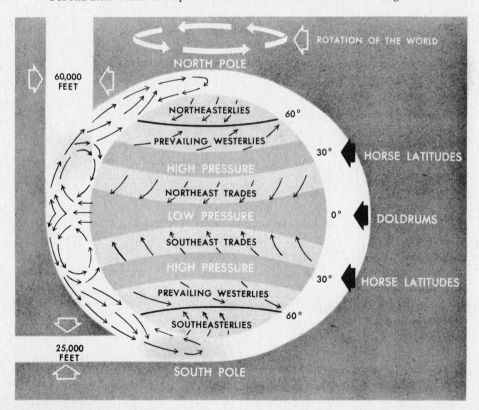

ROTATION OF THE WORLD

NORTH POLE

60,000 FEET

NORTHEASTERLIES   60°

PREVAILING WESTERLIES

30°   HORSE LATITUDES

HIGH PRESSURE

NORTHEAST TRADES

LOW PRESSURE   0°   DOLDRUMS

SOUTHEAST TRADES

HIGH PRESSURE

30°   HORSE LATITUDES

PREVAILING WESTERLIES

60°

SOUTHEASTERLIES

25,000 FEET

SOUTH POLE

FIGURE 2.3b.—Simplified diagram of the general circulation of the atmosphere.

Hemisphere and toward the left in the Southern Hemisphere. At some distance above the surface of the earth, the wind tends to blow along the isobars, being called the **geostrophic wind** if the isobars are straight (great circles), and **gradient wind** if they are curved. Near the surface of the earth, friction tends to divert the wind from the isobars toward the center of low pressure. At sea, where friction is less than on land, the wind follows the isobars more closely.

A simplified diagram of the general circulation pattern is shown in figure 2.3b. Figures 2.3c and 2.3d give a generalized picture of the world's pressure distribution and wind systems as actually observed.

The change in pressure with horizontal distance is called **pressure gradient.** It is maximum along a normal (perpendicular) to the isobars. A force results which is called **pressure gradient force** and is always directed from high to low pressure. Speed of the wind as illustrated in figures 2.3c and 2.3d is approximately proportional to this pressure gradient.

2.4. The Doldrums.—The belt of low pressure at the surface near the equator occupies a position approximately midway between high pressure belts at about latitude 30° to 35° on each side. Except for significant intradiurnal changes, the atmospheric pressure along the equatorial low is almost uniform. With minimal pressure gradient, wind speeds are light and directions are variable. Hot, sultry days are common. The sky is often overcast, and showers and thundershowers are relatively frequent; in these disturbed areas, brief periods of strong wind occur.

The area involved is a thin belt near the equator, the eastern part in both the Atlantic and Pacific being wider than the western part. However, both the position and extent of the belt vary with longitude and season. During all seasons in the Northern Hemisphere, the belt is centered in the eastern Atlantic and Pacific; however, there are wide excursions of the doldrums regions at longitudes with considerable landmass. On the average, the position is at 5°N, frequently called the **meteorological equator.**

2.5. The trade winds at the surface blow from the belts of high pressure toward the equatorial belts of low pressure. Because of the rotation of the earth, the moving air is deflected toward the west. Therefore, the trade winds in the Northern Hemisphere are from the northeast and are called the **northeast trades,** while those in the Southern Hemisphere are from the southeast and are called the **southeast trades.** The trade-wind directions are best defined over eastern ocean areas.

The trade winds are generally considered among the most constant of winds, blowing for days or even weeks with little change of direction or speed. However, at times they weaken or shift direction, and there are regions where the general pattern is disrupted. A notable example is found in the island groups of the South Pacific, where the trades are practically nonexistent during January and February. Their best development is attained in the South Atlantic and in the South Indian Ocean. In general, they are stronger during the winter than during the summer season.

In July and August, when the belt of equatorial low pressure moves to a position some distance north of the equator, the southeast trades blow across the equator, into the Northern Hemisphere, where the earth's rotation diverts them toward the right, causing them to be southerly and southwesterly winds. The "southwest monsoons" of the African and Central American coasts have their origin partly in such diverted southeast trades.

Cyclones (art. 2.12) from the middle latitudes rarely enter the regions of the trade winds, although tropical cyclones (ch. III) originate within these areas.

2.6. The horse latitudes.—Along the poleward side of each trade-wind belt, and corresponding approximately with the belt of high pressure in each hemisphere, is another region with weak pressure gradients and correspondingly light, variable winds.

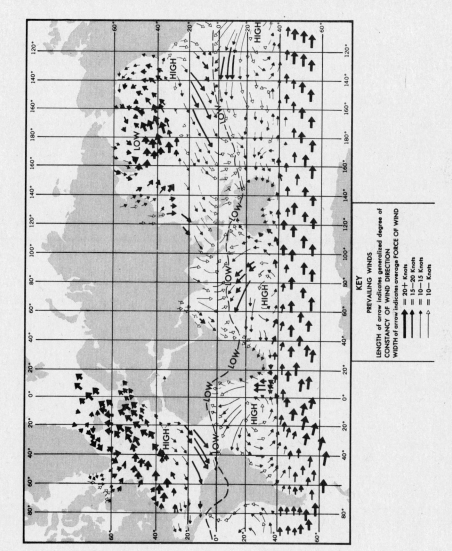

FIGURE 2.3c.—Generalized pattern of actual surface winds in January and February.

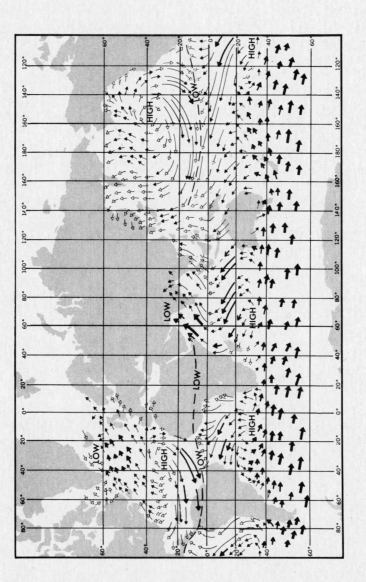

FIGURE 2.3d.—Generalized pattern of actual surface winds in July and August (see key with figure 2.3c).

These are called the **horse latitudes.** The weather is generally good although low clouds are common. Compared to the doldrums, periods of stagnation in the horse latitudes are less persistent, being of a more intermittent nature. The difference is due primarily to the fact that rising currents of warm air in the equatorial low carry large amounts of moisture which condenses as the air cools at higher levels, while in the horse latitudes the air is apparently descending and becoming less humid as it is warmed at lower heights.

2.7. The prevailing westerlies.—On the poleward side of the high pressure belt in each hemisphere the atmospheric pressure again diminishes. The currents of air set in motion along these gradients toward the poles are diverted by the earth's rotation toward the east, becoming southwesterly winds in the Northern Hemisphere and northwesterly in the Southern Hemisphere. These two wind systems are known as the **prevailing westerlies** of the temperate zones.

In the Northern Hemisphere this relatively simple pattern is distorted considerably by secondary wind circulations, due primarily to the presence of large landmasses. In the North Atlantic, between latitudes 40° and 50°, winds blow from some direction between south and northwest during 74 percent of the time, being somewhat more persistent in winter than in summer. They are stronger in winter, too, averaging about 25 knots (Beaufort 6) as compared with 14 knots (Beaufort 4) in the summer.

In the Southern Hemisphere the westerlies blow throughout the year with a steadiness approaching that of the trade winds (art. 2.5). The speed, though variable, is generally between 17 and 27 knots (Beaufort 5 and 6). Latitudes 40°S to 50°S (or 55°S) where these boisterous winds occur, are called the **roaring forties.** These winds are strongest at about latitude 50°S.

The greater speed and persistence of the westerlies in the Southern Hemisphere are due to the difference in the atmospheric pressure pattern, and its variations, from that of the Northern Hemisphere. In the comparatively landless Southern Hemisphere, the average yearly atmospheric pressure diminishes much more rapidly on the poleward side of the high pressure belt, and has fewer irregularities due to continental interference, than in the Northern Hemisphere.

2.8. Winds of polar regions.—Partly because of the low temperatures near the geographical poles of the earth, the surface pressure tends to remain higher than in surrounding regions. Consequently, the winds blow outward from the poles, and are deflected westward by the rotation of the earth, to become **northeasterlies** in the Arctic, and **southeasterlies** in the Antarctic. Where the polar easterlies meet the prevailing westerlies, near 50°N and 50°S on the average, a discontinuity in temperature and wind exists. This discontinuity is called the **polar front.** Here the warmer low-latitude air ascends over the colder polar air creating a zone of cloudiness and precipitation.

In the Arctic, the general circulation is greatly modified by surrounding landmasses. Winds over the Arctic Ocean are somewhat variable, and strong surface winds are rarely encountered.

In the Antarctic, on the other hand, a high central landmass is surrounded by water, a condition which augments, rather than diminishes, the general circulation. The high pressure, although weaker than in the horse latitudes, is stronger than in the Arctic, and of great persistence especially in eastern Antarctica. The cold air from the plateau areas moves outward and downward toward the sea and is deflected toward the west by the earth's rotation. The katabatic winds (art. 2.13) remain strong throughout the year, frequently attaining hurricane force near the base of the mountains. These are some of the strongest surface winds encountered anywhere in the world, with the possible exception of those in well-developed tropical cyclones (ch. III).

2.9. Modifications of the general circulation.—The general circulation of the atmosphere as described in articles 2.3–2.8 is greatly modified by various conditions.

The high pressure in the horse latitudes is not uniformly distributed around the belts, but tends to be accentuated at several points, as shown in figures 2.3b and 2.3c. These **semipermanent highs** remain at about the same places with great persistence.

Semipermanent lows also occur in various places, the most prominent ones being west of Iceland, and over the Aleutians (winter only) in the Northern Hemisphere, and in the Ross Sea and Weddell Sea in the antarctic areas. The regions occupied by these semipermanent lows are sometimes called the graveyards of the lows, since many lows move directly into these areas and lose their identity as they merge with and reinforce the semipermanent lows. The low pressure in these areas is maintained largely by the migratory lows which stall there, with topography also important, especially in Antarctica.

Another modifying influence is land, which undergoes greater temperature changes than does the sea. During the summer, a continent is warmer than its adjacent oceans. Therefore, low pressures tend to prevail over the land. If a climatological belt of high pressure encounters a continent, its pattern is distorted or interrupted, whereas a belt of low pressure is intensified over the same area. In winter, the opposite effect takes place, belts of high pressure being intensified over land and those of low pressure being weakened.

The most striking example of a wind system produced by the alternate heating and cooling of a landmass is the **monsoons** (seasonal wind) of the China Sea and Indian Ocean. A portion of this effect is shown in figures 2.9a and 2.9b. In the summer (fig. 2.9a), low pressure prevails over the warm continent of Asia, and relatively higher pressure prevails over the adjacent sea. Between these two systems the wind blows in a nearly steady direction. The lower portion of the pattern is in the Southern Hemisphere, extending to about 10° south latitude. Here the rotation of the earth causes a deflection to the left, resulting in southeasterly winds. As they cross the equator, the deflection is in the opposite direction, causing them to curve toward the right, becoming southwesterly winds. In the winter (fig. 2.9b), the positions of high and low pressure areas are interchanged, and the direction of flow is reversed.

In the China Sea the summer monsoon blows from the southwest, usually from May to September. The strong winds are accompanied by heavy squalls and thunderstorms, the rainfall being much heavier than during the winter monsoon. As the season advances, squalls and rain become less frequent. In some places the wind be-

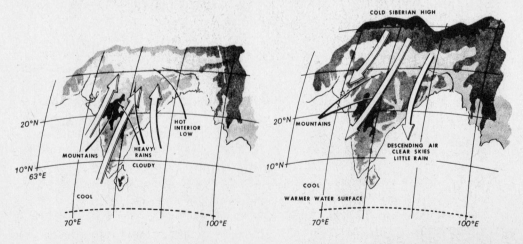

FIGURE 2.9a.—The summer monsoon.   FIGURE 2.9b.—The winter monsoon.

comes a light breeze which is unsteady in direction, or stops altogether, while in other places it continues almost undiminished, with changes in direction or calms being infrequent. The winter monsoon blows from the northeast, usually from October to April. It blows with a steadiness similar to that of the trade winds, often attaining the speed of a moderate gale (28–33 knots). Skies are generally clear during this season, and there is relatively little rain.

The general circulation is further modified by winds of cyclonic origin (art. 2.12), and various local winds (art. 2.13).

2.10. Airmasses.—Because of large differences in physical characteristics of the earth's surface, particularly the oceanic and continental contrasts, the air overlying these surfaces acquires differing values of temperature and moisture. The processes of radiation and convection in the lower portions of the troposphere act in differing characteristic manners for a number of well-defined regions of the earth. The air overlying these regions acquires characteristics common to the particular area, but contrasting to those of other areas. Each distinctive part of the atmosphere, within which common characteristics prevail over a reasonably large area, is called an **airmass.**

Airmasses are named according to their source regions. Four such regions are generally recognized: (1) *equatorial* (*E*), the doldrum area between the north and south trades; (2) *tropical* (*T*), the trade wind and lower temperate regions; (3) *polar* (*P*), the higher temperate latitudes; and (4) *arctic or antarctic* (*A*), the north or south polar regions of ice and snow. This classification is a general indication of relative temperature, as well as latitude of origin.

Airmasses are further classified as maritime (*m*) or continental (*c*), depending upon whether they form over water or land. This classification is an indication of the relative moisture content of the airmass. Tropical air, then, might be designated maritime tropical (*mT*) or continental tropical (*cT*). Similarly, polar air may be either maritime polar (*mP*) or continental polar (*cP*). Arctic/antarctic air, due to the predominance of landmasses and ice fields in the high latitudes, is rarely maritime arctic (*mA*). Equatorial air is found exclusively over the ocean surface and is designated neither (*cE*) nor (*mE*), but simply (*E*).

A third classification sometimes applied to tropical and polar airmasses indicates whether the airmass is *warm* (*w*) or *cold* (*k*) relative to the underlying surface. Thus, the symbol *mTw* indicates maritime tropical air which is warmer than the underlying surface, and *cPk* indicates continental polar air which is colder than the underlying surface. The *w* and *k* classifications are primarily indications of stability (i.e., change of temperature with increasing height). If the air is cold relative to the surface, the lower portion of the airmass is being heated, resulting in instability (temperature markedly decreases with increasing height) as the warmer air tends to rise by convection. Conversely, if the air is warm relative to the surface, the lower portion of the airmass is cooled, tending to remain close to the surface. This is a stable condition (temperature increases with increasing height).

Two other types of airmasses are sometimes recognized. These are *monsoon* (*M*), a transitional form between *cP* and *E*; and *superior* (*S*), a special type formed in the free atmosphere by the sinking and consequent warming of air aloft.

2.11. Fronts.—As airmasses move within the general circulation, they travel from their source regions and invade other areas dominated by air having different characteristics. Such a process leads to a zone of separation between the two airmasses. The gradients of thermal and moisture properties are maximized in the zone. Since the zone or discontinuity is so thin as to approach a sheet when viewed on a small scale map, it is called a **frontal surface.** The intersection of a frontal surface and a horizontal

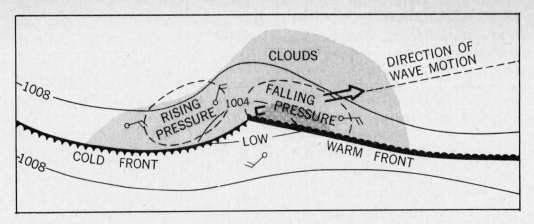

FIGURE 2.11a.—First stage in the development of a frontal wave (top view).

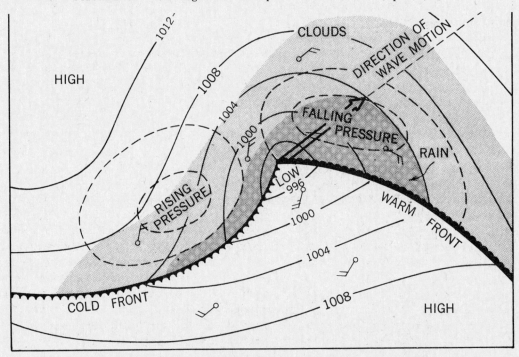

FIGURE 2.11b.—A fully developed frontal wave (top view).

plane in a line is called a **front,** although the term "front" is commonly used as a short expression for "frontal surface" when this will not introduce an ambiguity.

Indicative of the differences in the motion of adjacent airmasses, the front takes a wave like character, hence the term "frontal wave."

Before the formation of frontal waves, the isobars (lines of equal atmospheric pressure) tend to run parallel to the fronts. As a wave is formed, the pattern is distorted somewhat, as shown in figure 2.11a. In this illustration, colder air is north of warmer air. In figures 2.11a-2.11d isobars are drawn at 4-millibar intervals.

The wave tends to travel in the direction of the general circulation, which in the temperate latitudes is usually in a general easterly and slightly poleward direction.

Along the leading edge of the wave, warmer air is replacing colder air. This is called the **warm front.** The trailing edge is the **cold front,** where colder air is underrunning and displacing warmer air.

The warm air, being less dense, tends to ride up over the colder air it is replacing. The slope is gentle, varying between 1:100 and 1:300. Partly because of the replacement of cold, dense air with warm, light air, the pressure decreases. Since the slope is gentle, the upper part of a warm frontal surface may be many hundreds of miles ahead of the surface portion. The decreasing pressure, indicated by a "falling barometer," is often an indication of the approach of such a wave. In a slow-moving, well-developed wave, the barometer may begin to fall several *days* before the wave arrives. Thus, the amount and nature of the change of atmospheric pressure between observations, called **pressure tendency,** is of assistance in predicting the approach of such a system.

The advancing cold air, being more dense, tends to cut under the warmer air at the cold front, lifting it to greater heights. The slope here is such that the upper-air portion of the cold front is behind the surface position relative to its motion. The slope generally ranges from 1:25 to 1:100, being steeper than the warm front. After a cold front has passed, the pressure increases—a "rising barometer."

In the first stages, these effects are not marked, but as the wave continues to grow, they become more pronounced, as shown in figure 2.11b. As the amplitude of the wave increases, pressure near the center usually decreases, and the "low" is said to "deepen." As it deepens, its forward speed generally decreases.

The approach of a well-developed warm front (i.e., when the warm air is mT) is usually heralded not only by falling pressure, but also by a more-or-less regular sequence of clouds. First, cirrus appear. These give way successively to cirrostratus, altostratus, altocumulus, and nimbostratus. Brief showers may precede the steady rain accompanying the nimbostratus.

As the warm front passes, the temperature rises, the wind shifts clockwise (in the Northern Hemisphere), and the steady rain stops. Drizzle may fall from low-lying stratus clouds, or there may be fog for some time after the wind shift. During passage of the **warm sector** between the warm front and the cold front, there is little change in temperature or pressure. However, if the wave is still growing and the low deepening, the pressure might slowly decrease. In the warm sector the skies are generally clear or partly cloudy, with cumulus or stratocumulus clouds most frequent. The warm air is usually moist, and haze or fog may often be present.

As the faster moving, steeper cold front passes, the wind shifts clockwise in the Northern Hemisphere (counterclockwise in the Southern Hemisphere), the temperature falls rapidly, and there are often brief and sometimes violent showers, frequently accompanied by thunder and lightning. Clouds are usually of the convective type. A cold front usually coincides with a well-defined **wind-shift line** (a line along which the wind shifts abruptly from southerly or southwesterly to northerly or northwesterly in the Northern Hemisphere and from northerly or northwesterly to southerly or south-westerly in the Southern Hemisphere). At sea a series of brief showers accompanied by strong, shifting winds may occur along or some distance (up to 200 miles) ahead of a cold front. These are called **squalls** (in common nautical use the term squall may be additionally applied to any severe local storm accompanied by gusty winds, precipitation, thunder, and lightning), and the line along which they occur is called a **squall line.**

Because of its greater speed and steeper slope, which may approach or even exceed the vertical near the earth's surface (due to friction), a cold front and its associated weather passes more quickly than a warm front. After a cold front passes, the pressure rises, often quite rapidly, the visibility usually improves, and the clouds tend to diminish.

As the wave progresses and the cold front approaches the slower moving warm front, the low becomes deeper and the warm sector becomes smaller. This is shown in figure 2.11c.

Finally, the faster moving cold front overtakes the warm front (fig. 2.11d), resulting in an **occluded front** at the surface, and an **upper front** aloft (fig. 2.11e). When the two parts of the cold airmass meet, the warmer portion tends to rise above the colder part. The warm air continues to rise until the entire frontal system dissipates. As the warmer air is replaced by colder air, the pressure gradually rises, a process called "filling." This usually occurs within a few days after an occluded front forms. Finally, there results a **cold low,** or simply a low pressure system across which little or no gradient in temperature and moisture can be found.

The sequence of weather associated with a low depends greatly upon location with respect to the path of the center. That described above assumes that the observer is so located that he encounters each part of the system. If he is poleward of the path of the center of the low, the abrupt weather changes associated with the passage of fronts are not experienced. Instead, the change from the weather characteristically found ahead of a warm front to that behind a cold front takes place gradually, the exact sequence being dictated somewhat by distance from the center, as well as severity and age of the low.

Although each low follows generally the pattern given above, no two are ever exactly alike. Other centers of low pressure and high pressure and the airmasses associated with them, even though they may be 1,000 miles or more away, influence the formation and motion of individual low centers and their accompanying weather. Particularly, a high stalls or diverts a low. This is true of temporary highs as well as semipermanent highs.

2.12. Cyclones and anticyclones.—An area of relatively low pressure, generally circular, is called a **cyclone.** Its counterpart for high pressure is called an **anticyclone.** These terms are used particularly in connection with the winds associated with such centers. Wind tends to blow from an area of high pressure to one of low pressure, but due to rotation of the earth, they are deflected toward the right in the Northern Hemisphere and toward the left in the Southern Hemisphere (art. 2.3).

Because of the rotation of the earth, therefore, the circulation tends to be counterclockwise around areas of low pressure and clockwise around areas of high pressure in the Northern Hemisphere (figs. 2.11c and 2.11d), the speed being proportional to the spacing of isobars. In the Southern Hemisphere, the direction of circulation is reversed. Based upon this condition, a general rule (**Buys Ballot's Law**) can be stated thus:

If an observer in the Northern Hemisphere faces the surface wind, the center of low pressure is toward his right, somewhat behind him; and the center of high pressure is toward his left and somewhat in front of him.

If an observer in the Southern Hemisphere faces the surface wind, the center of low pressure is toward his left and somewhat behind him; and the center of high pressure is toward his right and somewhat in front of him.

In a general way, these relationships apply in the case of the general distribution of pressure, as well as to temporary local pressure systems.

The reason for the wind shift along a front is that the isobars have an abrupt change of direction along these lines, as shown in figures 2.11a-2.11d. Since the direction of the wind is directly related to the direction of isobars, any change in the latter results in a shift in the wind direction.

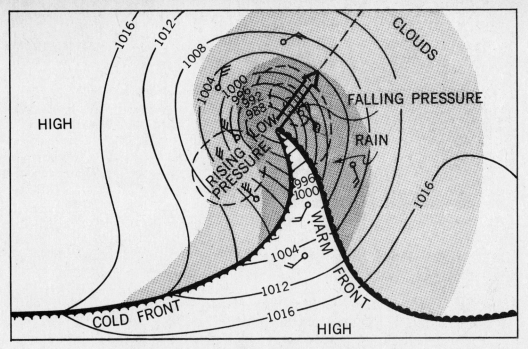

FIGURE 2.11c.—A frontal wave nearing occlusion (top view).

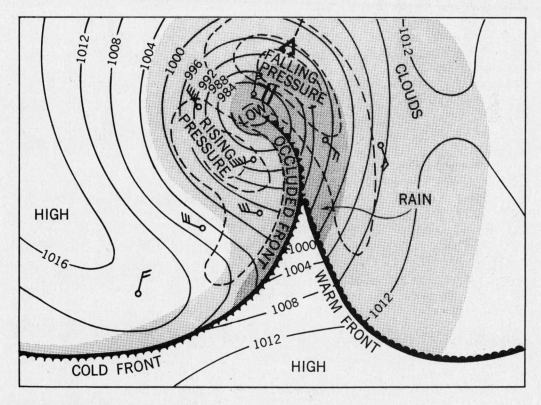

FIGURE 2.11d.—An occluded front (top view).

FIGURE 2.11e.—An occluded front (cross section).

In the Northern Hemisphere, the wind shifts toward the *right* (clockwise or veering) when either a warm or cold front passes. In the Southern Hemisphere, the shift is toward the *left* (counterclockwise or backing). When an observer is on the poleward side of the path of a frontal wave, wind shifts are reversed (i.e., backing in the Northern Hemisphere and veering in the Southern Hemisphere).

In an anticyclone, successive isobars are relatively far apart, resulting in light winds. In a cyclone, the isobars are more closely spaced. With a steeper pressure gradient, the winds are stronger.

Since an anticyclonic area is a region of outflowing winds, air is drawn into it from aloft. Descending air is warmed, and as air becomes warmer, its capacity for holding uncondensed moisture increases. Therefore, clouds tend to dissipate. Clear skies are characteristic of an anticyclone, although scattered clouds and showers are sometimes encountered.

In contrast, a cyclonic area is one of converging winds. The resulting upward movement of air results in cooling, a condition favorable to the formation of clouds and precipitation. More or less continuous rain and generally stormy weather are usually associated with a cyclone.

Between the two hemispheric belts of high pressure associated with the horse latitudes, called subtropical anticyclones (art. 2.6), cyclones form only occasionally over certain areas at sea, generally in summer and fall only. These **tropical cyclones** are usually quite violent, being known under various names according to their location. They are discussed in chapter III.

In the areas of the prevailing westerlies (art. 2.7) in temperate latitudes, migratory cyclones (**lows**) and anticyclones (**highs**) are a common occurrence. These are sometimes called **extratropical cyclones** and **extratropical anticyclones** to distinguish them from the more violent tropical cyclones. Formation occurs over sea and land. The lows intensify as they move poleward; the highs weaken as they move equatorward. In their early stages, cyclones are elongated, as shown in figure 2.11a, but as their life cycle proceeds, they become more nearly circular (figs. 2.11b–2.11d).

2.13. Local winds.—In addition to the winds of the general circulation (arts. 2.3–2.8) and those associated with migratory cyclones and anticyclones (art. 2.12), there are numerous local winds which influence the weather in various places.

The most common of these are the **land** and **sea breezes,** caused by alternate heating and cooling of land adjacent to water. The effect is similar to that which causes the monsoons (art. 2.9), but on a much smaller scale, and over shorter periods. By day the land is warmer than the water, and by night it is cooler. This effect occurs along many coasts during the summer. Between about 0900 and 1100 the temperature of the land becomes greater than that of the adjacent water. The lower levels of air over the land are warmed, and the air rises, drawing in cooler air from the sea. This is the **sea breeze.** Late in the afternoon, when the sun is low in the sky, the temperature of the two surfaces equalizes and the breeze stops. After sunset, as the land cools below the sea temperature, the air above it is also cooled. The contracting cool air becomes more dense, increasing the pressure near the surface. This results in an outflow of winds to the sea. This is the **land breeze,** which blows during the night and dies away near sunrise. Since the atmospheric pressure changes associated with this cycle are not great, the accompanying winds generally do not exceed gentle to moderate breezes. The circulation is usually of limited extent, reaching a distance of perhaps 20 miles inland, and not more than 5 or 6 miles offshore, and to a height of a few hundred feet. In the doldrums and subtropics, this process is repeated with great regularity throughout most of the year. As the latitude increases, it becomes less prominent, being masked by winds of migratory cyclones and anticyclones (art. 2.12). However, the effect often may be present to reinforce, retard, or deflect stronger prevailing winds.

Varying conditions of topography produce a large variety of local winds throughout the world. Winds tend to follow valleys, and to be deflected from high banks and shores. In mountain areas wind flows in response to temperature distribution and gravity. An **anabatic wind** is one that blows up an incline, usually as a result of surface heating. A **katabatic wind** is one which blows down an incline. There are two types, *foehn* and *fall wind*.

A dry wind with a downward component, warm for the season, is called a **foehn.** The foehn occurs when horizontally moving air encounters a mountain barrier. As it blows upward to clear the barrier, it is cooled below the dew point, resulting in loss of moisture by cloud formation and perhaps rain. As the air continues to rise, its rate of cooling is reduced because the condensing water vapor gives off heat to the surrounding atmosphere. After crossing the mountain barrier, the air flows downward along the leeward slope, being warmed by compression as it descends to lower levels. Thus, since it loses less heat on the ascent than it gains during descent, and since it loses moisture during ascent, it arrives at the bottom of the mountains as very warm, dry air. This accounts for the warm, arid regions along the eastern side of the Rocky Mountains and in similar areas. In the Rocky Mountain region this wind is known by the name **chinook.** It may occur at any season of the year, at any hour of the day or night, and have any speed from a gentle breeze to a gale. It may last for several days, or for a very short period. Its effect is most marked in winter, when it may cause the temperature to rise as much as 20° F to 30° F within 15 minutes, and cause snow and ice to melt within a few hours. On the west coast of the United States, a foehn wind, given the name **Santa Ana,** blows through a pass and down a valley by that name in Southern California. This wind may blow with such force that it endangers small craft immediately off the coast.

A cold wind blowing down an incline is called a **fall wind.** Although it is warmed somewhat during descent, as is the foehn, it remains cold relative to the surrounding air. It occurs when cold air is dammed up in great quantity on the windward side of a mountain and then spills over suddenly, usually as an overwhelming surge down the other side. It is usually quite violent, sometimes reaching hurricane force. A different name for this type wind is given at each place where it is common. The **tehuantepecer** of the Mexican and Central American coast, the **pampero** of the Argentine coast, the **mistral** of the western Mediterranean, and the **bora** of the eastern Mediterranean are examples of this type wind.

Many other local winds common to certain areas have been given distinctive names.

A **blizzard** is a violent, intensely cold wind laden with snow mostly or entirely picked up from the ground, although the term is often used popularly to refer to any heavy snowfall accompanied by strong wind. A **dust whirl** is a rotating column of air about 100 to 300 feet in height, carrying dust, leaves, and other light material. This wind, which is similar to a waterspout at sea (art. 2.24), is given various local names such as **dust devil** in southwestern United States and **desert devil** in South Africa. A **gust** is a sudden, brief increase in wind speed followed by a slackening, or the violent wind or squall that accompanies a thunderstorm. A puff of wind or a light breeze affecting a small area, such as would cause patches of ripples on the surface of water, is called a **cat's paw**.

2.14. **Fog** is a cloud (art. 1.13) whose base is low enough to restrict visibility. Fog is composed of droplets of water, or ice crystals (**ice fog**) formed by condensation or crystallization of water vapor in the air.

Radiation fog forms over low-lying land on clear, calm nights. As the land radiates heat and becomes cooler, it cools the air immediately above the surface. This causes a **temperature inversion** to form, the temperature for some distance upward *increasing* with height. If the air is cooled to its dew point (art. 1.12), fog forms. Often, cooler and more dense air drains down surrounding slopes to heighten the effect. Radiation fog is often quite shallow, and is usually densest at the surface. After sunrise the fog may "lift," as shown in figure 2.14, and gradually dissipate, usually being entirely gone by noon. At sea the temperature of the water undergoes little change between day and night, and so radiation fog is seldom encountered more than 10 miles from shore.

Advection fog forms when warm, moist air blows over a colder surface and is cooled below its dew point. This type, most commonly encountered at sea, may be quite dense and often persists over relatively long periods. Advection fog is common over cold ocean currents. If the wind is strong enough to thoroughly mix the air, condensation may take place at some distance above the surface of the earth, forming low stratus clouds (art. 1.13) rather than fog.

Off the coast of California, seasonal winds create an offshore current which displaces the warm surface water, causing an upwelling of colder water. Moist Pacific air is transported along the coast in the same wind system and is cooled by the relatively cold water. Advection fog results. In the coastal valleys, fog is sometimes formed when moist air blown inland during the afternoon is cooled by radiation during the night.

When very cold air moves over warmer water, wisps of visible water vapor may rise from the surface as the water "steams." In extreme cases this **frost smoke**, or **arctic sea smoke**, may rise to a height of several hundred feet, the portion near the surface constituting a dense fog which obscures the horizon and surface objects, but usually leaves the sky relatively clear.

Haze consists of fine dust or salt particles in the air, too small to be individually apparent, but in sufficient number to reduce horizontal visibility and cast a bluish or yellowish veil over the landscape, subduing its colors and making objects appear indistinct. This is sometimes called **dry haze** to distinguish it from **damp haze,** which consists of small water droplets or moist particles in the air, smaller and more scattered than light fog. In international meteorological practice, the term "haze" is used to refer to a condition of atmospheric obscurity caused by dust and smoke.

Mist is synonymous with **drizzle** in the United States but is often considered as intermediate between haze and fog in its properties.

A mixture of smoke and fog is called **smog**.

RADIATION FOG

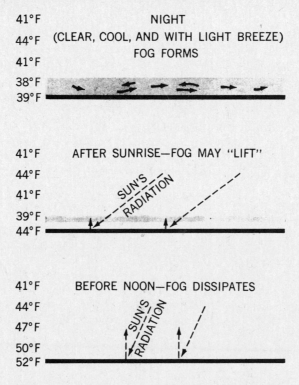

FIGURE 2.14.—Formation and dissipation of radiation fog.

2.15. Mirage.—Light is refracted as it passes through the atmosphere. When refraction is normal, objects appear slightly elevated, and the visible horizon is farther from the observer than it otherwise would be. Since the effects are uniformly progressive, they are not apparent to the observer. When refraction is not normal, some form of **mirage** may occur. A mirage is an optical phenomenon in which objects appear distorted, displaced (raised or lowered), magnified, multiplied, or inverted due to varying atmospheric refraction which occurs when a layer of air near the earth's surface differs greatly in density from surrounding air. This may occur when there is a rapid and sometimes irregular change of temperature or humidity with height.

If there is a temperature inversion (increase of temperature with height), particularly if accompanied by a rapid decrease in humidity, the refraction is greater than normal. Objects appear elevated, and the visible horizon is farther away. Objects which are normally below the horizon become visible. This is called **looming.** If the upper portion of an object is raised much more than the bottom part, the object appears taller than usual, an effect called **towering.** If the lower part of an object is raised more than the upper part, the object appears shorter, an effect called **stooping.** When the refraction is greater than normal, a **superior mirage** may occur. An inverted image is seen above the object, and sometimes an erect image appears over the inverted one, with the bases of the two images touching. Greater than normal refraction usually occurs when the water is much colder than the air above it.

If the temperature decrease with height is much greater than normal, refraction is less than normal, or may even cause bending in the opposite direction. Objects appear lower than normal, and the visible horizon is closer to the observer. This is called **sinking.** Towering or stooping may occur if conditions are suitable. When the refraction is reversed, an **inferior mirage** may occur. A ship or an island appears to be floating in the air above a shimmering horizon, possibly with an inverted image beneath it. Conditions suitable to the formation of an inferior mirage occur when the surface is much warmer than the air above it. This usually requires a heated landmass, and therefore is more common near the coast than at sea.

When refraction is not uniformly progressive, objects may appear distorted, taking an almost endless variety of shapes. The sun when near the horizon is one of the objects most noticeably affected. A **fata morgana** is a complex mirage characterized by marked distortion, generally in the vertical. It may cause objects to appear towering, magnified, and at times even multiplied.

2.16. Sky coloring.—White light is composed of light of all colors. Color is related to wavelength, the visible spectrum varying from about 0.000038 to 0.000076 centimeters. The characteristics of each color are related to its wavelength (or frequency). Thus, the shorter the wavelength, the greater the amount of bending when light is refracted. It is this principle that permits the separation of light from celestial bodies into a **spectrum** ranging from red, through orange, yellow, green, and blue, to violet, with long-wave infrared (black light) being slightly outside the visible range at one end and short-wave ultraviolet being slightly outside the visible range at the other end. Light of shorter wavelength is scattered and diffracted more than that of longer wavelength.

Light from the sun and moon is white, containing all colors. As it enters the earth's atmosphere, a certain amount of it is scattered. The blue and violet, being of shorter wavelength than other colors, are scattered most. Most of the violet light is absorbed in the atmosphere. Thus, the scattered blue light is most apparent, and the sky appears blue. At great heights, above most of the atmosphere, it appears black.

When the sun is near the horizon, its light passes through more of the atmosphere than when higher in the sky, resulting in greater scattering and absorption of blue and green light, so that a larger percentage of the red and orange light penetrates to the observer. For this reason the sun and moon appear redder at this time, and when this light falls upon clouds, they appear colored. This accounts for the colors at sunset and sunrise. As the setting sun approaches the horizon, the sunset colors first appear as faint tints of yellow and orange. As the sun continues to set, the colors deepen. Contrasts occur, due principally to difference in height of clouds. As the sun sets, the clouds become a deeper red, first the lower clouds and then the higher ones, and finally they fade to a gray.

When there is a large quantity of smoke, dust, or other material in the sky, unusual effects may be observed. If the material in the atmosphere is of suitable substance and quantity to absorb the longer wave red, orange, and yellow radiations, the sky may have a greenish tint, and even the sun or moon may appear green. If the green light, too, is absorbed, the sun or moon may appear blue. A **green moon** or **blue moon** is most likely to occur when the sun is slightly below the horizon and the longer wavelength light from the sun is absorbed, resulting in green or blue light being cast upon the atmosphere in front of the moon. The effect is most apparent if the moon is on the same side of the sky as the sun.

2.17. Rainbows.—The familiar arc of concentric colored bands seen when the sun shines on rain, mist, spray, etc., is caused by refraction, internal reflection, and diffraction of sunlight by the drops of water. The center of the arc is a point 180° from the sun, in the direction of a line from the sun, through the observer. The radius of the brightest rainbow is 42°. The colors are visible because of the difference in the amount of refraction of the different colors making up white light, the light being spread out to form a spectrum (art. 2.16). Red is on the outer side and blue and violet on the inner side, with orange, yellow, and green between, in that order from red.

Sometimes a secondary rainbow is seen outside the primary one, at a radius of about 50°. The order of colors of this rainbow is reversed. On rare occasions a faint rainbow is seen on the same side as the sun. The radius of this rainbow and the order of colors are the same as those of the primary rainbow.

A similar arc formed by light from the moon (a lunar rainbow) is called a **moonbow.** The colors are usually very faint. A faint, white arc of about 39° radius is occasionally seen in fog opposite the sun. This is called a **fogbow,** although its origin is controversial, some considering it a halo (art. 2.18).

2.18. Halos.—Refraction, or a combination of refraction and reflection, of light by ice crystals in the atmosphere (cirrostratus clouds, art. 1.13) may cause a **halo** to appear. The most common form is a ring of light of radius 22° or 46° with the sun or moon at the center. Occasionally a faint, white circle with a radius of 90° appears around the sun. This is called a **Hevelian halo.** It is probably caused by refraction and internal reflection of the sun's light by bipyramidal ice crystals. A halo formed by refraction is usually faintly colored like a rainbow (art. 2.17), with red nearest the celestial body, and blue farthest from it.

A brilliant rainbow-colored arc of about a quarter of a circle with its center at the zenith, and the bottom of the arc about 46° above the sun, is called a **circumzenithal arc.** Red is on the outside of the arc, nearest the sun. It is produced by the refraction and dispersion of the sun's light striking the top of prismatic ice crystals in the atmosphere. It usually lasts for only about 5 minutes, but may be so brilliant as to be mistaken for an unusually bright rainbow. A similar arc formed 46° *below* the sun, with red on the upper side, is called a **circumhorizontal arc.** Any arc tangent to a heliocentric halo (one surrounding the sun) is called a **tangent arc.** As the sun increases in elevation, such arcs tangent to the halo of 22° gradually bend their ends toward each other. If they meet, the elongated curve enclosing the circular halo is called a **circumscribed halo.** The inner edge is red.

A halo consisting of a faint, white circle through the sun and parallel to the horizon is called a **parhelic circle.** A similar one through the moon is called a **paraselenic circle.** They are produced by reflection of sunlight or moonlight from vertical faces of ice crystals.

A **parhelion** (plural *parhelia*) is a form of halo consisting of an image of the sun at the same altitude and some distance from it, usually 22°, but occasionally 46°. A similar phenomenon occurring at an angular distance of 120° (sometimes 90° or 140°) from the sun is called a **paranthelion.** One at an angular distance of 180°, a rare occurrence, is called an **anthelion,** although this term is also used to refer to a luminous, colored ring or **glory** sometimes seen around the shadow of one's head on a cloud or fog bank. A parhelion is popularly called a **mock sun** or **sun dog.** Similar phenomena in relation to the moon are called **paraselene** (popularly a **mock moon** or **moon dog**), **parantiselene,** and **antiselene.** The term *parhelion* should not be confused with *perihelion*, that orbital point nearest the sun when the sun is the center of attraction.

A **sun pillar** is a glittering shaft of white or reddish light occasionally seen extending above and below the sun, usually when the sun is near the horizon. A phenomenon similar to a sun pillar, but observed in connection with the moon, is called a **moon pillar.** A rare form of halo in which horizontal and vertical shafts of light intersect at the sun is called a **sun cross.** It is probably due to the simultaneous occurrence of a sun pillar and a parhelic circle.

2.19. Corona.—When the sun or moon is seen through altostratus clouds (art. 1.13), its outline is indistinct, and it appears surrounded by a glow of light called a **corona**. This is somewhat similar in appearance to the corona seen around the sun during a solar eclipse. When the effect is due to clouds, however, the glow may be accompanied by one or more rainbow-colored rings of small radii, with the celestial body at the center. These can be distinguished from a halo by their much smaller radii and also by the fact that the order of the colors is reversed, red being on the inside, nearest the body, in the case of the halo, and on the outside, away from the body, in the case of the corona.

A corona is caused by diffraction of light by tiny droplets of water. The radius of a corona is inversely proportional to the size of the water droplets. A large corona indicates small droplets. If a corona decreases in size, the water droplets are becoming larger and the air more humid. This may be an indication of an approaching rainstorm.

The glow portion of a corona is called an **aureole**.

2.20. The green flash.—As light from the sun passes through the atmosphere, it is refracted. Since the amount of bending is slightly different for each color, separate images of the sun are formed in each color of the spectrum. The effect is similar to that of imperfect color printing in which the various colors are slightly out of register. However, the difference is so slight that the effect is not usually noticeable. At the horizon, where refraction is maximum, the greatest difference, which occurs between violet at one end of the spectrum and red at the other, is about 10 seconds of arc. At latitudes of the United States, about 0.7 second of time is needed for the sun to change altitude by this amount when it is near the horizon. The red image, being bent least by refraction, is first to set and last to rise. The shorter wave blue and violet colors are scattered most by the atmosphere, giving it its characteristic blue color (art. 2.16). Thus, as the sun sets, the green image may be the last of the colored images to drop out of sight. If the red, orange, and yellow images are below the horizon, and the blue and violet light is scattered and absorbed, the upper rim of the green image is the only part seen, and the sun appears green. This is the **green flash.** The shade of green varies, and occasionally the blue image is seen, either separately or following the green flash (at sunset). On rare occasions the violet image is also seen. These colors may also be seen at sunrise, but in reverse order. They are occasionally seen when the sun disappears behind a cloud or other obstruction.

The phenomenon is not observed at each sunrise or sunset, but under suitable conditions is far more common than generally supposed. Conditions favorable to observation of the green flash are a sharp horizon, clear atmosphere, a temperature inversion (art. 2.14), and an attentive observer. Since these conditions are more frequently met when the horizon is formed by the sea than by land, the phenomenon is more common at sea. With a sharp sea horizon and clear atmosphere, an attentive observer may see the green flash at as many as 50 percent of sunsets and sunrises, although a telescope may be needed for some of the observations.

Duration of the green flash (including the time of blue and violet flashes) of as long as 10 seconds has been reported, but such length is rare. Usually it lasts for a period of about ½ second to 2½ seconds with about 1¼ seconds being average. This variability is probably due primarily to changes in the index of refraction of the air near the horizon.

Under favorable conditions, a momentary green flash has been observed at the setting of Venus and Jupiter. A telescope improves the chances of seeing such a flash from a planet, but is not a necessity.

2.21. Crepuscular rays are beams of light from the sun passing through openings in the clouds, and made visible by illumination of dust in the atmosphere along their paths. Actually, the rays are virtually parallel, but because of perspective, appear to diverge. Those appearing to extend downward are popularly called **backstays of the sun,** or **sun drawing water.** Those extending upward and across the sky, appearing to converge toward a point 180° from the sun, are called **anticrepuscular rays.**

2.22. The atmosphere and radio waves.—Radio waves traveling through the atmosphere exhibit many of the properties of light, being refracted, reflected, diffracted, and scattered.

2.23. Atmospheric electricity.—Various conditions induce the formation of electrical charges in the atmosphere. When this occurs, there is often a difference of electron charge between various parts of the atmosphere, and between the atmosphere and earth or terrestrial objects. When this difference exceeds a certain minimum value, depending upon the conditions, the static electricity is discharged, resulting in phenomena such as lightning or St. Elmo's fire.

Lightning is the discharge of electricity from one part of a thundercloud (art. 1.13) to another, from one such cloud to another, or between such a cloud and the earth or a terrestrial object.

Enormous electrical stresses build up within thunderclouds, and between such clouds and the earth. At some point the resistance of the intervening air is overcome. At first the process is a progressive one, probably starting as a brush discharge (St. Elmo's fire) and growing by ionization. The breakdown follows an irregular path along the line of least resistance. A hundred or more individual discharges may be necessary to complete the path between points of opposite polarity. When this "leader stroke" reaches its destination, a heavy "main stroke" immediately follows in the opposite direction. This main stroke is the visible lightning, which may be tinted any color, depending upon the nature of the gases through which it passes. The illumination is due to the high degree of ionization of the air, which causes many of the atoms to be in excited states and emit radiation.

Thunder, the noise that accompanies lightning, is caused by the heating and ionizing of the air by lightning, which results in rapid expansion of the air along its path and the sending out of a compression wave. Thunder may be heard at a distance of as much as 15 miles, but generally does not carry that far. The elapsed time between the flash of lightning and reception of the accompanying sound of thunder is an indication of the distance, because of the difference in travel time of light and sound. Since the former is comparatively instantaneous, and the speed of sound is about 1,117 feet per second, the approximate distance in nautical miles is equal to the elapsed time in seconds, divided by 5.5. If the thunder accompanying lightning cannot be heard due to its distance, the lightning is called **heat lightning,** a phenomenon not unusual during continental "hot spells."

St. Elmo's fire is a luminous discharge of electricity from pointed objects such as the masts and yardarms of ships, lightning rods, steeples, mountain tops, blades of grass, human hair, arms, etc., when there is a considerable difference in the electrical charge between the object and the air. It appears most frequently during a storm. An object from which St. Elmo's fire emanates is in danger of being struck by lightning, since this type discharge may be the initial phase of the leader stroke. Throughout history those who have not understood St. Elmo's fire have regarded it with superstitious awe, considering it a supernatural manifestation. This view is reflected in the name **corposant** (from "corpo santo," meaning "body of a saint") sometimes given this phenomenon.

The **aurora** is a luminous glow appearing in varied forms in the thin atmosphere high above the earth in middle and high latitudes due to radiation emissions from gases in the high atmosphere.

2.24. Waterspouts.—A waterspout is a small, whirling storm over the ocean or inland waters. Its chief characteristic is a funnel-shaped cloud; when fully developed it extends from the surface of the water to the base of a cumulus type cloud (fig. 2.24). The water in a spout is mostly confined to its lower portion, and may be either salt spray drawn up by the sea surface, or freshwater resulting from condensation due to the lowered pressure in the center of the vortex creating the spout. The air in waterspouts may rotate clockwise or counterclockwise, depending on the manner of formation. They are found most frequently in tropical regions, but are not uncommon in higher latitudes.

There are two types of waterspouts: those derived from violent convective storms over land moving seaward, called **tornadoes,** and those formed over the sea and which are associated with fair or foul weather. The latter type is most common, lasts a maximum of 1 hour, and has variable strength. Many waterspouts are no stronger than dust whirlwinds, which they resemble; at other times they are strong enough to destroy small craft or to cause damage to larger vessels, although modern ocean-going vessels have little to fear from this type.

Waterspouts vary in diameter from a few feet to several hundred feet, and in height from a few hundred feet to several thousand feet. Sometimes they assume fantastic shapes; in early stages of development an hour glass shape between cloud and sea is common. Since a waterspout is often inclined to the vertical, its actual length may be much greater than indicated by its height.

2.25. Deck ice.—Ships traveling through regions where the air temperature is below freezing may acquire thick deposits of ice as a result of salt spray freezing on the rigging or deck areas (fig. 2.25). This accumulation of ice is called **ice accretion.** Also, precipitation may freeze to the superstructure and exposed areas of the vessel, increasing the load of ice.

On small vessels in heavy seas and freezing weather, deck ice may accumulate very rapidly and increase the topside weight to such an extent as to reduce seriously the stability of the vessel.

2.26. Forecasting weather.—The prediction of weather at some future time is based upon an understanding of weather processes, and observations of present conditions. Thus, one learns that when there is a certain sequence of cloud types (art. 1.13), rain usually can be expected to follow within a certain period. If the sky is cloudless, more heat will be received from the sun by day, and more heat will be radiated outward from the warm earth by night than if the sky is overcast. If the wind is in such a direction that warm, moist air will be transported over a colder surface, fog can be expected. A falling barometer indicates the approach of a "low," probably accompanied by stormy weather. Thus, before meteorology passed from an "art" to "science," many individuals learned to interpret certain atmospheric phenomena in terms of future weather, and to make reasonably accurate forecasts for short periods into the future.

With the establishment of weather observation stations, continuous and accurate weather information became available. As such observations expanded, and communication facilities improved, knowledge of simultaneous conditions over wider areas became available. This made possible the collection of these "synoptic" reports at civilian and military forecast centers.

The individual observations are made at government-operated stations on shore, and aboard vessels at sea. Observations aboard merchant ships at sea are made and transmitted on a voluntary and cooperative basis. The various national meteorological services supply shipmasters with blank forms, printed instructions, and other materials essential to the making, recording, and interpreting of observations. Any shipmaster

FIGURE 2.24.—Waterspouts.

can render a particularly valuable service by reporting all unusual or non-normal weather occurrences.

Symbols and numbers are used to indicate on a **synoptic chart,** popularly called a **weather map,** the conditions at each observation station. Isobars are drawn through lines of equal atmospheric pressure, fronts are located and symbolically marked (fig. 2.26), areas of precipitation and fog are indicated, etc.

Ordinarily, weather maps for surface observations are prepared every 6 (sometimes 3) hours. In addition, synoptic charts for selected heights are prepared every 12 (sometimes 6) hours. Knowledge of conditions aloft is of value in establishing the three-dimensional structure and motion of the atmosphere as input to the forecast.

FIGURE 2.25.—Deck ice.

With the advent of the digital computer, highly sophisticated numerical models have been developed to analyze and prognosticate weather patterns. The civil and military weather centers prepare and disseminate vast numbers of weather charts (analyses and prognoses) daily to assist local forecasters in their efforts to provide users with accurate, predicted weather parameters. It must be remembered that in any area, the accuracy of forecasted parameters decreases with the length of the forecast period. Thus, a 12-hour forecast is likely to be more reliable than a 24-hour forecast. Long term forecasts for 2 weeks or a month in advance are limited to general statements. For example, a prediction is made as to which areas will have temperatures above or below normal, and how precipitation will compare with normal, but no attempt is made to state that rainfall will occur at a certain time and place.

Forecasts are issued for various areas. The national meteorological services of most maritime nations, including the United States, issue forecasts for ocean areas and warnings of the approach of storms. The efforts of the various nations are coordinated through the World Meteorological Organization.

2.27. Dissemination of weather information is carried out in a number of ways. Forecasts are widely broadcast by commercial and government radio stations, and printed in newspapers. Shipping authorities on land are kept informed by telegraph and telephone. Visual storm warnings are displayed in various ports, and storm warnings are broadcast by radio.

Through the use of codes, a simplified version of synoptic weather charts is transmitted to various stations ashore and afloat. Rapid transmission of completed maps has been made possible by the development of facsimile transmitters and receivers. This system is based upon detailed scanning , by a photoelectric detector, of properly illuminated black and white copy. The varying degrees of light intensity are converted to electric energy which is transmitted to the receiver and converted back to a black and white presentation.

Complete information on dissemination of weather information by radio is provided in International Meteorological Codes 1972 and *World Wide Synoptic Broadcasts*,

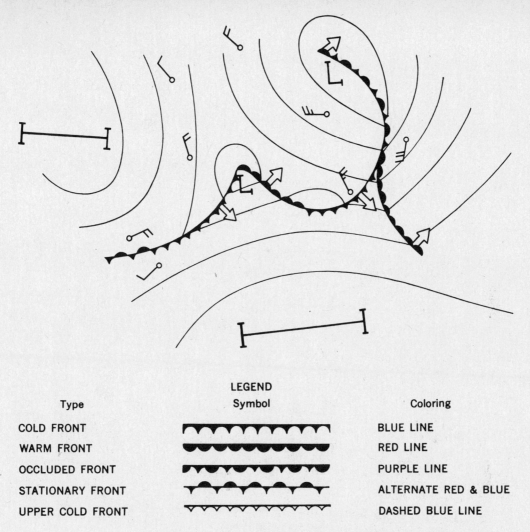

LEGEND

Type	Symbol	Coloring
COLD FRONT		BLUE LINE
WARM FRONT		RED LINE
OCCLUDED FRONT		PURPLE LINE
STATIONARY FRONT		ALTERNATE RED & BLUE
UPPER COLD FRONT		DASHED BLUE LINE

FIGURE 2.26.—Designation of fronts on weather maps.

NAVAIR 50–1P–11, published by The Naval Weather Service Command. This publication lists broadcast schedules and weather codes. Information on day and night visual storm warnings is given in the various volumes of sailing directions and coast pilots.

2.28. Interpreting the weather.—The factors which determine weather are numerous and varied. Ever-increasing knowledge regarding them makes possible a continually improving weather service. However, the ability to forecast is acquired through study and long practice, and therefore the services of a trained meteorologist should be utilized whenever available.

The value of a forecast is increased if one has access to the information upon which it is based, and understands the principles and processes involved. It is sometimes as important to know the various types of weather that *might* be experienced as it is to know which of several possibilities is *most likely* to occur.

At sea, reporting stations are unevenly distributed, sometimes leaving relatively large areas with incomplete reports, or none at all. Under these conditions, the locations of highs, lows, fronts, etc., are imperfectly known, and their very existence may even be in doubt. At such times the mariner who can interpret the observations made from his own vessel may be able to predict weather for the next several hours more reliably than a trained meteorologist some distance away with incomplete information for the area of concern.

Knowledge of the various relationships given in chapters I, II, and III is of value, but only the more elementary principles are presented. Further information can be obtained from meteorological publications such as those listed at the ends of the weather chapters. The information obtained from these references will provide a background for proper interpretation of individual weather experiences. If one uses every opportunity to observe and interpret weather sequences, knowledge, and skill can be developed that will serve as a valuable supplement to information given in weather broadcasts, or to supply information for areas not covered by such broadcasts.

2.29. Influencing the weather.—Meteorological activities are devoted primarily to understanding weather processes and predicting future weather. However, as knowledge regarding cause-and-effect relationships increases, the possibility of being able to induce certain results by artificially producing the necessary conditions becomes greater. The most promising results to date have been in inducing or increasing precipitation on a local scale by "seeding" supercooled clouds with powdered dry ice or silver iodide smoke. The effectiveness of this procedure on a larger scale is still controversial. Experiments in decreasing the intensity of severe tropical cyclones (i.e., hurricanes, typhoons), have been carried out but an operational method is still many years away.

References

American Meteorological Society. *Compendium of Meteorology*. Boston, American Meteorological Society, 1951.

American Meteorological Society. *Glossary of Meteorology*. Boston, American Meteorological Society, 1959 (with corrections, 1970).

Anderson, R. K. et al. *Applications of Meteorological Satellite Data in Analysis and Forecasting*, ESSA TR NESS51. Washington, D.C., National Oceanic and Atmospheric Administration, National Environmental Satellite Service, 1974.

Atkinson, G. D. *Forecasting Guide to Tropical Meteorology*, TR 240, Air Weather Service, U. S. Air Force, 1971.

Berry, F. A. Jr., E. Bollay, and N. R. Beers. *Handbook of Meteorology*. New York, McGraw-Hill, 1945.

Byers, H. R. *General Meteorology*, 4th ed. New York, McGraw-Hill, 1974.

Donn, W. L. *Meteorology*, 4th ed. New York, McGraw-Hill, 1975.

Haltiner, G. J., and F. L. Martin. *Dynamical and Physical Meteorology*, New York, McGraw-Hill, 1957.

Kotsch, W. J. *Weather for the Mariner*, Annapolis, Maryland, Naval Institute Press, 1970.

Neuberger, H., and J. Cahir. *Principles of Climatology, A Manual on Earth Science*. New York, Holt, Rinehart and Winston, Inc., 1969.

Petterssen, S. *Introduction to Meteorology*, 3rd ed. New York, McGraw-Hill, 1969.

Riehl, H. *Tropical Meteorology*. New York, McGraw-Hill, 1954.

Trewartha, G. T. *An Introduction to Climate*, 4th ed. New York, McGraw-Hill, 1968.

U. S. Department of Commerce. *Mariners Weather Log*. National Oceanic and Atmospheric Administration, Environmental Data Service. (periodical).

U. S. Department of the Navy. *Numerical Environmental Products Manual*. NAVAIR 50–1G–522. Naval Weather Service Command, 1975.

U. S. Department of the Navy. *U. S. Navy Marine Climatic Atlas of the World*, vol. VIII, The World. NAVAIR 50–1C–54, Naval Weather Service Command, 1969.

U. S. Department of the Navy. *Aerographer's Mate 1 and C*, NAVEDTRA 10362–B. Naval Education and Training Command, 1974.

U. S. Department of the Navy. *Aerographer's Mate 3 and 2*, NAVEDTRA 10363–E, 1976.

Williams, J., J. J. Higginson, and J. D. Rohrbough. *Sea and Air, The Marine Environment*. Annapolis, Maryland, Naval Institute Press, 1973.

World Meteorological Organization. *The Preparation and Use of Weather Maps by Mariners*, TN72. Secretariat, World Meteorological Organization, Geneva, Switzerland, 1966.

CHAPTER III

TROPICAL CYCLONES

3.1. Introduction.—A tropical cyclone is a cyclone (art. 2.13) originating in the Tropics or subtropics. Although it generally resembles the extratropical cyclone originating in higher latitudes, there are important differences, the principal one being the concentration of a large amount of energy into a relatively small area. Tropical cyclones are infrequent in comparison with middle- and high-latitude storms, but they have a record of destruction far exceeding that of any other type of storm. Because of their fury, and the fact that they are predominantly oceanic, they merit the special attention of all mariners, whether professional or amateur.

Rarely does the mariner who has experienced a fully developed tropical cyclone at sea wish to encounter a second one. He has learned the wisdom of avoiding them if possible. The uninitiated may be misled by the deceptively small size of a tropical cyclone as it appears on a weather map, and by the fine weather experienced only a few hundred miles from the reported center of such a storm. The rapidity with which the weather can deteriorate with approach of the storm, and the violence of the fully developed tropical cyclone, are difficult to visualize if they have not been experienced.

On his second voyage to the New World, Columbus encountered a tropical storm. Although his vessels suffered no damage, this experience proved valuable during his fourth voyage when his vessels were threatened by a fully developed hurricane. Columbus read the signs of an approaching storm from the appearance of a southeasterly swell, the direction of the high cirrus clouds, and the hazy appearance of the atmosphere. He directed his vessels to shelter. The commander of another group, who did not heed the signs, lost most of his ships; more than 500 men in their crews perished.

3.2. Definitions.—Tropical cyclone is the general term for cyclones originating in the Tropics or subtropics. These cyclones are classified by form and intensity as follows:

Tropical disturbance is a discrete system of apparently organized convection—generally 100 to 300 miles in diameter—having a nonfrontal migratory character, and having maintained its identity for 24 hours or more. It may or may not be associated with a detectable perturbation of the wind field. It has no strong winds and no closed isobars i.e., isobars that completely enclose the low. (In successive stages of intensification, the tropical cyclone may be classified as a tropical disturbance, tropical depression, tropical storm, and hurricane or typhoon.)

Tropical depression has one or more closed isobars and some rotary circulation at the surface. The highest sustained (1-minute mean) surface wind speed is 33 knots.

Tropical storm has closed isobars and a distinct rotary circulation. The highest sustained (1-minute mean) surface wind speed is 34 to 63 knots.

Hurricane or **typhoon** has closed isobars, a strong and very pronounced rotary circulation, and a sustained (1-minute mean) surface wind speed of 64 knots or higher.

3.3. Areas of occurence.—Tropical cyclones occur almost entirely in six rather distinct areas, four in the Northern Hemisphere and two in the Southern Hemisphere as shown in figure 3.3. The name by which the tropical cyclone is commonly known varies somewhat with the locality, as follows:

North Atlantic. A tropical cyclone with winds of 64 knots or greater is called a **hurricane.**

Eastern North Pacific. The name **hurricane** is used as in the North Atlantic.

Western North Pacific. A fully developed storm with winds of 64 knots or greater is called a **typhoon** or, locally in the Philippines, a **baguio.**

North Indian Ocean. A tropical cyclone with winds of 34 knots or greater is called a **cyclonic storm.**

South Indian Ocean. A tropical cyclone with winds of 34 knots or greater is called a **cyclone.**

Southwest Pacific and Australian Area. The name **cyclone** is used as in the South Indian Ocean. A severe tropical cyclone originating in the Timor Sea and moving southwest and then southeast across the interior of northwestern Australia is called a **willy-willy.**

Tropical cyclones have not been observed in the South Atlantic or in the South Pacific east of 140°W.

3.4. Origin, season, and frequency of occurence of the tropical cyclones in the six areas are as follows:

North Atlantic tropical cyclones can affect the entire North Atlantic Ocean in any month. However, they are mostly a threat south of about 35°N from June through November; August, September, and October are the months of highest incidence (tab. 3.4) about 9 or 10 tropical cyclones (tropical storms and hurricanes form each season; 5 or 6 reach hurricane intensity (winds of 64 knots and higher). A few hurricanes have generated winds estimated as high as 200 knots. Early- and late-season storms usually develop west of 50°W; during August and September, this spawning ground extends to the Cape Verde Islands. These storms usually move westward or westnorthwestward at speeds of less than 15 knots in the lower latitudes. After moving into the northern Caribbean or Greater Antilles regions, they will usually either move toward the Gulf of Mexico or recurve and accelerate in the North Atlantic. Some will recurve after reaching the Gulf of Mexico, while others will continue westward to landfall (fig. 3.4).

Eastern North Pacific season is from June through October, although a storm can form in any month. An average of 15 tropical cyclones (tropical storms and hurricanes) form each year with about 6 reaching hurricane strength. The most intense storms are often the early-and late-season ones; these form close to the coast and far south. Midseason storms form anywhere in a wide band from the Mexican-Central American coast to the Hawaiian Islands. August and September are the months of highest incidence. These storms differ from their North Atlantic counterparts in that they are usually smaller in size. However, they can be just as intense.

Western North Pacific. More tropical cyclones form in the tropical western North Pacific than anywhere else in the world. More than 25 (tropical storms and typhoons) develop each year, and about 18 become typhoons. These typhoons are the largest and most intense tropical cyclones in the world. Each year an average of five generate maximum winds over 130 knots; circulations covering more than 600 miles in diameter are not uncommon. Most of these storms form east of the Philippines, and move across the Pacific toward the Philippines, Japan, and China; a few storms form in the South China Sea. The season extends from April through December. However, tropical cyclones are more common in the off-season months in this area than anywhere else. The peak of the season is July through October, when nearly 70 percent of all typhoons develop. There is a noticeable seasonal shift in storm tracks in this region. From July

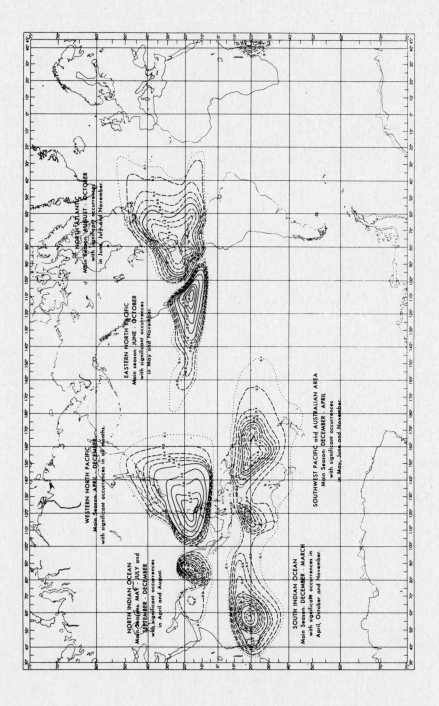

FIGURE 3.3.—Areas in which tropical cyclones occur. The average number of tropical cyclones per 5° square has been analyzed for this figure. The main season for intense tropical storm activity is also shown for each major basin.

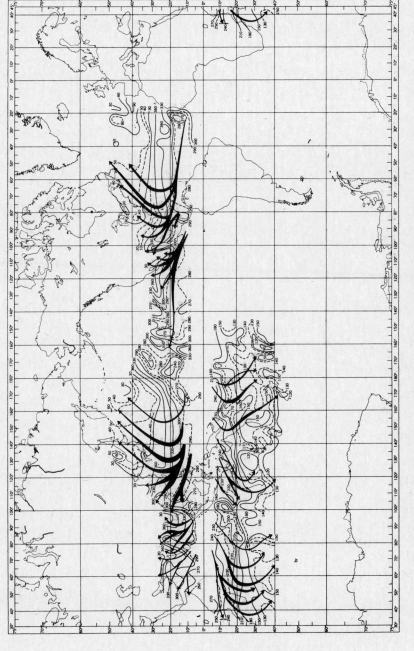

FIGURE 3.4.—Storm tracks. The width of the arrow indicates the approximate frequency of storms; the wider the arrow the higher the frequency. Isolines on the base map show the resultant direction toward which storms moved. Data for the entire year has been summarized for this figure.

AREA AND STAGE	JAN	FEB	MAR	APR	MAY	JUN	JUL	AUG	SEP	OCT	NOV	DEC	ANNUAL
NORTH ATLANTIC													
TROPICAL STORMS	*	*	*	*	0.1	0.4	0.3	1.0	1.5	1.2	0.4	*	4.2
HURRICANES	*	*	*	*	*	0.3	0.4	1.5	2.7	1.3	0.3	*	5.2
TROPICAL STORMS AND HURRICANES	*	*	*	*	0.2	0.7	0.8	2.5	4.3	2.5	0.7	0.1	9.4
EASTERN NORTH PACIFIC													
TROPICAL STORMS	*	*	*	*	*	1.5	2.8	2.3	2.3	1.2	0.3	*	9.3
HURRICANES	*	*	*	*	0.3	0.6	0.9	2.0	1.8	1.0	*	*	5.8
TROPICAL STORMS AND HURRICANES	*	*	*	*	0.3	2.0	3.6	4.5	4.1	2.2	0.3	*	15.2
WESTERN NORTH PACIFIC													
TROPICAL STORMS	0.2	0.3	0.3	0.2	0.4	0.5	1.2	1.8	1.5	1.0	0.8	0.6	7.5
TYPHOONS	0.3	0.2	0.2	0.7	0.9	1.2	2.7	4.0	4.1	3.3	2.1	0.7	17.8
TROPICAL STORMS AND TYPHOONS	0.4	0.4	0.5	0.9	1.3	1.8	3.9	5.8	5.6	4.3	2.9	1.3	25.3
SOUTHWEST PACIFIC AND AUSTRALIAN AREA													
TROPICAL STORMS	2.7	2.8	2.4	1.3	0.3	0.2	*	*	*	0.1	0.4	1.5	10.9
HURRICANES	0.7	1.1	1.3	0.3	*	*	0.1	0.1	*	*	0.3	0.5	3.8
TROPICAL STORMS AND HURRICANES	3.4	4.1	3.7	1.7	0.3	0.2	0.1	0.1	*	0.1	0.7	2.0	14.8
SOUTHWEST INDIAN OCEAN													
TROPICAL STORMS	2.0	2.2	1.7	0.6	0.2	*	*	*	*	0.3	0.3	0.8	7.4
HURRICANES	1.3	1.1	0.8	0.4	*	*	*	*	*	*	*	0.5	3.8
TROPICAL STORMS AND HURRICANES	3.2	3.3	2.5	1.1	0.2	*	*	*	*	0.3	0.4	1.4	11.2
NORTH INDIAN OCEAN													
TROPICAL STORMS	0.1	*	*	0.1	0.3	0.5	0.5	0.4	0.4	0.6	0.5	0.3	3.5
CYCLONES [1]	*	*	*	0.1	0.5	0.2	0.1	*	0.1	0.4	0.6	0.2	2.2
TROPICAL STORMS AND CYCLONES [1]	0.1	*	0.1	0.3	0.7	0.7	0.6	0.4	0.5	1.0	1.1	0.5	5.7

* Less than .05   [1] Winds ⩾ 48 Kts.

Monthly values cannot be combined because single storms overlapping two months were counted once in each month and once in the annual.

TABLE 3.4.—Monthly and annual average number of storms per year for each area.

through September, storms move north of the Philippines and recurve, while early- and late-season typhoons move on a more westerly track through the Philippines before recurving (fig. 3.4).

North Indian Ocean tropical cyclones develop in the Bay of Bengal and Arabian Sea during the spring and fall. Tropical cyclones in this area form between latitudes 8°N and 15°N, except from June through September, when the little activity that does occur is confined north of about 15°N. These storms are usually short-lived and weak; however, winds of 130 knots have been encountered. They often develop as perturbations along the Intertropical Convergence Zone (ITCZ); this inhibits summertime development since the ITCZ (art. 3.5) is usually over land during this monsoon season. However, it is sometimes displaced southward, and when this occurs, storms will form over the monsoon-flooded plains of Bengal. On the average, six cyclonic storms form each year. These include two storms that generate winds of 48 knots or greater. Another 10 tropical cyclones never develop beyond tropical depressions. The Bay of Bengal is the area of highest incidence. However, it is not unusual for a storm to move across southern India and reintensify in the Arabian Sea. This is particularly true during October— the month of highest incidence during the tropical cyclone season. It is also during this period that torrential rains from these storms dumped over already rain-soaked areas cause disastrous floods.

South Indian Ocean. Over the waters west of 100°E to the east African coast, an average of 11 tropical cyclones (tropical storms and hurricanes) form each season, and about 4 reach hurricane intensity. The season is from December through March, although it is possible for a storm to form in any month. Tropical cyclones in this region usually form south of 10°S. The latitude of recurvature usually migrates from about 20°S in January to around 15°S in April. After crossing 30°S, these storms sometimes become intense extratropical lows.

Southwest Pacific and Australian Area. These tropical waters spawn an annual average of 15 tropical cyclones (tropical storms and hurricanes), 4 of which reach hurricane intensity. The season extends from about December through April, although storms can form in any month. Activity is widespread in January and February, and it is in these months that tropical cyclones are most likely to affect Fiji, Samoa, and the other eastern islands.

Tropical cyclones usually form in the waters from 105°E to 160°W, between 5° and 20°S. Storms affecting northern and western Australia often develop in the Timor or Arafura Sea, while those that affect the east coast form in the Coral Sea. These storms are often small, but can develop winds in excess of 130 knots. New Zealand is sometimes reached by decaying Coral Sea storms; occasionally, it is reached by an intense hurricane. In general, tropical cyclones in this region move southwestward and then recurve southeastward (fig. 3.4).

3.5. Hurricane formation was once believed to result from an intensification of convective forces which produce the cumulonimbus towers of the doldrums. This view of hurricane generation held that surface heating caused warm moist air to ascend convectively to levels where condensation produced cumulonimbus clouds, which, after an inexplicable drop in atmospheric pressure, coalesced and were spun into a cyclonic motion by Coriolis force.

This hypothesis left much to be desired. Although some hurricanes develop from disturbances beginning in the doldrums (art. 2.5), very few reach maturity in that region. Also, the high incidence of seemingly ideal convective situations does not match the low incidence of Atlantic hurricanes. Finally, the hypothesis did not explain the drop in atmospheric pressure, so essential to development of hurricane-force winds.

There is still no exact understanding of the triggering mechanism involved in hurricane generation, the balance of conditions needed to generate hurricane circulation, and the relationships between large- and small-scale atmospheric processes. But scientists today, treating the hurricane system as an atmospheric heat engine, present a more comprehensive and convincing view.

They begin with a starter mechanism in which either internal or external forces intensify the initial disturbance. The initial disturbance becomes a region into which low-level air from the surrounding area begins to flow, accelerating the convection already occurring inside the disturbance. The vertical circulation becomes increasingly well organized as water vapor in the ascending moist layer is condensed (releasing large amounts of heat energy to drive the wind system) and as the system is swept into a counterclockwise cyclonic spiral. But this incipient hurricane would soon fill up because of inflow at lower levels unless the chimney in which converging air surges upward is provided the exhaust mechanism of high-altitude winds.

These high-altitude winds (fig. 3.5) pump ascending air out of the cyclonic system into a high-altitude anticyclone, which transports the air well away from the disturbance before sinking occurs. Thus, a large scale vertical circulation is set up in which low-level air is spiraled up the cyclonic twisting of the disturbance, and, after a trajectory over the sea, returned to lower altitudes some distance from the storm. This pumping action—and the heat released by the ascending air—may account for the sudden drop of atmospheric pressure at the surface, which produces the steep pressure gradient along which winds reach hurricane proportions.

It is believed that the interaction of low-level and high-altitude wind systems determines the intensity the hurricane will attain. If less air is pumped out than converges at low levels, the system will fill and die out. If more is pumped out than flows in, the circulation will be sustained and will intensify.

Research has shown that any process which increases the rate of low-level inflow is favorable for hurricane development, provided the inflowing air carries sufficient heat and moisture to fuel the hurricane's power system. It has also been shown that air above the developing disturbance at altitudes between 20,000 and 40,000 feet increases 1° to 3° in temperature about 24 hours before the disturbance develops into a hurricane. But it is not known whether low-level inflow and high-level warming *cause* hurricanes. They could very well be measurable symptoms of another effect which actually triggers the storm's increase to hurricane intensity.

The view of hurricanes as atmospheric engines is necessarily a general one. The exact role of each contributor is not completely understood. The engine seems to be both inefficient and unreliable; a myriad of delicate conditions must be satisfied for the atmosphere to produce a hurricane. Their relative infrequency indicates that many a potentially healthy hurricane ends early as a misfiring dud of a disturbance, somewhere over the sea.

3.6. Portrait of a hurricane.—In the early life of the hurricane, the spiral covers an area averaging 100 miles in diameter with winds of 64 knots and greater, and spreads gale-force winds over a 400-mile diameter. The cyclonic spiral (fig. 3.6) is marked by heavy cloud bands from which torrential rains fall, separated by areas of light rain or no rain at all. These spiral bands ascend in decks of cumulus and cumulonimbus clouds to the convective limit of cloud formation, where condensing water vapor is swept off as ice-crystal wisps of cirrus clouds. Thunderstorm electrical activity is observed in these bands, both as lightning and as tiny electrostatic discharge.

In the lower few thousand feet, air flows in through the cyclone, and is drawn upward through ascending columns of air near the center. The size and intensity decrease with altitude, the cyclonic circulation being gradually replaced above 40,000 feet by an anticyclonic circulation centered hundreds of miles away—the enormous high-altitude pump which is the exhaust system of the hurricane heat engine.

At lower levels, where the hurricane is more intense, winds on the rim of the storm follow a wide pattern, like the slower currents around the edge of a whirlpool; and, like those currents, these winds accelerate as they approach the center of the vortex. The

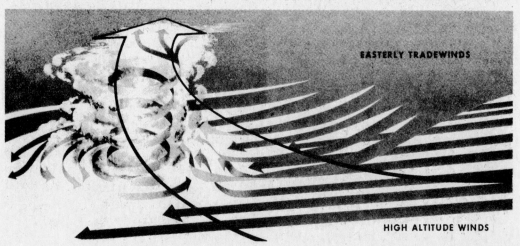

FIGURE 3.5.—Pumping action of high altitude winds.

outer band has light winds at the rim of the storm, perhaps no more than 25 knots; within 30 miles of the center, winds may have velocities exceeding 130 knots. The inner band is the region of maximum wind velocity, where the storm's worst winds are felt, and where ascending air is chimneyed upward, releasing heat to drive the storm. In most hurricanes, these winds reach 85 knots and more than 170 knots in the more memorable ones.

In the hurricane, winds flow toward the low pressure in the warm, comparatively calm core. There, converging air is whirled upward by convection, the mechanical thrusting of other converging air, and the pumping action of high-altitude circulations. This spiral is marked by the thick cloud walls curling inward toward the storm center, releasing heavy precipitation and enormous quantities of heat energy. At the center, surrounded by a band in which this strong vertical circulation is greatest, is the **eye** of the hurricane.

The eye, like the spiral rainbands, is unique to the hurricane; no other atmospheric phenomenon has this calm core. On the average, eye diameter is about 14 miles, although diameters of 25 miles are not unusual. From the heated tower of maximum winds and cumulonimbus clouds, winds diminish rapidly to something less than 15 miles per hour in the eye; at the opposite wall, winds increase again, but come from the opposite direction because of the cyclonic circulation of the storm. This transformation of storm into comparative calm, and calm into violence from another quarter is spectacular. The eye's abrupt existence in the midst of opaque rainsqualls and hurricane winds, the intermittent bursts of blue sky and sunlight through light clouds in the core of the cyclone, and the galleried cumulus and cumulonimbus clouds are unforgettable.

That is how an average hurricane is structured. But every hurricane is individual, and the more or less orderly circulation described here omits the extreme variability and instability within the storm system. Pressure and temperature gradients fluctuate wildly across the storm as the hurricane maintains its erratic life in the face of forces

FIGURE 3.6.—Cutaway view of a hurricane greatly exaggerated in vertical dimension. Actual hurricanes are less than 50,000 feet high and may have a diameter of several hundred miles.

which will ultimately destroy it. If it is an August storm, its average life expectancy is 12 days; if a July or November storm, it lives an average of 8 days.

3.7. Life of a tropical cyclone.—Reports from ships in the vicinity of an **easterly wave** (a westward-moving trough of low pressure embedded in deep easterlies) indicate that the atmospheric pressure in the region has fallen more than 5 millibars in the past 24 hours. This is cause for alarm because in the Tropics pressure varies little; the normal diurnal pressure change is only about 3 millibars. Satellite pictures indicate thickening middle and high clouds, squalls are reported ahead of the easterly wave, and wind reports indicate a cyclonic circulation is forming. The former easterly wave—now classified a *tropical disturbance*—is moving westward at 10 knots under the canopy of a large high-pressure system aloft. Sea surface temperatures in the vicinity are in the mid-80°F range.

Within 48 hours winds increase to 25 knots near the center of definite circulation, and central pressure has dropped below 1000 millibars. The disturbance is now classified as a *tropical depression*. Soon the circulation extends out to 100 miles and upward to 20,000 feet. Winds near the center increase to gale force, central pressure falls below 990 millibars, and towering cumulonimbus clouds shield a developing eye; a *tropical storm* has developed.

Satellite photographs now reveal a tightly organized tropical cyclone, and reconnaissance reports indicate maximum winds of 80 knots around a central pressure of 980 millibars; a *hurricane* has developed. A ship to the right (left in the Southern Hemisphere) of the hurricane's center (looking toward the direction of storm movement) reports a 30-foot sea. The hurricane is fast maturing; it continues eastward.

A few days later the hurricane reaches its peak. The satellite photographs a text-book picture (fig. 3.7), as 120-knot winds roar around a 940-millibar pressure center; hurricane-force winds extend 50 miles in all directions, and seas are reported up to 40 feet. There is no further deepening now, but the hurricane begins to expand. In 2 days, gales extend out to 200 miles, and hurricane winds out to 75 miles. Then the hurricane slows and begins to recurve; this turning marks the beginning of the end.

The hurricane accelerates, and, upon reaching the temperate latitudes, it begins to lose its tropical characteristics. The circulation continues to expand, but now cold air is intruding (cold air, cold water, dry air aloft, and land aid in the decay of a tropical cyclone). The warm core survives for a few more days before the transformation to a large extratropical low-pressure system is complete.

Not all tropical cyclones follow this ideal pattern. Most falter in the early stages, some dissipate over land, and others remain potent for several weeks.

The lowest-sea-level pressure ever recorded was 877 millibars in typhoon Ida, on September 24, 1958. The observation was taken by a reconnaissance aircraft dropsonde some 750 miles east of Luzon, Philippines. This observation was obtained again in typhoon Nora on October 6, 1973. The lowest barometric reading of record for the United States is 892.3 millibars obtained during a hurricane at Lower Matecumbe Key, Florida in September 1935. In hurricane Camille in 1969, a 905 millibar pressure was measured by reconnaissance aircraft. During a 1927 typhoon, the S. S. *Sapoeroea* recorded a pressure of 886.6 millibars, the lowest sea-level pressure reported from a ship. Pressure has been observed to drop more than 33 millibars per hour, with a pressure gradient amounting to a change of 3.7 millibars per mile.

3.8. The marine weather broadcast is the most important tool the mariner has for avoiding the tropical cyclone. This broadcast, covering all tropical areas, provides information about the tropical cyclone's present location, maximum winds and seas, and future condition.

FIGURE 3.7.—Satellite photograph of hurricane.

The U. S. Navy, the National Oceanic and Atmospheric Administration, and the U. S. Air Force have developed a highly effective surveillance system for the tropical cyclone areas of the world. Routine and special weather reports (from land stations, ships at sea, aircraft; daytime weather satellite reports; radar reports from land stations; special reports from ships at sea; and the specially instrumented weather reconnaissance aircraft of National Oceanic and Atmospheric Administration and the U. S. Air Force) enable accurate detection, location, and tracking of tropical cyclones. International cooperation is good. In addition to improved satellites permitting nighttime surveillance, data buoys provide another new source of information for the protection of the mariner.

The tropical warning services have three principal functions:

1. the collection and analysis of the necessary observational data;

2. the preparation of timely and accurate forecasts and warnings; and

3. the rapid and efficient distribution of advisories, warnings, and all other pertinent information.

To provide timely and accurate information and warnings regarding tropical cyclones, the oceans have been divided into overlapping geographical areas of responsibility.

For detailed information on the areas of responsibility of the countries participating in the international forecasting and warning program, and radio aids, refer to *Worldwide Marine Weather Broadcasts*, published jointly by the Naval Weather Service Command and the National Weather Service.

Although the areas of forecasting responsibility are fairly well defined for the Department of Defense, the international and domestic civilian system provides many overlaps and is dependent upon qualitative factors. For example, when a tropical storm or hurricane is traveling westward and crosses 35°W longitude, the continued issuance of forecasts and warnings to the general public, shipping interests, etc., becomes the responsibility of the National Hurricane Center of the National Weather Service at Miami, Florida. When a tropical storm or hurricane crosses 35°W longitude traveling from west to east, the National Hurricane Center ceases to issue formal public advisories, but will issue marine bulletins on any dangerous tropical cyclone in the North Atlantic, if it is of importance or constitutes a threat to shipping and other interests. These advisories are included in National Weather Service Marine Bulletins broadcast to ships four times daily at 0030, 0600, 1230, and 1830 GMT, over radio station NAM Norfolk, Virginia. Special advisories may be issued at any time.

In the eastern Pacific (east of longitude 140°W), responsibility for the issuance of tropical storm and hurricane advisories and warnings for the general public, merchant shipping, and other interests rests with the National Weather Service Eastern Pacific Hurricane Center, San Francisco, California. The Department of Defense responsibility rests with the U. S. Navy's Fleet Weather Central, Pearl Harbor, Hawaii. Formal advisories and warnings are issued at 0300, 0900, 1500, and 2100 GMT, and are included in the marine bulletins broadcast by radio stations KPH, KMI, KFS, NMC, ELH, DOE, NMQ, and KOU.

In the central Pacific (between the 180th meridian and longitude 140°W), the civilian responsibility rests with the National Weather Service Central Pacific Hurricane Center, Honolulu Hawaii. Department of Defense responsibility rests with the U. S. Navy's Fleet Weather Central, Pearl Harbor. Formal tropical storm and hurricane advisories and warnings are issued at 0300, 0900, 1500, and 2100 GMT, and are included in the marine bulletins broadcast by radio station KHK.

Tropical cyclone information messages generally contain position of the storm, intensity, direction and speed of movement, and a description of the area of strong winds. Also included is a forecast of future movement and intensity. When the storm is likely to affect any land area, details on when and where it will be felt, and data on tides, rain, floods, and maximum winds are also included. Figure 3.8 provides an example of a marine advisory issued by the National Hurricane Center.

The U. S. Navy's Fleet Weather Central in Guam, with its built-in Joint (Navy and Air Force) Typhoon Warning Center (JTWC), has a primary area of responsibility for all U. S. tropical storm and typhoon advisories and warnings from the 180th meridian westward to the mainland of Asia. A secondary area of responsibility extends westward to longitude 90°E. Whenever a tropical cyclone is observed in the western North

NOAA/NATIONAL HURRICANE CENTER MARINE ADVISORY NUM-
BER 13 HURRICANE LADY 0400Z SEPTEMBER 21 19--.

HURRICANE WARNINGS ARE DISPLAYED FROM KEY LARGO TO
CAPE KENNEDY. GALE WARNINGS ARE DISPLAYED FROM KEY
WEST TO JACKSONVILLE AND FROM FLORIDAY BAY TO CEDAR KEY.

HURRICANE CENTER LOCATED NEAR LATITUDE 25.5 NORTH
LONGITUDE 78.5 WEST AT 21/0400Z. POSITION EXCELLENT AC-
CURATE WITHIN 10 MILES BASED ON AIR FORCE RECONNAISSANCE
AND SYNOPTIC REPORTS.

PRESENT MOVEMENT TOWARD THE WEST NORTHWEST OR 285
DEGREES AT 10 KT. MAX SUSTAINED WINDS OF 100 KT NEAR
CENTER WITH GUSTS TO 160 KT.
MAX WINDS OVER INLAND AREAS 35 KT.
RAD OF 65 KT WINDS 90 NE 60 SE 80 SW 90 NW QUAD.
RAD OF 50 KT WINDS 120 NE 70 SE 90 SW 120 NW QUAD.
RAD OF 30 KT WINDS 210 NE 210 SE 210 SW 210 NW QUAD.
REPEAT CENTER LOCATED 25.5N 78.3W at 21/0400Z.

12 HOUR FORECAST VALID 21/1600Z LATITUDE 26.0N LONGI-
TUDE 80.5W.
MAX WINDS OF 100 KT NEAR CENTER WITH GUSTS TO 160 KT.
MAX WINDS OVER INLAND AREAS 65 KT.
RADIUS OF 50KT WINDS 120 NE 70 SE 90 SW 120 NW QUAD.
24 HOUR FORECAST VALID 22/0400Z LATITUDE 26.0N
LONGITUDE 83.0W.
MAX WINDS OF 75 KT NEAR CENTER WITH GUSTS TO 120 KT.
MAX WINDS OVER INLAND AREAS 45 KT.
RADIUS OF 50 KT WINDS 120 NE 120 SE 120 SW 120 NW QUAD.

STORM TIDE OF 9 TO 12 FT SOUTHEAST FLA COAST GREATER
MIAMI AREA TO THE PALM BEACHES.

NEXT ADVISORY AT 21/1000Z.

FIGURE 3.8.—Example of maritime advisory issued by National Hurricane Center.

Pacific area, serially numbered warnings, bearing an immediate precedence are broadcast from the Fleet Weather Central/JTWC at 0000, 0600, 1200, and 1800 GMT.

The responsibility for issuing gale and storm warnings for the Indian Ocean,
Arabian Sea, Bay of Bengal, Western Pacific, and South Pacific rests with many
countries. In general, warnings of approaching tropical cyclones which may be hazardous will include the following information: storm type, central pressure given in millibars, windspeed observed within the storm, storm location, speed and direction of
movement, the extent of the affected area, visibility, and the state of the sea, as well
as any other pertinent information received. All storm warning messages commence
with the international call sign "TTT."

These warnings are broadcast on prespecified radio frequency bands immediately
upon receipt of the information and at specific intervals thereafter. Generally, the
broadcast interval is every 6 to 8 hours, depending upon receipt of new information.

Bulletins and forecasts are excellent guides to the present and future behavior of
the tropical cyclone, and a plot should be kept of all positions.

3.9. The passage of a tropical cyclone at sea is an experience not soon to be
forgotten.

An early indication of the approach of such a storm is the presence of a long swell.
In the absence of a tropical cyclone, the crests of swell in the deep waters of the Atlantic
pass at the rate of perhaps eight per minute. Swell generated by a hurricane is about

twice as long, the crests passing at the rate of perhaps four per minute. Swell may be observed several days before arrival of the storm.

When the storm center is 500 to 1,000 miles away, the barometer usually rises a little, and the skies are relatively clear. Cumulus clouds, if present at all, are few in number and their vertical development appears suppressed. The barometer usually appears restless, **pumping** up and down a few hundredths of an inch.

As the tropical cyclone comes nearer, a cloud sequence begins which resembles that associated with the approach of a warm front in middle latitudes (art. 2.11). Snow-white, fibrous "mare's tails" (cirrus) appear when the storm is about 300 to 600 miles away. Usually these seem to converge, more or less, in the direction from which the storm is approaching. This convergence is particularly apparent at about the time of sunrise and sunset.

Shortly after the cirrus appears, but sometimes before, the barometer starts a long, slow fall. At first the fall is so gradual that it only appears to alter somewhat the normal daily cycle (two maxima and two minima in the Tropics). As the rate of fall increases, the daily pattern is completely lost in the more or less steady fall.

The cirrus becomes more confused and tangled, and then gradually gives way to a continuous veil of cirrostratus. Below this veil, altostratus forms, and then stratocumulus (art. 1.13). These clouds gradually become more dense, and as they do so, the weather becomes unsettled. A fine, mist-like rain begins to fall, interrupted from time to time by showers. The barometer has fallen perhaps a tenth of an inch.

As the fall becomes more rapid, the wind increases in gustiness, and its speed becomes greater, reaching a value of perhaps 22 to 40 knots (Beaufort 6–8). On the horizon appears a dark wall of heavy cumulonimbus (art. 1.13), the **bar** of the storm. Portions of this heavy cloud become detached from time to time and drift across the sky, accompanied by rainsqualls and wind of increasing speed. Between squalls, the cirrostratus can be seen through breaks in the stratocumulus.

As the bar approaches, the barometer falls more rapidly and wind speed increases. The seas, which have been gradually mounting, become tempestuous. Squall lines, one after the other, sweep past in ever increasing number and intensity.

With the arrival of the bar, the day becomes very dark, squalls become virtually continuous, and the barometer falls precipitously, with a rapid increase in wind speed. The center may still be 100 to 200 miles away in a fully developed tropical cyclone. As the center of the storm comes closer, the ever-stronger wind shrieks through the rigging and about the superstructure of the vessel. As the center approaches, rain falls in torrents. The wind fury increases. The seas become mountainous. The tops of huge waves are blown off to mingle with the rain and fill the air with water. Objects at a short distance are not visible. Even the largest and most seaworthy vessels become virtually unmanageable, and may sustain heavy damage. Less sturdy vessels may not survive. Navigation virtually stops as safety of the vessel becomes the prime consideration. The awesome fury of this condition can only be experienced. Words are inadequate to describe it.

If the eye of the storm passes over the vessel, the winds suddenly drop to a breeze as the wall of the eye passes. The rain stops, and the skies clear sufficiently to permit the sun to shine through holes in the comparatively thin cloud cover. Visibility improves. Mountainous seas approach from all sides, apparently in complete confusion. The barometer reaches its lowest point, which may be 1½ or 2 inches below normal in fully developed tropical cyclones. As the wall on the opposite side of the eye arrives, the full fury of the wind strikes as suddenly as it ceased, but from the opposite direction. The sequence of conditions that occurred during approach of the storm is reversed, and pass more quickly, as the various parts of the storm are not as wide in the rear of a storm as on its forward side.

Typical cloud formations associated with a hurricane are shown in figure 3.9.

3.10. Locating the center of a tropical cyclone.—If intelligent action is to be taken to avoid the full fury of a tropical cyclone, early determination of its location and direction of travel relative to the vessel is essential. The bulletins and forecasts are an excellent general guide, but they are not infallible and may be sufficiently in error to induce a mariner in a critical position to alter course so as to unwittingly increase the danger to his vessel. Often it is possible, using only those observations made aboard ship, to obtain a sufficiently close approximation to enable the vessel to maneuver to the best advantage.

As stated in article 3.9, the presence of an exceptionally long swell is usually the first visible indication of the existence of a tropical cyclone. In deep water it approaches from the general direction of origin (the position of the storm center *when the swell was generated*). However, in shoaling water this is a less reliable indication because the direction is changed by refraction, the crests being more nearly parallel to the bottom contours.

When the cirrus clouds appear, their point of convergence provides an indication of the direction of the storm center. If the storm is to pass well to one side of the observer, the point of convergence shifts slowly in the direction of storm movement. If the storm center will pass near the observer, this point remains steady. When the bar (art. 3.9) becomes visible, it appears to rest upon the horizon for several hours. The darkest part of this cloud is in the direction of the storm center. If the storm is to pass to one side, the bar appears to drift slowly along the horizon. If the storm is heading directly toward the observer, the position of the bar remains fixed. Once within the area of the dense, low clouds, one should observe their direction of movement, which is almost exactly along the isobars, with the center of the storm being 90° from the direction of cloud movement (left of direction of movement in the Northern Hemisphere, and right in the Southern Hemisphere).

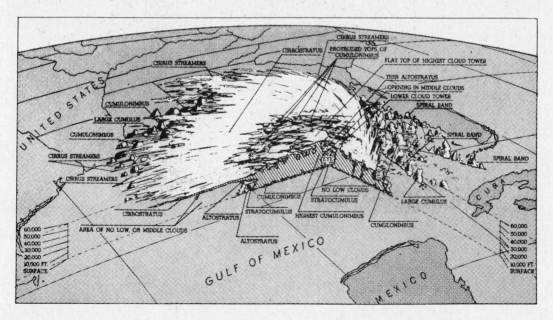

FIGURE 3.9.—Typical hurricane cloud formations.

The winds are probably the best guide to the direction of the center of a tropical cyclone. The circulation is cyclonic (art. 2.12), but because of the steep pressure gradient near the center, the winds there blow with greater violence and are more nearly circular than in extratropical cyclones.

According to Buys Ballot's law (art. 2.12) an observer who faces into the wind has the center of the low pressure on his right in the Northern Hemisphere, and on his left in the Southern Hemisphere, and in each case somewhat behind him. If the wind followed circular isobars exactly, the center would be exactly 8 points, or 90°, from dead ahead when facing into the wind. However, the track of the wind is usually inclined somewhat toward the center, so that the angle from dead ahead varies between perhaps 8 and 12 points (90° to 135°). The inclination varies in different parts of the same storm. It is least in front of the storm, and greatest in the rear, since the actual wind is the vector sum of that due to the pressure gradient and the motion of the storm along the track. A good average is perhaps 10 points in front, and 11 or 12 points in the rear. These values apply when the storm center is still several hundred miles away. Closer to the center, the wind blows more nearly along the isobars, the inclination being reduced by one or two points at the wall of the eye. Since wind direc-

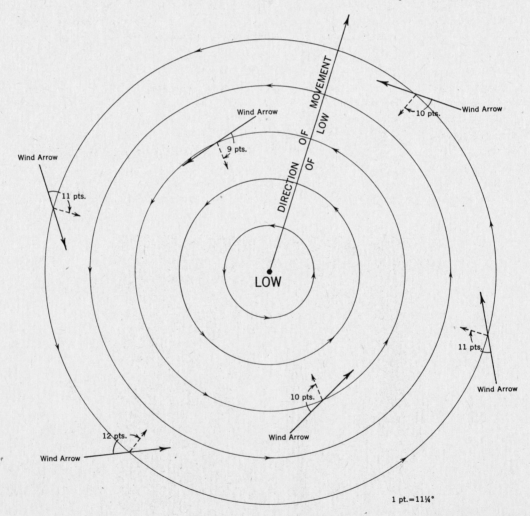

FIGURE 3.10a.—Approximate relationship of wind to isobars and storm center in the Northern Hemisphere.

tion usually shifts temporarily during a squall, its direction at this time should not be used for determining the position of the center. The approximate relationship of wind to isobars and storm center in the Northern Hemisphere is shown in figure 3.10a.

When the center is within radar range, it might be located by this equipment. However, since the radar return is predominantly from the rain, results can be deceptive, and other indications should not be neglected. Figure 3.10b shows a radar PPI presentation of a tropical cyclone. If the eye is out of range, the spiral bands (fig. 3.10b) may indicate its direction from the vessel. Tracking the eye or upwind portion of the spiral bands enables determining the direction and speed of movement; this should be done for at least 1 hour because the eye tends to oscillate. The tracking of individual cells, which tend to move tangentially around the eye, for 15 minutes or more, either at the end of the band or between bands, will provide an indication of the wind speed in that area of the storm.

Distance from the storm center is more difficult to determine than direction. Radar is perhaps the best guide. However, the rate of fall of the barometer is some indication.

3.11. Maneuvering to avoid the storm center.—The safest procedure with respect to tropical cyclones is to avoid them. If action is taken sufficiently early, this is simply a matter of setting a course that will take the vessel well to one side of the probable track of the storm, and then continuing to plot the positions of the storm center, as given in the weather bulletins, revising the course as needed.

However, such action is not always possible. If one finds himself within the storm area, the proper action to take depends in part upon his position relative to the storm center and its direction of travel. It is customary to divide the circular area of the storm into two parts. In the Northern Hemisphere, that part to the *right* of the storm track (facing in the direction *toward* which the storm is moving) is called the **dangerous semicircle.** It is considered dangerous because (1) the actual wind *speed* is greater than that due to the pressure gradient alone, since it is augmented by the forward motion of the storm, and (2) the *direction* of the wind and sea is such as to carry a vessel into the path of the storm (in the forward part of the semicircle). The part to the left

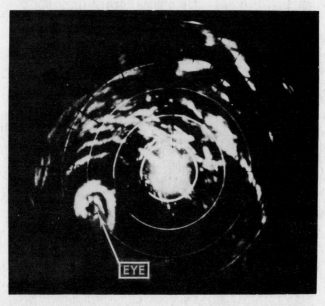

FIGURE 3.10b.—Radar PPI presentation of a tropical cyclone.

of the storm track is called the **navigable semicircle.** In this part, the wind is decreased by the forward motion of the storm, and the wind blows vessels away from the storm track (in the forward part). Because of the greater wind speed in the dangerous semicircle, the seas are higher than in the navigable semicircle. In the Southern Hemisphere, the dangerous semicircle is to the left of the storm track, and the navigable semicircle is to the right of the storm track.

A plot of successive positions of the storm center should indicate the semicircle in which a vessel is located. However, if this is based upon weather bulletins, it is not a reliable guide because of the lag between the observations upon which the bulletin is based and the time of reception of the bulletin, with the ever present possibility of a change in the direction of motion of the storm. The use of one's radar eliminates this lag, but the return is not always a true indication of the center. Perhaps the most reliable guide is the wind. Within the cyclonic circulation, a *veering* wind (one changing direction to the right in the Northern Hemisphere and to the left in the Southern Hemisphere) indicates the vessel is probably in the dangerous semicircle, and a *backing* wind (one changing in a direction opposite to a veering wind) indicates the vessel is probably in the navigable semicircle. However, if a vessel is underway, its motion should be considered. If it is outrunning the storm or pulling rapidly toward one side (which is not difficult during the early stages of a storm, when its speed is low), the opposite effect occurs. This should usually be accompanied by a rise in atmospheric pressure, but if motion of the vessel is nearly along an isobar, this may not be a reliable indication. If in doubt, the safest action is usually to stop long enough to determine definitely the semicircle. The loss in valuable time may be more than offset by the minimizing of the possibility of taking the wrong action and increasing the danger to the vessel. If the wind direction remains steady (for a vessel which is stopped), with increasing speed and falling barometer, the vessel is in or near the path of the storm. If it remains steady with decreasing speed and rising barometer, the vessel is on the storm track, behind the center.

The first action to take if one finds himself within the cyclonic circulation, is to determine the position of his vessel with respect to the storm center. While the vessel can still make considerable way through the water, a course should be selected to take it as far as possible from the center. If the vessel can move faster than the storm, it is a relatively simple matter to outrun the storm if sea room permits. But when the storm is faster, the solution is not as simple. In this case, the vessel, if ahead of the storm, will approach nearer to the the center. The problem is to select a course that will produce the greatest possible minimum distance. This is best determined by means of a relative movement plot, as shown in the following example solved on a maneuvering board.

Example.—A tropical cyclone is estimated to be moving in direction 320° at 19 knots. Its center bears 170°, at an estimated distance of 200 miles from a vessel which has a maximum speed of 12 knots.

Required.—(1) The course to steer at 12 knots to produce the greatest possible minimum distance between the vessel and the storm center.

(2) The distance of the storm center at nearest approach.

(3) Elapsed time until nearest approach.

Solution (fig. 3911).—Consider the vessel remaining at the center of the plot throughout the solution, as on a radar PPI.

(1) Plot point C at a distance of 200 miles (scale 20:1) in direction 170° from the center of the diagram, to locate the position of the storm center relative to the vessel. From the center of the diagram, draw RA, the speed vector of the storm center, in direction 320°, speed 19 knots (scale 2:1). From A draw a line tangent to the 12-knot speed circle (labeled 6 at scale 2:1) *on the side opposite the storm center.* From the center

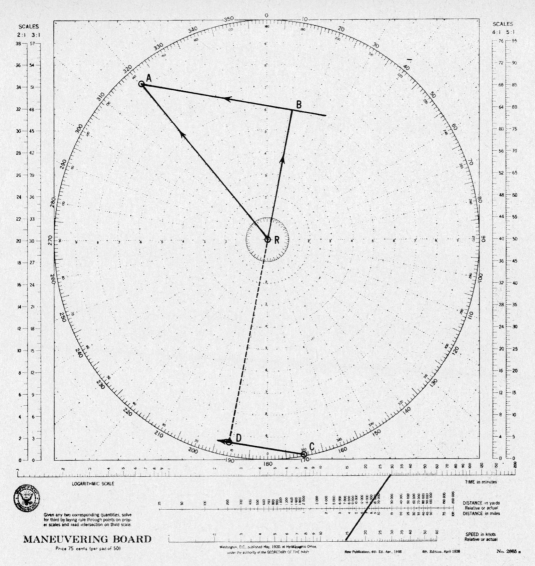

FIGURE 3.11.—Solution to determine course for avoiding storm center.

of the diagram draw a perpendicular to this tangent line, locating point *B*. The line *RB* is the required speed vector for the vessel. Its direction, 011°, is the required course.

(2) The path of the storm center *relative to the vessel*, will be along a line from *C* in the direction *BA*, if both storm and vessel maintain course and speed. The point of nearest approach will be at *D*, the foot of a perpendicular from the center of the diagram. This distance, at scale 20:1, is 187 miles.

(3) The length of the vector *BA* (14.8 knots) is the speed of the storm with respect to the vessel. Mark this on the lowest scale of the nomogram at the bottom of the diagram. The relative distance *CD* is 72 miles, by measurement. Mark this (scale 10:1) on the middle scale at the bottom of the diagram. Draw a line between the two points and extend it to intersect the top scale at 29.2 (292 at 10:1 scale). The elapsed time is therefore 292 minutes, or 4 hours 52 minutes, or 5 hours, approximately.

Answers.—(1) C 011°, (2) D 187 mi., (3) t 5^h (approximately).

The storm center will be dead astern at its nearest approach.

As a very general rule, for a vessel in the Northern Hemisphere, safety lies in placing the wind on the starboard bow in the dangerous semicircle and on the starboard quarter in the navigable semicircle. If on the storm track ahead of the storm, the wind should be put about 2 points on the starboard quarter until the vessel is well within the navigable semicircle, and the rule for that semicircle then followed. A study of figure 3.10a should indicate why these headings are desirable. In the Southern Hemisphere the same rules hold, but with respect to the port side. With a faster than average vessel, the wind can be brought a little farther aft in each case. However, as the speed of the storm increases along its track, the wind should be brought farther forward. If land interferes with what would otherwise be the best maneuver, the solution should be altered to fit the circumstances. If the speed of a vessel is greater than that of the storm, it is possible for the vessel, if behind the storm, to overtake it. In this case, the only action usually needed is to slow enough to let the storm pull ahead.

In all cases, one should be alert to changes in the direction of movement of the storm center, particularly in the area where the track normally curves toward the pole. If the storm maintains its direction and speed, the ship's course should be maintained as the wind shifts.

If it becomes necessary for a vessel to heave to, the characteristics of the vessel should be considered. A power vessel is concerned primarily with damage by direct action of the sea. A good general rule is to heave to with head to the sea in the dangerous semicircle or stern to the sea in the navigable semicircle. This will result in greatest amount of headway away from the storm center, and least amount of leeway toward it. If a vessel handles better with the sea astern or on the quarter, it may be placed in this position in the navigable semicircle or in the rear half of the dangerous semicircle, but *never* in the forward half of the dangerous semicircle. It has been reported that when the wind reaches hurricane speed and the seas become confused, some ships ride out the storm best if the engines are stopped, and the vessel is permitted to seek its own position. In this way, it is said, the ship rides *with* the storm instead of fighting *against* it.

In a sailing vessel, while attempting to avoid a storm center, one should steer courses as near as possible to those prescribed above for power vessels. However, if it becomes necessary for such a vessel to heave to, the wind is of greater concern than the sea. A good general rule always is to heave to on whichever tack permits the shifting wind to draw aft. In the Northern Hemisphere this is the starboard tack in the dangerous semicircle and the port tack in the navigable semicircle. In the Southern Hemisphere these are reversed.

While each storm requires its own analysis, and frequent or continual resurvey of the situation, the general rules for a steamer may be summarized as follows:

NORTHERN HEMISPHERE

Right or dangerous semicircle.—Bring the wind on the starboard bow (045° relative), hold course and make as much way as possible. If obliged to heave to, do so with head to the sea.

Left or navigable semicircle.—Bring the wind on the starboard quarter (135° relative), hold course and make as much way as possible. If obliged to heave to, do so with stern to the sea.

On storm track, ahead of center.—Bring the wind 2 points on the starboard quarter (about 160° relative), hold course and make as much way as possible. When well within the navigable semicircle, maneuver as indicated above.

On storm track, behind center.—Avoid the center by the best practicable course, keeping in mind the tendency of tropical cyclones to curve northward and eastward.

SOUTHERN HEMISPHERE

Left or dangerous semicircle.—Bring the wind on the port bow (315° relative), hold course and make as much way as possible. If obliged to heave to, do so with head to the sea.

Right or navigable semicircle.—Bring the wind on the port quarter (225° relative), hold course and make as much way as possible. If obliged to heave to, do so with stern to the sea.

On storm track, ahead of center.—Bring the wind 2 points on the port quarter (about 200° relative), hold course and make as much way as possible. When well within the navigable semicircle, maneuver as indicated above.

On storm track, behind center.—Avoid the center by the best practicable course, keeping in mind the tendency of tropical cyclones to curve southward and eastward.

Whenever a tropical cyclone is encountered, the wise procedure is to begin preparing the vessel for heavy weather in sufficient time to permit thorough preparation, so that damage may be minimized. One should be particularly careful to keep free surfaces of liquids to a minimum.

It is possible, particularly in temperate latitudes after the storm has recurved, that the dangerous semicircle is the left one in the Northern Hemisphere (right one in the Southern Hemisphere). This can occur if a large high lies north of the storm and causes a tightening of the pressure gradient in the region.

Typhoon Havens Handbook for the Western Pacific and Indian Oceans is published by the Naval Environmental Prediction Research Facility, Monterey, California, as an aid to commanders and commanding officers of ships in evaluating a typhoon situation and to assist them in deciding whether to sortie, to evade, or to remain in port to take shelter within a specific harbor.

3.12. Effects.—The high winds of a tropical cyclone inflict widespread damage when such a storm leaves the ocean and crosses land. Aids to navigation may be blown out of position or destroyed. Craft in harbors, unless they are properly secured, drag anchor or are blown against obstructions. Ashore, trees are blown over, houses are damaged, power lines are blown down, etc. The greatest damage usually occurs in the dangerous semicircle a short distance from the center, where the strongest winds occur. As the storm continues on across land, its fury subsides faster than it would if it had remained over water.

Wind instruments are usually incapable of measuring the 175- to 200-knot winds of the more intense hurricanes; if the instrument holds up, often the supporting structure gives way.

Wind gusts, which are usually 30 to 50 percent higher than sustained winds, add significantly to the destructiveness of the tropical cyclone. Many tropical cyclones that reach hurricane intensity develop winds of more than 90 knots sometime during their lives, but few develop winds of more than 130 knots.

Tropical cyclones have produced some of the world's heaviest rainfalls. While average amounts range from 6 to 10 inches, totals near 100 inches over a 4-day period have been observed. A 24-hour world's record of 73.62 inches fell at Reunion Island during a tropical cyclone in 1952. Forward movement of the storm and land topography have a considerable influence on rainfall totals. Torrential rains can occur when a storm moves against a mountain range; this is common in the Philippines and Japan, where even weak tropical depressions produce considerable rainfall. A 24-hour total of 46 inches was recorded in the Philippines during a typhoon in 1911. As hurricane Camille crossed southern Virginia's Blue Ridge Mountains in August of 1969, there was nearly

30 inches of rain in about 8 hours. This caused some of the most disastrous floods in the state's history.

Flooding is an extremely destructive by-product of the tropical cyclone's torrential rains. Whether an area will be flooded depends on the physical characteristics of the drainage basin, rate and accumulation of precipitation, and river stages at the time the rains begin. When heavy rains fall over flat terrain, the countryside may lie underwater for a month or so, and while buildings, furnishings, and underground powerlines may be damaged, there are usually few fatalities. In mountainous or hill country, disastrous flood's develop rapidly and can cause a great loss of life.

There have been occasional reports in tropical cyclones of waves greater than 40 feet in height, and numerous reports in the 30- to 40- foot category. However, in tropical cyclones, strong winds rarely persist for a sufficiently long time or over a large enough area to permit enormous wave heights to develop. The direction and speed of the wind changes more rapidly in tropical cyclones than in extratropical storms. Thus, the maximum duration and fetch length for any wind condition is often less in tropical cyclones than in extratropical storms of similar intensity, and the waves accompanying any given local wind conditions are generally not so high as those expected, with similar local wind conditions, in the high-latitude-type storms. In hurricane Camille, significant waves of 43 feet were recorded; an extreme wave height reached 72 feet.

Exceptional conditions may arise when waves of certain dimensions travel within the storm at a speed equal to the storm's speed, thus, in effect, extending the duration and fetch of the wave and significantly increasing its height. This occurs most often to the right of the track in the Northern Hemisphere (left of the track in the Southern Hemisphere). Another condition that may give rise to exceptional wave heights is the intersection of waves from two or more distinct directions. This may lead to a zone of confused seas in which the heights of some waves will equal the sum in each individual wave train. This process can occur in any quadrant of the storm and so it should not be assumed that the highest waves will always be encountered to the right of the storm track in the Northern Hemisphere (left of the track in the Southern Hemisphere).

When these waves move beyond the influence of the generating winds, they become swell. They are recognized by their smooth, undulating form, in contrast to the steep, ragged crests of the winds' waves. This swell, particularly that generated by the right side of the storm, can travel a thousand miles or more and may produce tides 3 or 4 feet above normal along several hundred miles of coastline.

When a tropical cyclone moves close to a coast, wind often causes a rapid rise in water level, and along with the falling pressure may produce a storm **surge.** This surge is usually confined to the right of the track in the Northern Hemisphere (left of the track in the Southern Hemisphere) and to a relatively small section of the coastline. It most often occurs with the approach of the storm, but in some cases, where a surge moves into a long channel, the effect may be delayed. Occasionally, the greatest rise in water is observed on the opposite side of the track, when northerly winds funnel into a partially landlocked harbor. The surge could be 3 feet or less, or it could be 20 feet or more, depending on the combination of all the factors involved.

There have been reports of a "hurricane wave," described as a "wall of water," which moves rapidly toward the coastline. Authenticated cases are rare, but some of the world's greatest natural disasters have occurred as a result of this wave, which may be just a rapidly rising and abnormally high storm surge. In India, such a disaster occurred in 1876, between Calcutta and Chittagong, and drowned more than 100,000 persons.

Along the coast, particularly, greater damage may be inflicted by water than by the wind. There are at least four sources of water damage. First, the unusually high seas generated by the storm winds pound against shore installations and craft in their way. Second, the continued blowing of the wind toward land causes the water level to increase perhaps 3 to 10 feet above its normal level. This **storm tide,** which may begin when the storm center is 500 miles or even farther from the shore, gradually increases until the storm passes. The highest storm tides are caused by a slow-moving tropical cyclone of large diameter, because both of these effects result in greater duration of wind in the same direction. The effect is greatest in a partly enclosed body of water, such as the Gulf of Mexico, where the concave coastline does not readily permit the escape of water. It is least on small islands, which present little obstruction to the flow of water. Third, the furious winds which blow around the wall of the eye create a ridge of water called a **storm wave,** which strikes the coast and often inflicts heavy damage. The effect is similar to that of a **seismic sea wave,** caused by an earthquake in the ocean floor. Both of these waves are popularly called **tidal waves.** Storm waves of 20 feet or more have occurred. About 3 or 4 feet of this is due to the decrease of atmospheric pressure, and the rest to winds. Like the damage caused by wind, that due to high seas, the storm surge and tide, and the storm wave is greatest in the dangerous semicircle, near the center. The fourth source of water damage is the heavy rain that accompanies a tropical cyclone. This causes floods that add to the damage caused in other ways.

There have been many instances of tornadoes occurring within the circulation of tropical cyclones. Most of these have been associated with tropical cyclones of the North Atlantic Ocean and have occurred in the West Indies and along the gulf and Atlantic coasts of the United States. They are usually observed in the forward semicircle or along the advancing periphery of the storm. These tornadoes are usually short-lived and less intense than those that occur in the midwestern United States.

When proceeding along a shore recently visited by a tropical cyclone, a navigator should remember that time is required to restore aids to navigation which have been blown out of position or destroyed. In some instances the aid may remain but its light, sound apparatus, or radiobeacon may be inoperative. Landmarks may have been damaged or destroyed.

References

Australia Bureau of Meteorology. *Occurrence of Tropical Depressions and Cyclones in the Northeastern and Northwestern Australian Regions* (annual summary 1957–1962). Maribyrnong, Department of Supply, Central Drawing Office.

Australia Bureau of Meteorology. *Tropical Cyclones in the Northern Australian Regions* (annual summary 1962–1969). Maribyrnong, Department of Supply, Central Drawing Office.

Brand, S., and J. W. Blelloch. *Typhoon Havens Handbook for the Western Pacific and Indian Oceans.* Technical Paper 5–76, Naval Environmental Prediction Research Facility, Monterey, California, 1976.

Chin, P. C. *Tropical Cyclones in the Western Pacific and China Sea Area from 1884 to 1953.* Hong Kong, Royal Observatory, 1958.

Crutcher, H. L. and R. G. Quayle. *Mariners Worldwide Climatic Guide to Tropical Storms at Sea,* Naval Weather Service Command, U. S. Govt. Print. Off., 1974.

Cry, G. W. "Tropical Cyclones of the North Atlantic Ocean: Tracks and Frequencies of Hurricanes and Tropical Storms, 1871–1963," Technical Paper No. 55, Washington, U. S. Govt. Print. Off., 1965.

DeAngelis, R. M. "North Pacific Hurricanes: Timid or Treacherous?" *Mariners Weather Log,* vol. 11, No. 6 (November 1967), pp. 193–200.

Dunn, G. E. and B. I. Miller. *Atlantic Hurricanes.* Louisiana State University Press, 1960.

Environmental Science Services Administration. *Hurricane the greatest storm on earth*. Washington, U. S. Govt. Print. Off., 1967.

Harding, E. T. and Wm. J. Kotsch. *Heavy Weather Guide*. Annapolis, U. S. Naval Institute, 1965.

Harris, D. L. "Wave Patterns in Tropical Cyclones." *Mariners Weather Log*, vol. 6, No. 5 (September 1962) pp. 156–160.

Harris, D. L. "Characteristics of the Hurricane Storm Surge," Technical Paper No. 48. Washington, U. S. Govt. Print. Off., 1963.

Hodge, W. T. "North Pacific Typhoons Where and When Are They Most Frequent?" *Mariners Weather Log*, vol. 9, No. 3 (May 1965), pp. 73–76.

India Meteorological Department. *Tracks of Storms and Depressions in the Bay of Bengal and the Arabian Sea, 1877–1960*. New Delhi, India Meteorological Department, 1964.

Joint Typhoon Warning Center. *Annual Typhoon Reports 1959–1973*. Guam, U. S. Fleet Weather Central/Joint Typhoon Warning Center.

Mauritius Meteorological Department. *Annual Report of the Meteorological Department 1950–1966*. Port Louis, Mauritius Government Printer.

New Zealand Meteorological Service. *Annual Meteorological Summary 1957–1966*. Laucala Bay, Suva, Fiji Government Press.

Nimitz, Chester W., et. al. "Typhoon Doctrine." U. S. Naval Institute Proceedings, vol. 82, No. 1 (January 1956), pp. 83–93.

U. S. Weather Bureau. *Climatological and Oceanographic Atlas for Mariners*, Vol. I North Atlantic Ocean. Washington, U. S. Govt. Print. Off., 1959.

U. S. Weather Bureau. *Climatological and Oceanographic Atlas for Mariners*, Vol. II North Pacific Ocean. Washington, U. S. Govt. Print. Off., 1961.

CHAPTER IV

THE OCEANS

4.1. Introduction.—Oceanography is the application of the sciences to the phenomena of the oceans. It includes a study of their forms; physical, chemical, geological, and biological features; and phenomena. Thus, it embraces the widely separated fields of geography, geology, chemistry, physics, and biology. Many subdivisions of these sciences, such as sedimentation, ecology (biological relationship between organisms and their environment), bacteriology, biochemistry, hydrodynamics, acoustics, and optics, have been extensively studied in the oceans.

The oceans cover 70.8 percent of the surface of the earth. The Atlantic covers 16.2 percent, the Pacific 32.4 percent (3.2 percent more than the land area of the entire earth), the Indian Ocean 14.4 percent, and marginal and adjacent areas (of which the largest is the Arctic Ocean) 7.8 percent. Their extent alone makes them an important subject for study. However, greater incentive lies in their use for transportation, their influence upon weather and climate, and their potentiality as a source of power, food, freshwater, and mineral and organic substances.

4.2. History of oceanography.—The earliest studies of the oceans were concerned principally with problems of navigation. Information concerning tides, currents, soundings, ice, and distances between ports was needed as ocean commerce increased. According to Posidonius, a depth of 1,000 fathoms had been measured in the Sea of Sardinia as early as the second century BC. About the middle of the 19th century, the Darwinian theories of evolution gave a great impetus to the collection of marine organisms, since it is believed by some that all terrestrial forms have evolved from oceanic ancestors. Later, the serious depletion of many fisheries called for investigation of the relation of the economically valuable organisms to the physical characteristics of their environment, especially in northwestern Europe and off Japan. Still later, the growing use of the oceans in warfare, particularly after the development of the submarine, required that much effort be expended in problems of detection and attack, resulting in the study of many previously neglected scientific aspects of the sea.

Oceanographic exploration. Exploration of the seas was primarily geographical until the 19th century, although the accumulated observations of seafarers, as recorded in the early charts and sailing directions, often included data on tides, currents, and other oceanographic phenomena. The great voyages of discovery, particularly those beginning in 1768 with Captain Cook, and continued by such commanders as La Pérouse, Bellingshausen, and Wilkes, included scientists in their complements. However, scientific work on the oceans at this period was severely limited by lack of suitable instruments for probing conditions below the surface. Meanwhile, Lieutenant Matthew Fontaine Maury, USN, working in the forerunner of the U. S. Navy Hydrographic Office in Washington, developed to a high degree of perfection the analysis of log-book observations. His first results, published in 1848, were of great importance to ship operations in the recommendation of favorable sailing routes, and they stimulated international cooperation in the fields of oceanography and marine meteorology.

In the rapid advances in technology after 1850, oceanographic instrumentation problems were not neglected, with the result that the British Navy in 1872–76 was able to send HMS *Challenger* around the world on the first purely deep-sea oceanographic expedition ever attempted. Her bottom samples, as analyzed by Sir John Murray, laid the foundation of geological oceanography, and 77 of her seawater samples, analyzed by C. R. Dittmar, proved for the first time that various constituents of the salts in seawater are everywhere in virtually the same proportions.

Since that time, the coastal waters and fishing banks of many nations have been extensively studied, and numerous vessels of various nationalities have conducted work on the high seas. Notable among these have been the American *Albatross* from 1882 to 1920; the Austrian *Pola* in the Mediterranean and Red Seas between 1890 and 1896; the Danish *Dana*, which during its voyages of 1920–22 discovered the breeding place of the European eels in the Sargasso Sea; the American *Carnegie* in 1927–29; the German *Meteor* in the Atlantic from 1928 to 1938; and the British *Discovery II* in the Antarctic between 1930 and 1939. Notable also were the drifts of the Norwegian vessels *Fram* and *Maud* in the arctic ice pack from 1893 to 1896 and 1918 to 1925, respectively; the attempt by Sir George Hubert Wilkins to operate under the ice in the British submarine *Nautilus* in 1931; and the Russian station set up at the North Pole in 1937, which made observations from the drifting pack ice.

At the same time, investigations pursued ashore provided the theoretical basis for the explanation of ocean currents, under the leadership of Helland-Hansen in Norway and Ekman and the Bjerknes in Sweden, while Martin Knudsen in Denmark worked out the precise details of the relationship between chlorinity, salinity, and density, enabling the theories to be verified by field observations.

During World War II, basic investigations were interrupted while work on purely military applications of oceanography was carried out. Deep-sea expeditions were renewed by the Swedish *Albatross* after the war, followed by the Danish *Galathea*, the second British *Challenger* (built in 1931), and *Discovery II* in the Antarctic, and vessels of the American Scripps Institution in the Pacific. Oceanographic work was carried out by Americans and Russians in the Arctic.

4.3. Origin of the oceans.—Although many leading geologists still disagree with the conclusion that the structure of the continents is fundamentally different from that of the oceans, there is a growing body of evidence in support of the theory that the rocks underlying the ocean floors are more dense than those underlying the continents. According to this theory, all the earth's crust floats on a central liquid core, and the portions that make up the continents, being lighter, float with a higher freeboard. Thus, the thinner areas, composed of heavier rock, form natural basins where water has collected.

The shape of the oceans is constantly changing due to **continental drift.** The surface of the earth may be conceived as consisting of several "plates." These plates are joined along **fracture** or **fault lines.** There is constant and measurable movement of these plates.

The origin of the water in the oceans is also controversial. Although some geologists have postulated that all the water existed as vapor in the atmosphere of the primeval earth, and that it fell in great torrents of rain as soon as the earth cooled sufficiently, another school holds that the atmosphere of the original hot earth was lost, and that the water gradually accumulated as it was given off in steam by volcanoes or worked to the surface in hot springs.

Most of the water on the earth's crust is now in the oceans—about 328,000,000 cubic statute miles, or about 85 percent of the total. The mean depth of the ocean is 2,075 fathoms, and the total area is 139,000,000 square statute miles.

4.4. Oceanographic chemistry may be divided into three main parts: the chemistry of (1) seawater, (2) marine sediments, and (3) organisms living in the sea. The first is of particular interest to the navigator.

Chemical properties of seawater are determined by analyzing samples of water obtained at various places and depths. Samples from below the surface are obtained by means of metal bottles designed for this purpose. The open bottles are attached at suitable intervals to a wire lowered into the sea. When they reach the desired depths, a metal ring or **messenger** is dropped down the wire. When the messenger arrives at the first bottle, it causes the bottle to close, trapping a sample of the water at that depth, and releasing a second messenger which travels on down the wire. The process is repeated at each bottle until all are closed, when they are hauled up and each bottle detached as it comes within reach. Of the various types devised, the **Nansen bottle** is the most widely known. It is equipped with a removable frame for attaching a thermometer.

4.5. Physical properties of seawater are dependent primarily upon salinity, temperature, and pressure. However, factors like motion of the water and the amount of suspended matter affect such properties as color and transparency, conduction of heat, absorption of radiation, etc.

4.6. Salinity is the amount of dissolved solid material in the water when carbonate has been converted to oxide, bromide and iodide to chloride, and organic material oxidized. It is usually expressed as parts per thousand (by weight), under certain standard conditions. This is not the same as **chlorinity,** which is equal approximately to the amount of chlorine, with bromides and oxides converted to chloride. (Actually the chlorine content is about 1.00045 times the chlorinity as determined by standard procedures.) The two have been found to be related empirically by the formula:

$$\text{salinity} = 0.03 + 1.805 \times \text{chlorinity}.$$

Historically the determination of salinity was a slow and difficult process, while chlorinity could be determined easily and accurately by titration with silver nitrate. It was customary to determine chlorinity and compute salinity by the formula given above. By this process, salinity could be determined with an error not exceeding 0.02 parts per thousand. Salinity can now be measured directly using a **salinometer** which measures changes in conductivity. Salinity generally varies between about 33 and 37 parts per thousand, the average being about 35 parts per thousand. However, when the water has been diluted, as near the mouth of a river or after a heavy rainfall, the salinity is somewhat less; and in areas of excessive evaporation, the salinity may be as high as 40 parts per thousand. In certain confined bodies of water, notably the Great Salt Lake in Utah, and the Dead Sea in Asia Minor, the salinity is several times this maximum. Chlorinity accounts for about 55 percent of salinity, the average being about 19 parts per thousand.

4.7. Temperature in the ocean varies widely, both horizontally and with depth. Maximum values of about 90°F are encountered at the surface in the Persian Gulf in summer, and the lowest possible values of about 28°F (the usual minimum freezing point of seawater) occur in polar regions and near the ocean bottom everywhere, including the Tropics. Pub. No. 225, *World Atlas of Sea Surface Temperatures*, shows in detail the average sea surface temperatures for each month. The following tabulation gives the percentage distribution of temperatures for the world for the months of February and August, as derived from this source:

Surface temperature °F	Percentage of area of ocean	
	February	August
<35	12. 0	13. 1
35–40	6. 5	3. 3
40–45	4. 0	3. 0
45–50	4. 5	5. 0
50–55	4. 0	6. 5
55–60	5. 0	6. 0
60–65	5. 5	6. 3
65–70	8. 0	7. 0
70–75	10. 0	10. 4
75–80	17. 5	16. 5
80–85	23. 0	22. 7
85–90	0. 0	0. 2

The vertical distribution of temperature in the sea nearly everywhere shows a decrease of temperature with depth. Since colder water is denser (assuming the same salinity), it sinks below warmer water. This results in a temperature distribution just opposite to that of the earth's crust, where temperature increases with depth below the surface of the ground.

In general, in the sea there is usually a mixed layer of isothermal water below the surface, where the temperature is the same as that of the surface. This layer is caused by two physical processes: wind mixing, and convective overturning as surface water cools and becomes more dense. The layer is best developed in the Arctic and Antarctic regions and seas like the Baltic and Sea of Japan during the winter, where it may extend to the bottom of the ocean. In the Tropics, the wind-mixed layer may exist to a depth of 125 meters. The layer may exist throughout the year. Below this layer is a zone of rapid temperature decrease, called the **thermocline,** to the temperature of the deep oceans. At a depth greater than 200 fathoms, the temperature everywhere is below 60°F, and in the deeper layers, fed by cooled waters that have sunk from the surface in the Arctic and Antarctic, temperatures as low as 28°.5F exist.

In the colder regions the cooling creates the convective overturning and isothermal water in the winter; but in the summer a seasonal thermocline is created as the upper water becomes warmer.

A typical curve of temperature at various depths is shown in figure 9.3a. Temperature at any desired depth can be determined by means of a **reversing thermometer** attached to a Nansen bottle (art. 4.4). When the bottle closes, the thermometer measures the temperature to within 0°.04F, thus providing a reading for a particular time and point. Instruments with **thermistors** (devices that utilize the change in conductivity of a semiconductor with change in temperature) are commonly used to measure temperature. The STD (salinity-temperature-depth) is an instrument that provides continuous signals as it is lowered from the vessel; temperature is determined by means of a thermistor, salinity by conductivity, and depth by pressure. Continuous records of temperature were first obtained by an instrument called a **bathythermograph,** invented by Spilhaus in 1938. This device functioned to a depth of 75 meters.

The mechanical bathythermograph has been replaced almost entirely by the **expendable bathythermograph (XBT),** which uses a thermistor. The XBT is connected to the vessel by a fine wire. The wire is coiled inside the probe and as the probe free-falls in the ocean, the wire plays out. Depth is determined by elapsed time and a known sink rate. Depth range is determined by the amount of wire stored in the probe; the most common model has a depth range of 500 meters. At the end of the drop, the wire breaks and the probe falls to the ocean bottom. One instrument of this type is dropped from an aircraft, the data being relayed to the aircraft from a buoy to which the wire of the XBT is attached.

4.8. Pressure.—In oceanographic work, pressure is generally expressed in units of the centimeter-gram-second system. The basic unit of this system is 1 dyne per square centimeter. This is a very small unit, one million constituting a practical unit called a bar, which is nearly equal to 1 atmosphere. Atmospheric pressure is often expressed in terms of **millibars,** 1,000 of these being equal to 1 bar. In oceanographic work, water pressure is commonly expressed in terms of **decibars,** 10 of these being equal to 1 bar. One decibar is equal to nearly 1½ pounds per square inch. This unit is convenient because it is very nearly the pressure exerted by 1 meter of water. Thus, the pressure in decibars is approximately the same as the depth in meters, the unit of depth customarily used in oceanographic research. In terms more familiar to the mariner, the pressure at various depths is as follows:

Depth in fathoms	Pressure in pounds per square inch
1,000	2,680
2,000	5,390
3,000	8,100
4,000	10,810
5,000	13,520

The increase in pressure with depth is nearly constant because water is only slightly compressible.

Although virtually all of the physical properties of seawater are affected to a measurable extent by pressure, the effect is not as great as those of salinity and temperature. Pressure is of particular importance to submarines, directly because of the stress it induces on the materials of the craft, and indirectly because of its effect upon buoyancy.

4.9. Density is mass per unit volume. Oceanographers use the centimeter-gram-second system, in which density is expressed as grams per cubic centimeter. The ratio of the density of a substance to that of a standard substance under stated conditions is called **specific gravity.** By definition, the density of distilled water at 4°C(39°.2 F) is 1 gram per milliliter (approximately 1 gram per cubic centimeter). Therefore, if this is used as the standard, as it is in oceanographic work, density and specific gravity are virtually identical numerically.

The density of seawater depends upon salinity, temperature, and pressure. At constant temperature and pressure, density varies with salinity or, because of the relationship between this and chlorinity, with the chlorinity. A temperature of 32°F and atmospheric pressure are considered standard for density determination. The effects of thermal expansion and compressibility are used to determine the density at other temperatures and pressures. The density at a particular pressure affects the buoyancy of submarines. It is also important in its relation to ocean currents.

The greatest changes in density of seawater occur at the surface, where the water is subject to influences not present at depths. Here density is decreased by precipitation, run-off from land, melting of ice, or heating. When the surface water becomes less dense, it tends to float on top of the more dense water below. There is little tendency for the water to mix, and so the condition is one of stability. The density of surface water is increased by evaporation, formation of sea ice, and by cooling. If the surface water becomes more dense than that below, it causes convective mixing. The more dense surface water sinks and mixes with less dense water below. The resultant layer of water is of intermediate density. This process continues until the density of the mixed layer becomes less than that of the water below. The convective circulation established as part of this process can create very deep uniform mixed layers. If the surface water becomes sufficiently dense, it sinks all the way to the bottom. If this occurs in an area where horizontal flow is unobstructed, the water which has descended spreads to other regions, creating a dense bottom layer. Since the greatest increase in density occurs in polar regions, where the air is cold and great quantities of ice form, the cold, dense polar water sinks to the bottom and then spreads to lower latitudes. In the Arctic Ocean region, the cold, dense water is confined by the Bering Strait and the underwater ridge from Greenland to Iceland to Europe. In the Antarctic, however, there are no similar geographic restrictions and large quantities of very cold, dense water formed there flow to the north along the ocean bottom. This process has continued for a sufficiently long period of time that the entire ocean floor is covered with this dense water, thus explaining the layer of cold water at great depths in all the oceans.

In some respects, oceanographic processes are similar to those occuring in the atmosphere (ch. III). The convective circulation in the ocean is somewhat similar to that in the atmosphere. Water masses having nearly uniform characteristics are analogous to airmasses.

4.10. Compressibility.—Seawater is nearly incompressible, its coefficient of compressibility being only 0.000046 per bar under standard conditions. This value changes slightly with changes of temperature or salinity. The effect of compression is to force the molecules of the substance closer together, causing it to become more dense. Even though the compressibility is low, its total effect is considerable because of the amount of water involved. If the compressibility of seawater were zero, sea level would be about 90 feet higher than it now is.

4.11. Viscosity is resistance to flow. Seawater is slightly more viscous than freshwater. Its viscosity increases with greater salinity, but the effect is not nearly as marked as that occurring with decreasing temperature. The rate is not uniform, becoming greater as the temperature decreases. Because of the effect of temperature upon viscosity, an incompressible object might sink at a faster rate in warm surface water than in colder water below. However, for most objects, this effect may be more than offset by the compressibility of the object.

The actual relationships existing in the ocean are considerably more complicated than indicated by the simple explanation given above, because of turbulent motion within the sea. The disturbing effect is called **eddy viscosity.**

4.12. Specific heat is the amount of heat required to raise the temperature of a unit mass of a substance a stated amount. In oceanographic work, specific heat is stated, in centimeter-gram-second units, as the number of calories needed to raise 1 gram of the substance 1°C. Specific heat at constant pressure is usually the quantity desired when liquids are involved, but occasionally the specific heat at constant volume is required. The ratio of these two quantities has a direct relationship to the speed of sound in seawater.

The specific heat of seawater decreases slightly as salinity increases. However, it is much greater than that of land. The ocean is a giant sink and source for heat. It can absorb large quantities of heat with very little change in temperature. This is partly due to the high specific heat of water and partly due to mixing in the ocean that distributes the heat throughout a layer. Land has a lower specific heat and, in addition, all heat is lost or gained from a thin layer at the surface. This accounts for the greater temperature range of land and the atmosphere above it, resulting in monsoons (art. 2.10) and the familiar land and sea breezes of tropical and temperate regions (art. 2.14).

4.13. Thermal expansion.—One of the more interesting differences between salt- and freshwater relates to thermal expansion. Saltwater continues to become more dense as it cools to the freezing point; freshwater reaches maximum density at 4°C and then expands (becomes less dense) as the water cools to 0°C and freezes. This means that the convective mixing of freshwater stops at 4°C; freezing proceeds very rapidly beyond that point. The rate of expansion with increased temperature is greater in seawater than in freshwater. Thus, at temperature 15°C (59°F), and atmospheric pressure, the coefficient of thermal expansion is 0.000151 per degree Celsius for freshwater and 0.000214 per degree Celsius for water of 35 parts per thousand salinity. The coefficient of thermal expansion increases not only with greater salinity, but also with increased temperature and pressure. At 35 parts per thousand, the coefficient of surface water increases from 0.000051 per degree Celsius at 0°C (32°F) to 0.000334 per degree Celsius at 30°C (86°F). At a constant temperature of 0°C (32°F) and a salinity of 34.85 parts per thousand, the coefficient increases to 0.000276 per degree Celsius at a pressure of 10,000 decibars (at a depth of approximately 10,000 meters).

4.14. Thermal conductivity.—In water, as in other substances, one method of heat transfer is by conduction. Freshwater is a poor conductor of heat, having a coefficient of thermal conductivity of 0.00139 calories per second per centimeter per degree Celsius. For seawater it is slightly less but increases with greater temperature or pressure.

However, if turbulence is present, which it nearly always is to some extent in the ocean, the processes of heat transfer are altered. The effect of turbulence is to increase greatly the rate of heat transfer. The "eddy" coefficient used in place of the still-water coefficient is so many times larger, and so dependent upon the degree of turbulence that the effects of temperature and pressure are not important.

4.15. Electrical conductivity.—Water without impurities is a very poor conductor of electricity. However, when salt is in solution in water, the salt molecules are ionized and therefore are carriers of electricity. (What is commonly called freshwater has many impurities and is a good conductor of electricity; only *pure* distilled water is a poor conductor of electricity.) Hence, the electrical conductivity of seawater is directly proportional to the number of salt molecules in the water. For any given salinity, the conductivity increases with an increase in temperature.

4.16. Radioactivity.—Although the amount of radioactive material in seawater is very small, this material is present in marine sediments to a greater extent than in the rocks of the earth's crust. This is probably due to precipitation of radium or other radioactive material from the water. The radioactivity of the top layers of sediment is less than that of deeper layers. This may be due to absorption of radioactive material in the soft tissues of marine organisms.

4.17. Refractive index of seawater increases as salinity becomes greater, or as temperature decreases. Since it varies with frequency of the radiant energy, the "D line" of sodium is usually used as the standard for comparison.

4.18. Surface tension of water in dynes per square centimeter is approximately equal to $75.64-0.144T+0.0399Cl$, where T is temperature in degrees Celsius (centigrade) and Cl is the chlorinity of the water in parts per thousand. As indicated by the last term, the surface tension increases with chlorinity, and is therefore a little more for seawater than for freshwater. However, the presence of impurities causes it to be somewhat less than indicated by the formula.

4.19. Transparency of seawater varies with the number, size, and nature of particles suspended in the water, as well as with the nature and intensity of illumination. The rate of decrease of light energy with depth is called the "extinction coefficient." The earliest method of measuring transparency was by means of a **Secchi disk,**

a white disk 30 centimeters (a little less than 1 foot) in diameter. This was lowered into the sea, and the depth at which it disappeared was recorded. In coastal waters the depth varies from about 5 to 25 meters (16 to 82 feet). Offshore, the depth is usually about 45 to 60 meters (148 to 197 feet). The greatest recorded depth at which the disk has disappeared is 66 meters (217 feet), in the Sargasso Sea.

Although the Secchi disk still affords a simple method of measuring transparency, more exact methods have been devised.

4.20. Color.—The color of seawater varies considerably. Water of the Gulf Stream is a deep indigo blue, while a similar current off Japan was named Kuroshio (Black Stream) because of the dark color of its water. Along many coasts the water is green. In certain localities a brown or brownish-red water has been observed. Colors other than blue are caused by biological sources, such as plankton, or by suspended sediments from river runoff.

Offshore, some shade of blue is common, particularly in tropical or subtropical regions. It is due to scattering of sunlight by minute particles suspended in the water, or by molecules of the water itself. Because of its short wavelength, blue light is more effectively scattered than light of longer waves. Thus, the ocean appears blue for the same reason that the sky does (art. 2.17). The green color often seen near the coast is a mixture of the blue due to scattering of light and a stable soluble yellow pigment associated with phytoplankton (art. 4.24). Brown or brownish-red water receives its color from large quantities of certain types of **algae,** microscopic plants in the sea or from river runoff.

4.21. Marine geology is a branch of oceanography dealing with bottom relief, particularly the characteristics of ocean basins and the geological processes that brought them into being and tend to alter them, as well as with marine sediments.

4.22. Bottom relief.—Compared to land, relatively little is known of relief below the surface of the sea. Until recent years, the sea has proved an effective barrier to acquisition of knowledge of features below its surface. Although soundings of 1,000 fathoms were probably made as early as the second century BC (art. 4.2), the number of deep sea soundings by means of a weight lowered to the bottom had been relatively few. The process was a time-consuming one requiring special equipment. Several hours were needed for a single sounding. Since the development of an effective echo sounder in 1922, the number of deep sea soundings has greatly increased. Later, a recording echo sounder was developed to permit the continuous tracing of a **bottom profile.** This has assisted materially in the acquisition of knowledge of bottom relief. By this means, many underwater mountain ranges, and other features have been discovered. Although the main features are becoming known, a great many details are yet to be learned.

Along most of the coasts of the continents, the bottom slopes gradually downward to a depth of about 100 fathoms or somewhat less, where it falls away more rapidly to greater depths. This **continental shelf** (fig. 4.22a) averages about 30 miles in width, but varies from nothing to about 800 miles, the widest part being off the Siberian arctic coast. A similar shelf extending outward from an island or group of islands is called an **island shelf.** At the outer edge of the shelf, the steeper slope of 2° to 4° is called the **continental slope,** or the **island slope,** according to whether it surrounds a continent or group of islands. The shelf itself is not uniform, but has numerous hills, ridges, terraces, and canyons, the largest being comparable in size to the Grand Canyon.

The relief of the ocean floor is comparable to that of land. Both have steep, rugged mountains, deep canyons, rolling hills, plains, etc. Most of the ocean floor is considered to be made up of a number of more-or-less circular or oval depressions called **basins,** surrounded by walls (sills) of lesser depth.

Undersea features (figs. 4.22a and 4.22b) are defined as follows:

Archipelagic apron or *apron.*—A gentle slope with a generally smooth surface on the sea floor, particularly as found around groups of islands or seamounts.

Bank.—An elevation of the sea floor located on a shelf and over which the depth of water is relatively shallow but sufficient for safe surface navigation.

Basin.—A depression of variable extent and more-or-less circular or oval in form.

Borderland or *continental borderland.*—A region adjacent to a continent, normally occupied by or bordering a shelf, that is highly irregular with depths well in excess of those typical of a shelf.

Canyon.—A relatively narrow, deep depression with steep slopes, the bottom of which generally grades downward.

Cone.—See FAN.

Continental borderland.—See BORDERLAND.

Continental margin.—The zone separating the emergent continent from the deep sea bottom, generally consisting of the rise, slope, and shelf.

Continental rise.—A gentle slope rising toward the foot of the continental slope. See RISE.

Continental shelf.—See SHELF.

Cordillera.—An entire mountain system including all the subordinate ranges, interior plateaus, and basins.

Escarpment or *scarp.*—An elongated and comparatively steep slope of the sea floor, separating flat or gently sloping areas.

Fan or *cone.*—A gently sloping, fan-shaped feature normally located near the lower termination of a canyon.

Fracture zone.—An extensive linear zone of unusually irregular topography of the sea floor characterized by large seamounts, steep-sided or asymmetrical ridges, troughs, or escarpments.

Gap.—A depression cutting transversely across a ridge or rise.

Hill.—A small elevation rising generally less than 200 meters from the sea floor.

Hole.—A small depression of the sea floor.

Knoll.—An elevation rising less than 1,000 meters from the sea floor and of limited extent across the summit.

Levee.—An embankment bordering either one or both sides of a seachannel or the low-gradient seaward part of a canyon or valley.

Moat.—An annual depression that may not be continuous, located at the base of many seamounts or islands.

Mountains.—A well delineated subdivision of a large and complex positive feature, generally part of a cordillera.

Peak.—An individual pointed top on a ridge or a complex seamount.

Plain.—A flat, gently sloping or nearly level region of the sea floor.

Plateau.—A comparatively flat-topped elevation of the sea floor of considerable extent across the summit and usually rising more than 200 meters on at least one side.

Province.—A region composed of a group of similar bathymetric features whose characteristics are markedly in contrast with surrounding areas.

Range.—A series of ridges or seamounts, generally parallel.

Reef.—An offshore consolidated rock hazard to navigation with a least depth of 20 meters (or 10 fathoms) or less.

Ridge.—A long, narrow elevation of the sea floor with steep sides.

Rise.—A long, broad elevation that rises gently and generally smoothly from the sea floor.

Saddle.—A low part on a ridge or between seamounts.

Seachannel.—A long, narrow, U-shaped, or V-shaped, shallow depression of the sea floor, usually occurring on a gently sloping plain or fan.

Seamount.—An elevation rising 1,000 meters or more from the sea floor, and of limited extent across the summit.

Shelf or *continental shelf.*—A zone adjacent to a continent or around an island, and extending from the low waterline to the depth at which there is usually a marked increase of slope to greater depth.

Shoal.—An offshore hazard to navigation with a least depth of 20 meters (or 10 fathoms) or less, composed of unconsolidated material.

Sill.—The low part of the ridge or rise separating ocean basins from one another or from the adjacent sea floor.

Slope or *continental slope.*—The declivity seaward from a shelf into greater depth.

Spur.—A subordinate elevation, ridge, or rise projecting from a larger feature.

Tablemount or *Guyot.*—A seamount having a comparatively smooth, flat top.

Terrace or *bench.*—A bench-like feature bordering an undersea feature.

Trench.—A long, narrow and deep depression of the sea floor, with relatively steep sides.

Trough.—A long depression of the sea floor, normally wider and shallower than a trench.

Valley.—A relatively shallow, wide depression with gentle slopes, the bottom of which generally grades continuously downward. This term is used for features that do not have canyon-like characteristics in any significant part of their extent.

The term **deep** may be used for a very deep part of the ocean, generally that part deeper than 3,000 fathoms.

The average depth of water in the oceans is 2,075 fathoms (12,450 feet), as compared to an average height of land above the sea of about 2,750 feet. The greatest known depth is 35,800 feet, in the Marianas Trench in the Pacific. The highest known land is Mount Everest, 29,002 feet. About 23 percent of the ocean is shallower than 10,000 feet, about 76 percent is between 10,000 and 20,000 feet, and a little more than 1 percent is deeper than 20,000 feet.

4.23. Marine sediments.—The ocean floor is composed of material deposited there through the years. This material consists principally of (1) earth and rocks washed into the sea by streams and waves, (2) volcanic ashes and lava, and (3) the remains of marine organisms. Lesser amounts of land material are carried into the sea by glaciers, blown out to sea by wind, or deposited by chemical means. This latter process is responsible for the manganese nodules that cover some parts of the ocean floor. In the ocean, the material is transported by ocean currents, waves, and ice. Near shore the material is deposited at the rate of about 3 inches in 1,000 years, while in the deep water offshore the rate is only about half an inch in 1,000 years. Marine deposits in water deep enough to be relatively free from wave action are subject to little erosion. Recent studies have shown that some bottom currents are strong enough to move sediments. There are **turbidity currents,** similar to land slides, that move large masses of sediments. Turbidity currents have been known to rip apart large transoceanic cables on the ocean bottom. Because of this and the slow rate of deposit, marine sediments provide a better geological record than does the land.

Marine sediments are composed of individual particles of all sizes from the finest clay to large boulders. In general, the inorganic deposits near shore are relatively coarse (sand, gravel, shingle, etc.), while those in deep water are much finer (clay). In some areas the siliceous remains of marine organisms or the calcareous deposits (of either organic or inorganic origin) are sufficient to predominate on the ocean floor.

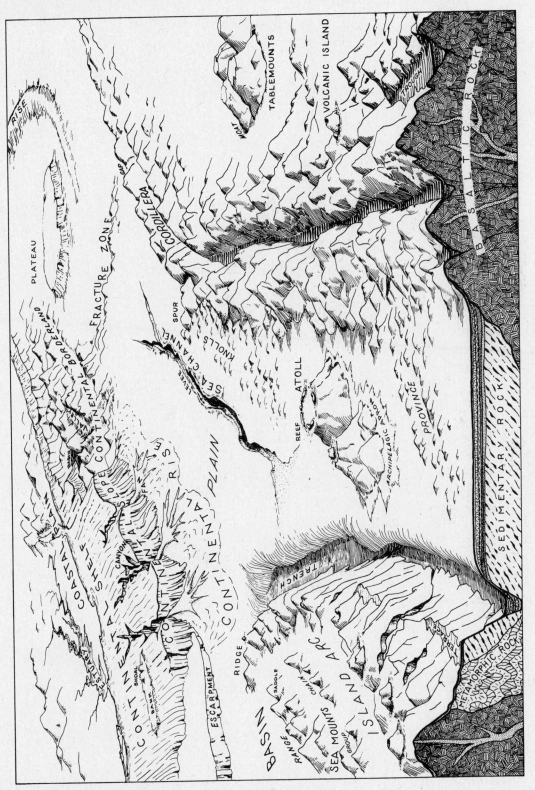

FIGURE 4.22a.—Ocean basin features.

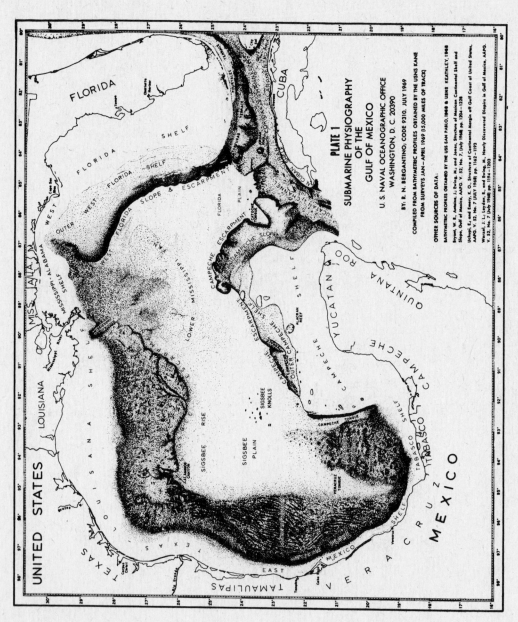

FIGURE 4.22b.—Ocean basin features.

A wide range of colors is found in marine sediments. The lighter colors (white or a pale tint) are usually associated with coarse-grained quartz or limestone deposits. Darker colors (red, blue, green, etc.) are usually found in mud having a predominance of some mineral substance, such as an oxide of iron or manganese. Black mud is often found in an area that is little disturbed, such as at the bottom of an inlet or in a depression without free access to other areas.

Marine sediments are studied primarily by means of bottom samples. Samples of surface deposits are obtained by means of a **snapper** (for mud, sand, etc.) or "dredge" (usually for rocky material). If a sample of material below the bottom surface is desired, a "coring" device is used. This device consists essentially of a tube driven into the bottom by weights or explosives. A sample obtained in this way preserves the natural order of the various layers. Samples of more than 100 feet in depth have been obtained by means of coring devices. The bottom sample obtained by the mariner, by arming his lead with tallow or soap, is an incomplete indication of bottom surface conditions. conditions.

4.24. Synoptic oceanography.—Bathythermograph and sea surface temperature observations are reported directly to the Fleet Numerical Weather Central, Monterey, California. These synoptic reports are then analyzed by computer to determine ocean thermal conditions at any point in the Northern Hemisphere.

References

Crease, J. "The Origin of Ocean Currents." *Journal of the Institute of Navigation* (British), vol. 5, no. 3 (July 1952).

Day, A., Rear Admiral. "Navigation and Hydrography." *Journal of the Institute of Navigation* (British), vol. 6, no. 1 (January 1953).

Deacon, G. E. R. "Oceanographical Research and Navigation." *Journal of the Institute of Navigation* (British), vol. 4, no. 3 (July 1951).

Defant, A. *Physical Oceanography.* (2 vols.) New York, Pergamon, 1961.

Marmer, H. A. *The Scope of Oceanography.* James Johnstone Memorial Volume. Liverpool, University Press of Liverpool, 1934.

National Research Council. *Physics of the Earth—Oceanography.* Bulletin no. 85, Chapter V. Washington, The National Academy of Sciences, 1932.

Satow, P. G. "Some Problems of Underwater Navigation." *Journal of the Institute of Navigation* (British), vol. 4, no. 3 (July 1951).

Shepard, F. P. *Submarine Geology.* New York, Harper, 1948.

Sverdrup, H. U., M. W. Johnson, and R. H. Fleming. *The Oceans, Their Physics, Chemistry and General Biology.* New York, Prentice-Hall, 1942.

CHAPTER V

TIDES AND TIDAL CURRENTS

General

5.1. The tidal phenomenon is the periodic motion of the waters of the sea due to differences in the attractive forces of various celestial bodies, principally the moon and sun, upon different parts of the rotating earth. It can be either a help or hindrance to the mariner—the water's rise and fall may at certain times provide enough depth to clear a bar and at others may prevent him from entering or leaving a harbor. The flow of the current may help his progress or hinder it, may set him toward dangers or away from them. By understanding this phenomenon and by making intelligent use of predictions published in tide and tidal current tables and of descriptions in sailing directions, the mariner can set his course and schedule his passage to make the tide serve him, or at least to avoid its dangers.

5.2. Tide and current.—In its rise and fall, the tide is accompanied by a periodic horizontal movement of the water called **tidal current.** The two movements, tide and tidal current, are intimately related, forming parts of the same phenomenon brought about by the tide-producing forces of the sun and moon, principally.

It is necessary, however, to distinguish clearly between tide and tidal current, for the relation between them is not a simple one nor is it everywhere the same. For the sake of clearness and to avoid misunderstanding, it is desirable that the mariner adopt the technical usage: **tide** for the vertical rise and fall of the water, and **tidal current** for the horizontal flow. The tide rises and falls, the tidal current floods and ebbs. In British usage, tidal current is called **tidal stream.**

5.3. Cause.—It is often said of science that the ability to predict a natural event is indicative of understanding. Since tides are the most accurately predictable oceanographic phenomena, one could easily assume that physical oceanographers truly understand them. Unfortunately, this is not true; significant gaps remain. An examination of the details of this apparent contradiction gives insight into one of the most exciting areas of oceanography—ocean tides.

To facilitate this examination, it will be desirable, first of all, to discuss the fundamental tide-generating forces and the theoretical equilibrium tide they try to produce. The principal tide-generating forces on the surface of the earth result from the differential gravitational forces of the moon and sun. The moon is the main tide-generating body. Due to its greater distance, the effect of the sun is only 46 percent of the effect due to the moon. After the theoretical equilibrium tide produced by the sun and moon is discussed in this article, the actual tide as observed in nature is described in article 5.4. Observed tides will differ considerably from the tides predicted by equilibrium theory since size, depth, and configuration of the basin or waterway, friction, landmasses, inertia of watermasses, Coriolis acceleration, and other factors are neglected in this theory. Nevertheless, equilibrium theory will be sufficient to describe the magnitude and distribution of the main tide-generating forces across the surface of the earth.

Tide-Generating Forces

Newton's universal law of gravitation governs both the orbits of celestial bodies and the tide-generating forces which occur on these bodies. The force of gravitational attraction between any two masses, m_1 and m_2, is given by

$$F = \frac{Gm_1m_2}{d^2},$$

where d is the distance between the two masses and G is a constant which depends upon the units employed. This law assumes that m_1 and m_2 are point masses. Newton was able to show that homogeneous spheres could be treated as point masses when determining their orbits. However, when computing differential gravitational forces, the actual dimensions of the masses must be taken into account.

Using the law of gravitation, it is found that the orbits of two point masses are conic sections about the barycenter of the two masses. If either one or both of the masses are homogenous spheres instead of point masses, the orbits are the same as the orbits which would result if all of the mass of the sphere were concentrated at a point at the center of the sphere. In the case of the earth-moon system, both the earth and the moon describe elliptical orbits about their barycenter if the simplifying assumption is made that both bodies are homogeneous spheres and the gravitational forces of the sun and other planets are neglected. The earth-moon barycenter is located at a distance of 0.74 R_E from the earth's center, where R_E is the radius of the earth. This is approximately three-fourths of the distance from the center of the earth to the surface of the earth along the line connecting the centers of the earth and moon (fig. 5.3a).

Thus, the center of mass of the earth describes a very small ellipse about the earth-moon barycenter whereas the center of mass of the moon describes a much larger ellipse about the same barycenter. If the gravitational forces of the other bodies of the solar system are neglected, Newton's law of gravitation also predicts that the earth-moon barycenter will describe an orbit which is approximately elliptical about the barycenter of the sun-earth-moon system. This barycentric point lies inside the sun (fig. 5.3b).

The *differences* in gravitational attraction of various celestial bodies, principally the moon and sun, upon different parts of the rotating and revolving earth are the fundamental tide-generating forces. These *differential gravitational* forces are described here by considering first the effects of the moon only. The results will be general and can be applied directly to describe the differential gravitational forces of the sun.

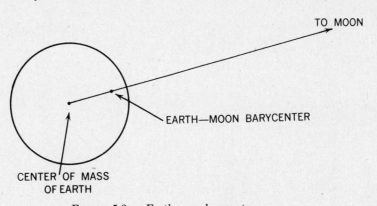

FIGURE 5.3a.—Earth-moon barycenter.

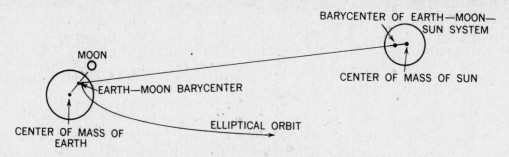

FIGURE 5.3b.—Orbit of earth-moon barycenter.

Earth-Moon System

When determining the orbit of the earth's center of mass about the earth-moon barycenter, the gravitational force exerted by the moon on the earth is given by

$$F = \frac{GM_E M_M}{d^2{}_M},$$

where M_E and M_M are the masses of the earth and moon, and d_M is the distance between their centers of mass. Acceleration and force are related by Newton's second law of motion, $F = ma$. The acceleration, a, of a mass, m, is then $a = F/m$. The terms "acceleration" and "force per unit mass" may be used interchangeably. Combining Newton's law of gravitation and his second law of motion, the acceleration of the earth's center of mass about the earth-moon barycenter is

$$a_c = \frac{GM_M}{d^2{}_M}.$$

In determining the direction and magnitude of tide-generating forces of the moon, the simplest case to treat is that of the sublunar point and its antipode on the earth. These two points are labeled P_1 and P_2 in figure 5.3c.

The acceleration at the point P_1 due to the gravitational attraction of the moon is

$$a_1 = \frac{GM_M}{(d_M - R_E)^2},$$

where R_E is the radius of the earth. The acceleration at the point P_2 due to the gravitational attraction of the moon is

$$a_2 = \frac{GM_M}{(d_M + R_E)^2}.$$

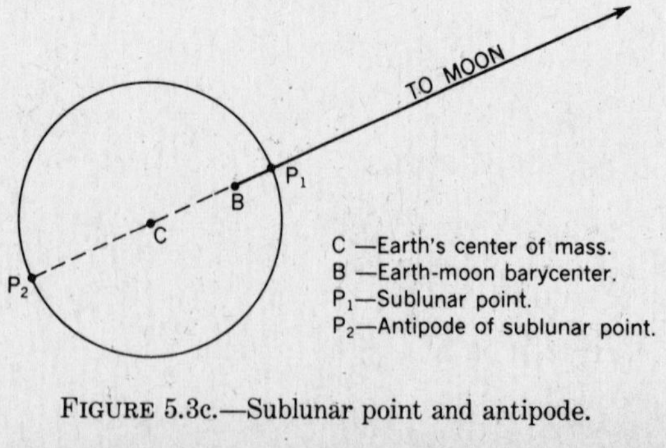

TO MOON

C —Earth's center of mass.
B —Earth-moon barycenter.
P_1—Sublunar point.
P_2—Antipode of sublunar point.

FIGURE 5.3c.—Sublunar point and antipode.

The differential acceleration at P_1, that is, the acceleration at P_1 relative to the acceleration of the center of the earth, is

$$a_1 - a_c = \frac{GM_M}{(d_M - R_E)^2} - \frac{GM_M}{d^2_M}.$$

With some simplification, this expression becomes

$$a_1 - a_c = \frac{2GM_M R_E}{d^3_M}.$$

This is the differential acceleration or differential force per unit mass at the sublunar point P_1. In a similar manner, the differential acceleration at the antipode, point P_2, is found to be

$$a_2 - a_c = \frac{-2GM_M R_E}{d^3_M}.$$

Both the accelerations and the resulting differential accelerations or differential forces per unit mass are shown in figure 5.3d.

Note that the differential gravitational force per unit mass at the antipode, point P_2, is negative. Thus, at both the sublunar point and the antipode, the moon's differential gravitational forces are vertical and directed away from the center of the earth.

A more complicated situation occurs when the moon's gravitational forces act on points of the earth's surface other than the sublunar point and the antipode. To find the differential accelerations at the sublunar point and the antipode, it was only necessary to take the algebraic difference between the accelerations at the surface and the acceleration of the earth's center. At other points on the surface, however, both the magnitude and direction of the moon's gravitational forces differ from that at the earth's center (fig. 5.3e). To obtain the differential accelerations at the other points, it is necessary to subtract the accelerations vectorially: $\vec{F}_D = \bar{a}_M - \bar{a}_C$, where $\vec{F}_D$ is the differential acceleration or differential force per unit mass at any point on the surface of the earth, $\bar{a}_C$ is the acceleration of the earth's center of mass, and $\bar{a}_M$ is the acceleration or force per unit mass at the point under consideration due to the moon's gravitational force acting at that point (fig. 5.3f).

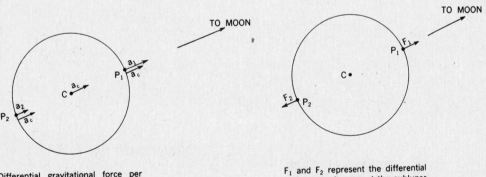

Differential gravitational force per unit mass at sublunar point P_1 is $a_1 - a_c$.

Differential gravitational force per unit mass at antipode, point P_2, is $a_2 - a_c$.

F_1 and F_2 represent the differential forces per unit mass at the sublunar point and the antipode, points P_1 and P_2, where
$F_1 = a_1 - a_c$
$F_2 = a_2 - a_c$

FIGURE 5.3d.—Differential gravitational forces per unit mass at sublunar point and antipode.

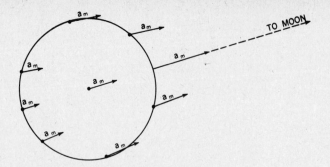

a $_m$ is the acceleration or force per unit mass at various points on the earth's surface due the moon's gravitational forces at these points.

FIGURE 5.3e.—Forces per unit mass on earth's surface due to moon's gravitational forces. Only the acceleration at the earth's center and the acceleration along one great circle through the sublunar point and the antipode are shown.

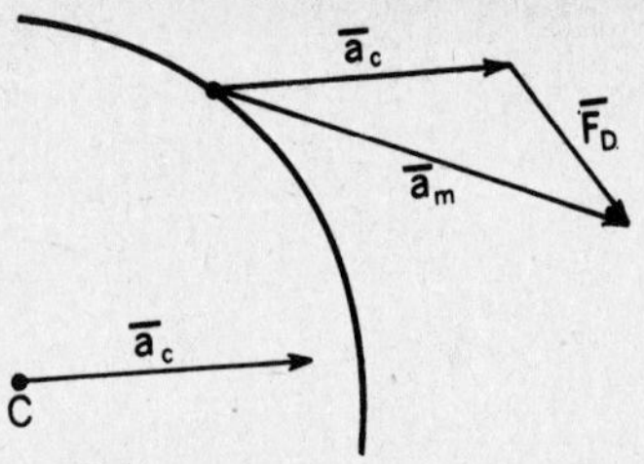

FIGURE 5.3f.—Differential force per unit mass, $\vec{F}_D$, is the vector difference $\bar{a}_M - \bar{a}_C$.

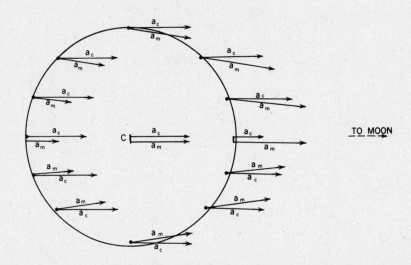

FIGURE 5.3g.—Accelerations due to the moon's gravitational forces, $\bar{a}_M$, compared to the acceleration of the center of the earth, $\bar{a}_C$. Comparisons are made at the earth's center and at various points along a great circle connecting the sublunar point and the antipode. The effects will be the same along all great circles. connecting these two points.

The relative effects of $\overline{a}_M$ and $\bar{a}_C$ at the center of the earth and at various points along one great circle through the sublunar point and the antipode are shown in figure 5.3g. The resultant differential forces per unit mass are shown in figure 5.3h.

If it is assumed that the entire surface of the earth is covered with a uniform layer of water, the differential forces may be resolved into components perpendicular and parallel to the surface of the earth (fig. 5.3i) to determine their effect.

The components of these differential forces which are perpendicular to the earth's surface have the effect of changing the weight of the mass on which they are acting. These vertical components do not contribute to the tidal effect. The horizontal components which are parallel to the earth's surface, although small, have the effect of moving the water in a horizontal direction towards the sublunar and antipodal points until an equilibrium position is found. The horizontal components of the differential forces are

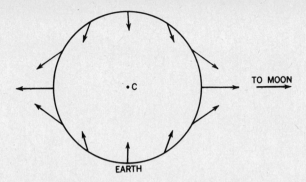

FIGURE 5.3h.—Differential forces along a great circle connecting the sublunar point and antipode.

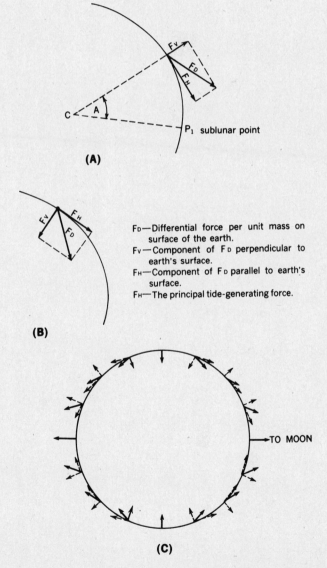

F_D—Differential force per unit mass on surface of the earth.
F_V—Component of F_D perpendicular to earth's surface.
F_H—Component of F_D parallel to earth's surface.
F_H—The principal tide-generating force.

FIGURE 5.3i.—Differential force resolved into horizontal and vertical components. (A) F_D directed out of surface. (B) F_D directed into surface. (C) Varying directions of F_D along a great circle through sublunar point.

the principal tide-generating forces. These are also called **tractive forces.** Figure 5.3j shows the tractive forces across the surface of the earth.

The magnitudes of the horizontal and vertical components are

$$F_H = \frac{3}{2} \frac{GM_M R_E}{d^3{}_M} \sin 2A$$

$$F_V = \frac{GM_M R_E}{d^3{}_M} (3 \cos^2 A - 1),$$

where A is the angle at the center of the earth between the line connecting the sublunar point and the antipode and the line from the center of the earth to the point under consideration (fig. 5.3i). Thus, it can be seen that the horizontal component which is the tide-generating force, is zero when the angle A is zero (sublunar point and antipode). It is also zero when the angle A is 90°. This corresponds approximately to the great circle connecting observers for which the moon is setting. The maximum value of F_H occurs when the angle A is 45°.

Equilibrium will be reached when a bulge of water has formed at the sublunar and antipodal points such that the tractive forces due to the moon's differential gravitational forces on the mass of water covering the surface of the earth are just balanced by the earth's gravitational attraction (fig. 5.3k).

Consider now the effects of the rotation of the earth which were previously neglected. If the declination of the moon is 0°, the bulges will lie on the equator of the earth. As the earth rotates, an observer at the equator will note that the moon transits approximately every 24 hours and 50 minutes. Since there are two bulges of water on the equator, one at the sublunar point and the other at the antipode, the observer will also see two high tides during this interval with one high tide occurring when the moon is overhead and another high tide 12 hours 25 minutes later when the observer is located at the antipode. He will also experience two low tides, one between each high tide. The range of these equilibrium tides at the equator will be less than 1 meter.

The heights of the two high tides should be equal at the equator. At points north or south of the equator, an observer would still experience two high and two low tides, but the heights of the high tides, although still equal, would not be as great as they are at the equator.

The effects of the declination of the moon are shown in figure 5.3l.

The preceding paragraphs addressed the tide-generating forces due to the differential gravitational forces of the moon on the earth. For the sublunar point, the force per unit mass was found to be

$$\frac{2GM_M R_E}{d^3{}_M}.$$

In a similar manner, the differential gravitational force per unit mass due to the sun at the subsolar point is found to be

$$\frac{2GM_S R_E}{d^3{}_S},$$

where M_S is the mass of the sun and d_S is the distance between the centers of mass of the sun and earth. To find the relative effects of the sun and moon, the ratio of the expressions can be used. This ratio is

$$\left(\frac{M_S}{M_M} \right) \left(\frac{d_M}{d_S} \right)^3.$$

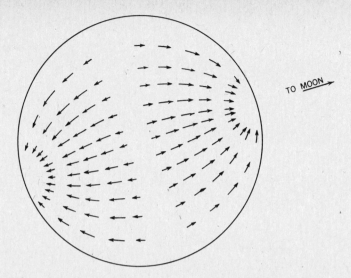

FIGURE 5.3j.—Tractive forces across the surface of the earth. Tractive forces are zero at the sublunar and antipodal points and along the great circle halfway between these two points. Tractive forces are maximum along the small circles located 45° from the sublunar point and the antipode.

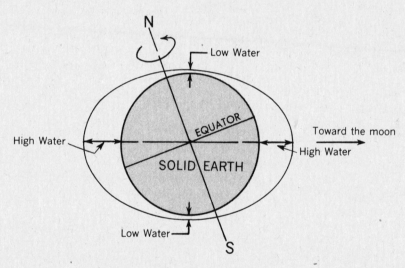

FIGURE 5.3k.—Theoretical equilibrium configuration due to moon's differential gravitational forces. One bulge of the water envelope is located at the sublunar point, the other bulge at the antipode.

The numerical value of this ratio is 0.46. Thus, the effect of the moon is approximately two and one-quarter times greater than the effect of the sun even though the moon's mass is but a fraction of the sun's. This is due to the fact that the differential forces vary inversely as the *cube* of the distance. Thus the moon's smaller mass is offset by its much shorter distance to the earth.

The preceding discussion pertaining to the effects of the moon is equally valid when discussing the effects of the sun, taking into account that the magnitude of the solar effects are smaller than the lunar effects. Hence, the tides will also vary according to the sun's declination and its varying distance from the earth. A second envelope of water representing the equilibrium tides due to the sun would resemble the envelope shown in figure 5.3k except that the heights of the high tides would be smaller.

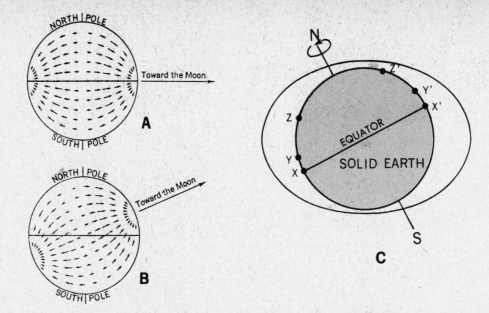

FIGURE 5.31.—Effects of the declination of the moon. (A) When the moon is in the plane of the equator, the forces are equal in magnitude at the two points on the same parallel of latitude and 180° apart in longitude. (B) When the moon is at north (or south) declination, the forces are unequal at such points and tend to cause an inequality in the two high waters and the two low waters of a day. (C) Observers at points X, Y, and Z experience one high tide when moon is on their meridian, then another high tide 12 hours 25 minutes later when at X', Y', and Z'. The second high tide is the same at X' as at X. High tides at Y' and Z' are lower than high tides at Y and Z.

Spring and Neap Tides

The combined lunar-solar effect is obtained by adding the sun's tractive forces vectorially to the moon's tractive forces. The resultant tidal bulge will be predominantly lunar with modifying solar effects upon both the height of the tide and the direction of the tidal bulge. Special cases of interest occur during the times of new and full moon (fig. 5.3m). With the earth, moon, and sun lying approximately on the same line, the tractive forces of the sun are acting in the same direction as the moon's tractive forces (modified by declination effects). The results are tides called *spring tides* whose ranges are greater than average.

Another case of interest occurs when the moon is at first and third quarters. At those times, the tractive forces of the sun are acting at approximately right angles to the moon's tractive forces (fig. 5.3m). The results are tides called *neap tides* whose ranges are less than average.

With the moon in positions between quadrature and new and full moon, the effect of the sun is to cause the tidal bulge to either lag or precede the moon (fig. 5.3n). These effects are called *priming* and *lagging* the tides.

Tide

5.4. General features.—Tide is the periodic rise and fall of the water accompanying the tidal phenomenon. At most places it occurs twice daily. The tide rises until it reaches a maximum height, called **high tide** or **high water,** and then falls to a minimum level called **low tide** or **low water.**

The rate of rise and fall is not uniform. From low water, the tide begins to rise slowly at first but at an increasing rate until it is about halfway to high water. The

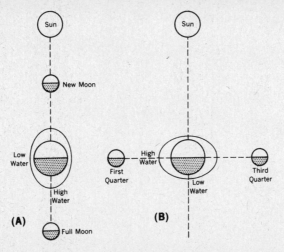

FIGURE 5.3m.—(A) Spring tides occur at times of new and full moon. Range of tide is greater than average since solar and lunar tractive forces act in same direction. (B) Neap tides occur at times of first and third quarters. Range of tide is less than average since solar and lunar tractive forces act at right angles.

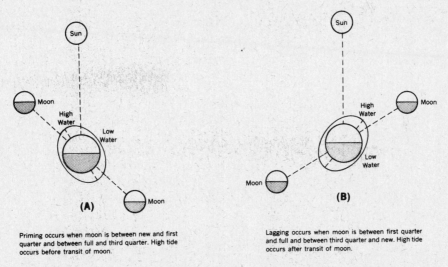

Priming occurs when moon is between new and first quarter and between full and third quarter. High tide occurs before transit of moon.

Lagging occurs when moon is between first quarter and full and between third quarter and new. High tide occurs after transit of moon.

FIGURE 5.3n.—Priming and lagging the tides.

rate of rise then decreases until high water is reached and the rise ceases. The falling tide behaves in a similar manner. The period at high or low water during which there is no sensible change of level is called **stand.** The difference in height between consecutive high and low waters is the **range.**

Figure 5.4 is a graphical representation of the rise and fall of the tide at New York during a 24-hour period. The tide curve has the general form of a sine curve.

5.5. Types of tide.—A body of water has a natural period of oscillation that is dependent upon its dimensions. None of the oceans appears to be a single oscillating body, but rather each one is made up of a number of oscillating basins. As such basins are acted upon by the tide-producing forces, some respond more readily to daily or diurnal forces, others to semidiurnal forces, and others almost equally to both. Hence, tides at a place are classified as one of three types—**semidiurnal, diurnal,** or **mixed**—according to the characteristics of the tidal pattern occurring at the place.

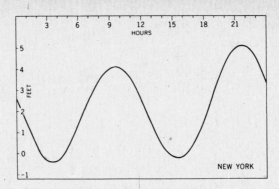

FIGURE 5.4.—The rise and fall of the tide at New York, shown graphically.

In the **semidiurnal** type of tide, there are two high and two low waters each tidal day, with relatively small inequality in the high and low water heights. Tides on the Atlantic coast of the United States are representative of the semidiurnal type, which is illustrated in figure 5.5a by the tide curve for Boston Harbor.

In the **diurnal** type of tide, only a single high and single low water occur each tidal day. Tides of the diurnal type occur along the northern shore of the Gulf of Mexico, in the Java Sea, the Gulf of Tonkin (off the Vietnam-China coast), and in a few other localities. The tide curve for Pei-Hai, China, illustrated in figure 5.5b, is an example of the diurnal type.

In the **mixed** type of tide, the diurnal and semidiurnal oscillations are both important factors and the tide is characterized by a large inequality in the high water heights, low water heights, or in both. There are usually two high and two low waters each day, but occasionally the tide may become diurnal. Such tides are prevalent along the Pacific coast of the United States and in many other parts of the world. Examples of mixed types of tide are shown in figure 5.5c. At Los Angeles, it is typical that the inequalities in the high and low waters are about the same. At Seattle the greater inequalities are typically in the low waters, while at Honolulu it is the high waters that have the greater inequalities.

5.6. Solar tide.—The natural period of oscillation of a body of water may accentuate either the solar or the lunar tidal oscillations. Though it is a general rule that the tides follow the moon, the relative importance of the solar effect varies in different areas. There are a few places, primarily in the South Pacific and the Indonesian areas, where the solar oscillation is the more important, and at those places the high and low waters occur at about the same time each day. At Port Adelaide, Australia (fig. 5.6), the solar and lunar semidiurnal oscillations are equal and nullify one another at neaps (art. 5.8).

5.7. Special effects.—As a progressive wave enters shallow water its speed is decreased. Since the trough is shallower than the crest, its retardation is greater, re-

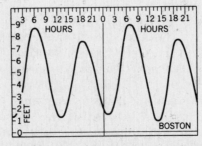

FIGURE 5.5a.—Semidiurnal type of tide.

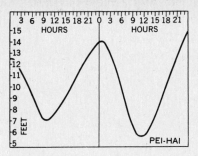

FIGURE 5.5b.—Diurnal type of tide.

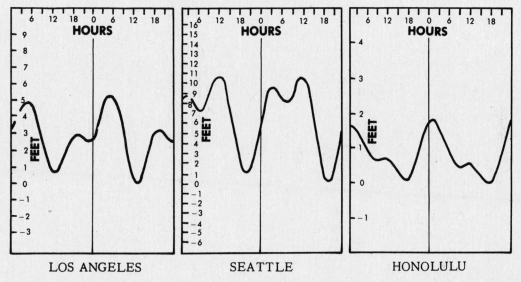

LOS ANGELES   SEATTLE   HONOLULU

FIGURE 5.5c.—Mixed type of tide.

sulting in a steepening of the wave front. Therefore, in many rivers, the duration of rise is considerably less than the duration of fall. In a few estuaries, the advance of the low water trough is so much retarded that the crest of the rising tide overtakes the low, and advances upstream as a churning, foaming wall of water called a **bore**. Bores that are large and dangerous at times of large tidal ranges may be mere ripples at those times of the month when the range is small. Examples occur in the Petitcodiac River in the Bay of Fundy, and at Haining, China, in the Tsientang Kaing. The tide tables indicate where bores occur.

Other special features are the **double low water** (as at Hoek Van Holland) and the **double high water** (as at Southampton, England). At such places there is often a slight fall or rise in the middle of the high or low water period. The practical effect is to create a longer period of stand at high or low tide. The tide tables direct attention to these and other peculiarities where they occur.

5.8. Variations in range.—Though the tide at a particular place can be classified as to type, it exhibits many variations during the month (fig. 5.6). The range of the tide varies in accordance with the intensity of the tide-producing force, though there may be a lag of a day or two (**age of tide**) between a particular astronomic cause and the tidal effect.

Thus, when the moon is at the point in its orbit nearest the earth (at *perigee*), the lunar semidiurnal range is increased and **perigean** tides occur; when the moon is farthest from the earth (at *apogee*), the smaller **apogean** tides occur. When the moon and sun are in line and pulling together, as at new and full moon, **spring** tides occur (the term

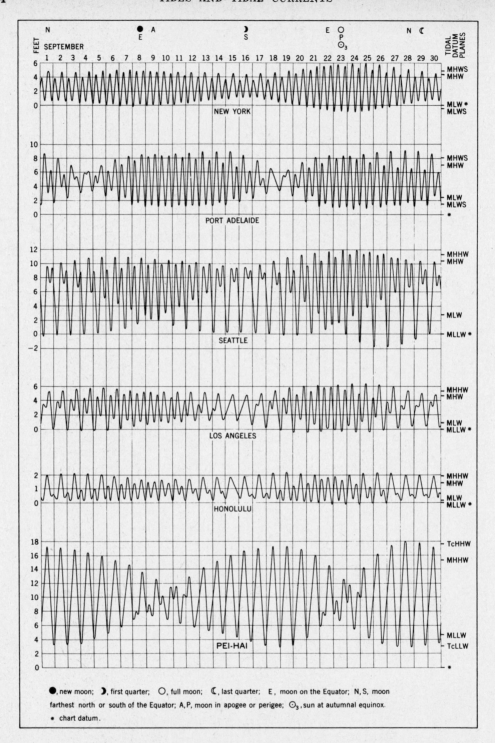

FIGURE 5.6.—Tidal variations at various places during a month.

spring has nothing to do with the season of year); when the moon and sun oppose each other, as at the quadratures, the smaller **neap** tides occur.

When certain of these phenomena coincide, the great **perigean spring** tides, the small **apogean neap** tides, etc., occur.

These are variations in the semidiurnal portion of the tide. Variations in the diurnal portion occur as the moon and sun change declination. When the moon is at its maximum semi-monthly declination (either north or south), **tropic** tides occur in which the diurnal effect is at a maximum; when it crosses the equator, the diurnal effect is a minimum and **equatorial** tides occur.

It should be noted that when the range of tide is increased, as at spring tides, there is more water available only at *high* tide; at *low* tide there is less, for the high waters rise higher and the low waters fall lower at these times. There is more water at neap low water than at spring low water. With tropic tides, there is usually more depth at one low water during the day than at the other. While it is desirable to know the meanings of these terms, the best way of determining the height of the tide at any place and time is to examine the tide predictions for the place as given in the tide tables. Figure 5.8 illustrates variations in the ranges and heights of tides in a locality such as the Indian Ocean where predicted and observed water levels are referenced to a chart sounding datum that will always cause them to be additive relative to the charted depth.

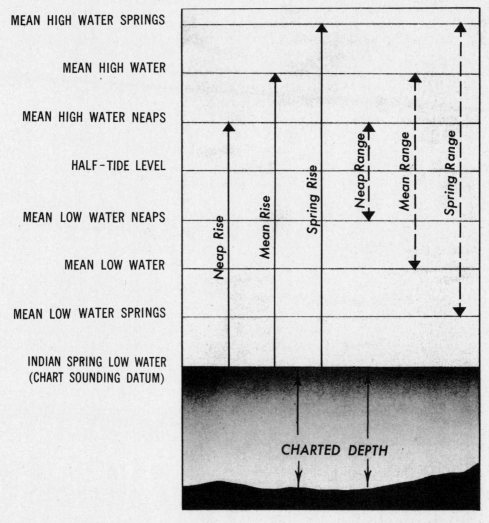

FIGURE 5.8.—Variations in the ranges and heights of tide in a locality where the chart sounding datum is Indian Spring Low Water.

5.9. Tidal cycles.—Tidal oscillations go through a number of cycles. The shortest cycle, completed in about 12 hours and 25 minutes for a semidiurnal tide, extends from any phase of the tide to the next recurrence of the same phase. During a **lunar day** (averaging 24 hours and 50 minutes) there are two highs and two lows (two of the shorter cycles) for a semidiurnal tide. The moon revolves around the earth with respect to the sun in a **synodical month** of about 29½ days, commonly called the **lunar month.** The effect of the phase variation is completed in one-half a synodical month or about 2 weeks as the moon varies from new to full or full to new. The effect of the moon's declination is also repeated in one-half of a **tropical month** of 27⅓ days or about each 2 weeks. The cycle involving the moon's distance requires an **anomalistic month** of about 27½ days. The sun's declination and distance cycles are respectively a half year and a year in length. An important lunar cycle, called the **nodal period,** is 18.6 years (usually expressed in round figures as 19 years). For a tidal value, particularly a range, to be considered a true mean, it must be either based upon observations extended over this period of time or adjusted to take account of variations known to occur during the cycle.

5.10. Time of tide.—Since the lunar tide-producing force has the greater effect in producing tides at most places, the tides "follow the moon." Because of the rotation of the earth, high water lags behind meridian passage (upper and lower) of the moon. The **tidal day,** which is also the **lunar day,** is the time between consecutive transits of the moon, or 24 hours and 50 minutes on the average. Where the tide is largely semidiurnal in type, the **lunitidal interval**—the interval between the moon's meridian transit and a particular phase of tide—is fairly constant throughout the month, varying somewhat with the tidal cycles. There are many places, however, where solar or diurnal oscillations are effective in upsetting this relationship, and the newer editions of charts of many countries now omit intervals because of the tendency to use them for prediction even though accurate predictions are available in tide tables. However, the lunitidal interval may be encountered. The interval generally given is the average elapsed time from the meridian transit (upper or lower) of the moon until the next high tide. This may be called **mean high water lunitidal interval** or **corrected** (or **mean**) **establishment.** The **establishment of the port, high water full and change** (HWF &C), or **vulgar** (or **common**) **establishment,** sometimes given, is the average interval on days of full or new moon, and approximates the mean high water lunitidal interval.

In the ocean, the tide may be of the nature of a progressive wave with the crest moving forward, a stationary or standing wave which oscillates in a seesaw fashion, or a combination of the two. Consequently, caution should be used in inferring the time of tide at a place from tidal data for nearby places. In a river or estuary, the tide enters from the sea and is usually sent upstream as a progressive wave, so that the tide occurs progressively later at various places upstream.

5.11. Tidal datums.—A **tidal datum** is a level from which heights and depths are measured. There are a number of such levels of reference that are important to the mariner. The relation of the tide each day during a month to these datums is shown, for certain places, in figure 5.6.

The most important level of reference to the mariner is the datum of soundings on charts. Since the tide rises and falls continually while soundings are being taken during a hydrographic survey, the tide should be observed during the survey so that soundings taken at all stages of the tide can be reduced to a common **chart sounding datum.** Soundings on charts show depths below a selected low water datum (occasionally mean sea level), and tide predictions in tide tables show heights above the same level. The depth of water available at any time is obtained by adding the height of the tide at the time in question to the charted depth, or by subtracting the predicted height if it is negative.

By international agreement, the level used as chart datum should be just low enough so that low waters do not go far below it. At most places, however, the level used is one determined from a mean of a number of low waters (usually over a 19-year period); therefore, some low waters can be expected to fall below it. The following are some of the datums in general use.

The highest low water datum in considerable use is **mean low water (MLW),** which is the average height of all low waters at a place. About half of the low waters fall below it. **Mean low water springs (MLWS),** usually shortened to **low water springs,** is the average level of the low waters that occur at the times of spring tides. **Mean lower low water (MLLW)** is the average height of the lower low waters of each tidal day. **Tropic lower low water (TcLLW)** is the average height of the lower low waters (or of the single daily low waters if the tide becomes diurnal) that occur when the moon is near maximum declination and the diurnal effect is most pronounced. This datum is not in common use as a tidal reference. **Indian spring low water (ISLW)** sometimes called **Indian tide plane** or **harmonic tide plane,** is a low water datum that includes the spring effect of the semi-diurnal portion of the tide and the tropic effect of the diurnal portion. It is about the level of lower low water of mixed tides at the time that the moon's maximum declination coincides with the time of new or full moon. **Mean lower low water springs** is the average level of the lower of the two low waters on the days of spring tides. Some still lower datums used on charts are determined from tide observations and some are determined arbitrarily and later referred to the tide. Most of them fall close to one or the other of the following two datums. **Lowest normal low water** is a datum that approximates the average height of monthly lowest low waters, discarding any tides disturbed by storms. **Lowest low water** is an extremely low datum. It conforms generally to the lowest tide observed, or even somewhat lower. Once a tidal datum is established, it is sometimes retained for an indefinite period, even though it might differ slightly from a better determination from later observations. When this occurs, the established datum may be called **low water datum, lower low water datum,** etc. These datums are used in a limited area and primarily for river and harbor engineering purposes. Examples are *Boston Harbor Low Water Datum* and *Columbia River Lower Low Water Datum*.

In some areas where there is little or no tide, such as the Baltic Sea, **mean sea level (MSL)** is used as chart datum. This is the average height of the surface of the sea for all stages of the tide over a 19-year period. This may differ slightly from **half-tide level,** which is the level midway between mean high water and mean low water.

Inconsistencies of terminology are found among charts of different countries and between charts issued at different times. For example, the spring effect as defined here is a feature of only the semidiurnal tide, yet it is sometimes used synonymously with tropic effect to refer to times of increased range of a diurnal tide. Such inconsistencies are being reduced through increased international cooperation.

Large-scale charts usually specify the datum of soundings and may contain a tide note giving mean heights of the tide at one or more places on the chart. These heights are intended merely as a rough guide to the change in depth to be expected under the specified conditions. They should not be used for the prediction of heights on any particular day. Such predictions should be obtained from *tide tables* (arts. 11.3–11.6).

5.12. High water datums.—Heights of land features are usually referred on nautical charts to a high water datum. The one used on charts of the United States, its territories, and possessions, and widely used elswehere, is **mean high water (MHW),**

which is the average height of all high waters over a 19-year period. Any other high water datum in use on charts is likely to be higher than this. Other high water datums are **mean high water springs** (**MHWS**), which is the average level of the high waters that occur at the time of spring tides; **mean higher high water** (**MHHW**), which is the average height of the higher high waters of each tidal day; and **tropic higher high water** (**TcHHW**), which is the average height of the higher high waters (or the single daily high waters if the tide becomes diurnal) that occur when the moon is near maximum declination and the diurnal effect is most pronounced. A reference merely to "high water" leaves some doubt as to the specific level referred to, for the height of high water varies from day to day. Where the range is large, the variation during a 2-week period may be considerable.

As there are periodic and apparent secular trends in sea level, a specific 19-year cycle (the **National Tidal Datum Epoch**) is issued for all United States datums. The National Tidal Datum Epoch officially adopted by the National Ocean Survey is presently 1941 through 1959. The Epoch will be reviewed for consideration for revision at 25-year intervals.

5.13. Observations and predictions.—Since the tide at different places responds differently to the tide-producing forces, the nature of the tide at any place can be determined most accurately by actual observation. The predictions in tide tables and the tidal data on nautical charts are based upon observations.

Tides are usually observed by means of a continuously recording gage. A year of observations is the minimum length desirable for determining the **harmonic constants** used in prediction. For establishing mean sea level and the long-time changes in the relative elevations of land and sea, as well as for other special uses, observations have been made over periods of 20, 30, and even 120 years at important locations. Observations for a month or less will establish the *type* of tide and suffice for comparison with a longer series of a similar type to determine tidal differences and constants.

Mathematically, the variations in the lunar and solar tide-producing forces, such as those due to changing phase, distance, and declination, are considered as separate constituent forces, and the **harmonic analysis** of observations reveals the response of each constituent of the tide to its corresponding force. At any one place this response remains constant and is shown for each constituent by **harmonic constants** which are in the form of a phase angle for the time relation and an amplitude for the height. Harmonic constants are used in making technical studies of the tide and predictions on computers and mechanical **tide predicting machines.** Most published tide predictions are made by computer.

5.14. Tide tables are published annually by most of the maritime nations of the world. They consist primarily of two parts. One contains predictions of the time and height of each high and low water for every day of the year for many important ports called **reference stations.** The other part contains tidal differences and ratios for thousands of other places, called **subordinate stations,** and specifies the reference station to which the differences are to be applied in order to obtain time and height of tide for any day at the subordinate station. The type of tide at a subordinate station is the same as at its reference station. The use of tide tables is explained in articles 11.3-11.6.

5.15. Meteorological effects.—The foregoing discussion of tide behavior assumes normal weather conditions. The level of the sea is affected by wind and atmospheric pressure. In general, onshore winds raise the level and offshore winds lower it, but the amount of change varies at different places. During periods of low atmospheric pressure, the water level tends to be higher than normal. For a stationary low, the increase in elevation can be found by the formula

$$R_0 = 0.0325(1010 - P),$$

in which R_0 is the increase in elevation in feet, and P is the atmospheric pressure in millibars. This is equal approximately to 1 centimeter per millibar depression, or 1 foot (13.6 inches) per inch depression. For a moving low, the increase in elevation is given by the formula

$$R = \frac{R_0}{1 - \frac{C^2}{gh}},$$

in which R is the increase in elevation in feet, R_0 is the increase in feet for a stationary low, C is the rate of motion of the low in feet per second, g is the acceleration due to gravity (32.2 feet per second per second), and h is the depth of water in feet.

Where the range of tide is very small, the meteorological effect may sometimes be greater than the normal tide.

Tidal Current

5.16. Tidal and nontidal currents.—Horizontal movement of the water is **current.** It may be classified as "tidal" and "nontidal." **Tidal current** is the periodic horizontal flow of water accompanying the rise and fall of the tide, and results from the same cause. **Nontidal current** is any current not due to the tidal movement. Nontidal currents include the permanent currents in the general circulatory system of the oceans as well as temporary currents arising from meteorological conditions. The current experienced at any time is usually a combination of tidal and nontidal currents.

In navigation, the effect of the tidal current is often of more importance than the changing depth due to the tide, and many mariners speak of "the tide," when they have in mind the flow of the tidal current.

5.17. General features.—Offshore, where the direction of flow is not restricted by any barriers, the tidal current is **rotary;** that is, it flows continuously, with the direction changing through all points of the compass during the tidal period. The tendency for the rotation in direction has its origin in the deflecting force of the earth's rotation, and unless modified by local conditions, the change is clockwise in the Northern Hemisphere and counterclockwise in the Southern Hemisphere. The speed usually varies throughout the tidal cycle, passing through two maximums in approximately opposite directions, and two minimums about halfway between the maximums in time and direction. Rotary currents can be depicted as in figure 5.17a, by a series of arrows representing the direction and speed of the current at each hour. This is sometimes called a **current rose.** Because of the elliptical pattern formed by the ends of the arrows, it is also referred to as a **current ellipse.**

In rivers or straits, or where the direction of flow is more or less restricted to certain channels, the tidal current is **reversing;** that is, it flows alternately in approximately opposite directions with an instant or short period of little or no current, called **slack water,** at each reversal of the current. During the flow in each direction, the speed varies from zero at the time of slack water to a maximum, called **strength of flood** or **ebb,** about midway between the slacks. Reversing currents can be indicated graphically, as in figure 5.17b, by arrows that represent the speed of the current at each hour. The flood is usually depicted above the slack waterline and the ebb below it. The tidal current curve formed by the ends of the arrows has the same characteristic sine form as the tide curve. (In illustrations for certain purposes, as in figures 5.18b and 5.20b, it is convenient to omit the arrows and show only the curve.)

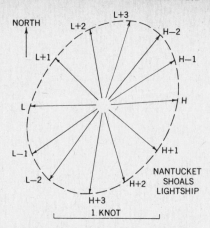

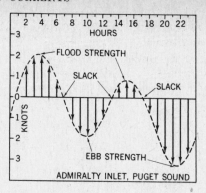

FIGURE 5.17b.—Reversing tidal current.
See figure 5.20b.

FIGURE 5.17a.—Rotary tidal current.
Times are hours before and after high
and low tide at Nantucket Shoals
Lightship. The bearing and length of
each arrow represents the hourly
direction and speed of the current.
See figure 5.20a.

A slight departure from the sine form is exhibited by the reversing current in a strait, such as East River, New York, that connects two tidal bodies of water. The tides at the two ends of a strait are seldom in phase or equal in range, and the current, called **hydraulic current,** is generated largely by the continuously changing difference in height of water at the two ends. The speed of a hydraulic current varies nearly as the square root of the difference in height. The speed reaches a maximum more quickly and remains at strength for a longer period than shown in figure 5.17b, and the period of weak current near the time of slack is considerably shortened.

The current *direction* or **set** is the direction *toward* which the current flows. The *speed* is sometimes called the **drift.** The term "velocity" is often used as the equivalent of "speed" when referring to current, although strictly "velocity" implies direction as well as speed. The term "strength" is also used to refer to speed, but more often to greatest speed between consecutive slack waters. The movement toward shore or up-stream is the **flood,** the movement away from shore or downstream is the **ebb.** In a purely semidiurnal type of current unaffected by nontidal flow, the flood and ebb each last about 6 hours and 13 minutes. But if there is either diurnal inequality or non-tidal flow, the durations of flood and ebb may be quite unequal.

5.18. Types of tidal current.—Tidal currents may be of the semidiurnal, diurnal, or **mixed** type; corresponding to a considerable degree to the type of tide at the place, but often with a stronger semidiurnal tendency.

The tidal currents in tidal estuaries along the Atlantic coast of the United States are examples of the semidiurnal type of reversing current. At Mobile Bay entrance they are almost purely diurnal. At most places, however, the type is mixed to a greater or lesser degree. At Tampa and Galveston entrances there is only one flood and one ebb each day when the moon is near its maximum declination, and two floods and two ebbs each day when the moon is near the equator. Along the Pacific coast of the United States there are generally two floods and two ebbs every day, but one of the floods or ebbs has a greater speed and longer duration than the other, the inequality varying with the declination of the moon. The inequalities in the current often differ considerably from place to place even within limited areas, such as adjacent passages in Puget Sound and various passages between the Aleutian Islands. Figure 5.18a shows several types of reversing current. Figure 5.18b shows how the flood disappears as the diurnal inequality increases at one station.

Offshore rotary currents that are purely semidiurnal repeat the elliptical pattern (fig. 5.17a) each tidal cycle of 12 hours and 25 minutes. If there is considerable diurnal inequality, the plotted hourly current arrows describe a set of two ellipses of different sizes during a period of 24 hours and 50 minutes, as shown in figure 5.18c, and the greater the diurnal inequality, the greater the difference between the sizes of the two ellipses. In a completely diurnal rotary current, the smaller ellipse disappears and only one ellipse is produced in 24 hours and 50 minutes.

5.19. Variations and cycles.—Tidal currents have periods and cycles similar to those of the tides (art. 5.9) and are subject to similar variations, but flood and ebb of the current do not necessarily occur at the same times as the rise and fall of the tide. The relationship is explained further in article 5.21.

The speed at strength increases and decreases during the 2-week period, month, and year with the variations in the range of tide. Thus, the stronger **spring** and **perigean currents** occur near the times of new and full moon and near the times of the moon's perigee, or at times of spring and perigean tides (art. 5.8); the weaker **neap** and **apogean**

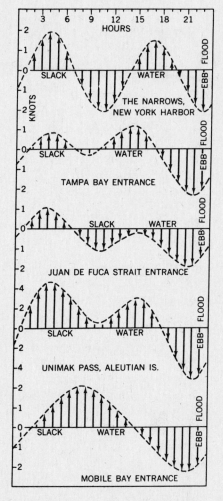

FIGURE 5.18a.—Several types of reversing current. The pattern changes gradually from day to day, passing through cycles somewhat similar to that shown for tides in figure 5.06.

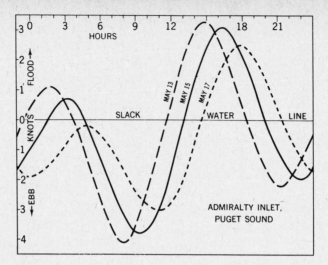

FIGURE 5.18b.—Changes in current of the mixed type. Note that each day as the inequality increases, the morning slacks draw together in time until on the 17th the morning flood disappears. On that day the current ebbs throughout the morning.

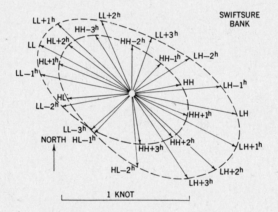

FIGURE 5.18c.—Rotary tidal current with diurnal inequality. Times are in hours referred to tides (higher high, lower low, lower high, and higher low) at Swiftsure Bank.

currents occur at the times of neap and apogean tides; **tropic currents** with increased diurnal speeds or with larger diurnal inequalities in speed occur at times of tropic tides; and **equatorial currents** with a minimum diurnal effect occur at times of equatorial tides; etc.

As with the tide, a *mean value* represents an average obtained from a 19-year series. Since a series of current observations is usually limited to a few days, and seldom covers more than a month or two, it is necessary to adjust the observed values, usually by comparison with tides at a nearby place, to obtain such a mean.

5.20. Effect of nontidal flow.—The current existing at any time is seldom purely tidal, but usually includes also a nontidal current that is due to drainage, oceanic circulation, wind, or other cause. The method in which tidal and nontidal currents combine is best explained graphically, as in figures 5.20a and 5.20b. The pattern of the tidal current remains unchanged, but the curve is shifted from the point or line from which the currents are measured in the direction of the nontidal current and by an amount equal to it. It is sometimes more convenient graphically merely to move the line or point of origin in the opposite direction.

Thus, the speed of the current flowing in the direction of the nontidal current is increased by an amount equal to the magnitude of the nontidal current, and the speed of the current flowing in the opposite direction is decreased by an equal amount. In figure 5.20a a nontidal current is represented in both direction and speed by the vector *AO*. Since this is greater than the speed of the tidal current in the opposite direction, the point *A* is outside the ellipse. The direction and speed of the combined tidal and nontidal currents at any time is represented by a vector from *A* to that point on the curve representing the given time, and can be scaled from the graph. The strongest and weakest currents may no longer be in the directions of the maximum and minimum of the tidal current. In a reversing current (fig. 5.20b), the effect is to advance the time of one slack and to retard the following one. If the speed of the nontidal current exceeds that of the reversing tidal current, the resultant current flows continuously in one direction without coming to a slack. In this case, the speed varies from a maximum to a minimum and back to a maximum in each tidal cycle. In figure 5.20b the horizontal line *A* represents slack water if only tidal currents are present. Line *B* represents the effect of a 0.5-knot nontidal ebb, and line *C* the effect of a 1.0-knot nontidal ebb. With the condition shown at *C* there is only one flood each tidal day. If the nontidal ebb were to increase to approximately 2 knots, there would be no flood, two maximum ebbs and two minimum ebbs occurring during a tidal day.

5.21. Relation between time of tidal current and time of tide.—At many places where current and tide are both semidiurnal, there is a definite relation between times of current and times of high and low water in the locality. Current atlases and notes on nautical charts often make use of this relationship by presenting for particular locations the direction and speed of the current at each succeeding hour after high and low water at a place for which tide predictions are available.

In localities where there is considerable diurnal inequality in tide or current, or where the type of current differs from the type of tide, the relationship is not constant, and it may be hazardous to try to predict the times of current from times of tide.

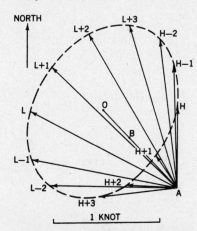

FIGURE 5.20a.—Effect of nontidal current on the rotary tidal current of figure 5.17a. If the nontidal current is northwest at 0.3 knot, it may be represented by *BO*, and all hourly directions and speeds will then be measured from *B*. If it is 1.0 knot, it will be represented by *AO* and the actual resultant hourly directions and speeds will be measured from *A*, as shown by the arrows.

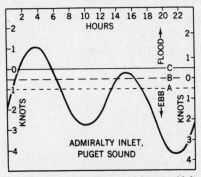

FIGURE 5.20.—Effect of nontidal current on the reversing tidal current of figure 5.17b. If the nontidal current is 0.5 knot in the ebb direction, the ebb is increased by moving the slack water line from position *A* up 0.5 knot to position *B*. Speeds will then be measured from this broken line as shown by the scale on the right, and times of slack are changed. If the nontidal current is 1.0 knot in the ebb direction, as shown by line *C*, the speeds are as shown on the left, and the current will not reverse to a flood in the afternoon; it will merely slacken at about 1500.

Note the current curve for Unimak Pass in the Aleutians in figure 5.18a. It shows the current as predicted in the tidal current tables. Predictions of high and low waters in the tide tables might have led one to expect the current to change from flood to ebb in the late morning, whereas actually the current continued to run flood with some strength at that time.

Since the relationship between times of tidal current and tide is not everywhere the same, and may be variable at the same place, one should exercise extreme caution in using general rules. The belief that slacks occur at local high and low tides and that the maximum flood and ebb occur when the tide is rising or falling most rapidly may be approximately true at the seaward entrance to, and in the upper reaches of, an inland tidal waterway. But generally this is not true in other parts of inland waterways. When an inland waterway is extensive or its entrance constricted, the slacks in some parts of the waterway often occur midway between the times of high and low tide. Usually in such waterways the relationship changes from place to place as one progresses upstream, slack water getting progressively closer in time to the local tide maximum until at the head of tidewater (the inland limit of water affected by a tide) the slacks occur at about the times of high and low tide.

5.22. Relation between speed of current and range of tide.—The variation in the speed of the tidal current from place to place is not necessarily consistent with the range of tide. It may be the reverse. For example, currents are weak in the Gulf of Maine where the tides are large, and strong near Nantucket Island and in Nantucket Sound where the tides are small.

At any one place, however, the speed of the current at strength of flood and ebb varies during the month in about the same proportion as the range of tide, and one can use this relationship to determine the relative strength of currents on any day.

5.23. Variation across an estuary.—In inland tidal waterways the *time* of tidal current varies across the channel from shore to shore. On the average, the current turns earlier near shore than in midstream, where the speed is greater. Differences of half an hour to an hour are not uncommon, but the difference varies and the relationship may be nullified by the effect of nontidal flow.

The *speed* of the current also varies across the channel, usually being greater in midstream or midchannel than near shore, but in a winding river or channel the strongest currents occur near the concave shore. Near the opposite (convex) shore the currents are weak or may eddy.

5.24. Variation with depth.—In tidal rivers the subsurface current acting on the lower portion of the hull may differ considerably from the surface current. An appreciable subsurface current may be present when the surface movement appears to be practically slack, and the subsurface current may even be flowing with appreciable speed in the opposite direction to the surface current.

In a tidal estuary, particularly in the lower reaches where there is considerable difference in density from top to bottom, flood usually begins earlier near the bottom than at the surface. The differences may be an hour or two or as little as a few minutes, depending upon the estuary, the location in the estuary, and freshet conditions. Even when the freshwater runoff becomes so great as to prevent the surface current from flooding, it may still flood below the surface. The difference in time of ebb from surface to bottom is normally small but subject to variation with time and location.

The ebb speed at strength usually decreases gradually from top to bottom, but the speed of flood at strength often is stronger at subsurface depths than at the surface.

5.25. Observations.—Observations of the current are made by means of a current meter or current pole and log line. In the past, most successful meters required a vessel and observers in continual attendance, as is necessary with the pole and line. Because of the difficulty and expense of such observations, they usually covered only a period of a day or two at a place. Observations of a month are the exception, and longer series were obtained only where ship and observers were available because of other duties, such as at lightships, where observations have been continued over a number of years.

Newer meters have been and are being developed that are suspended from a buoy and that record either in the buoy or send speed and direction impulses by radio to a base station on ship or land. With them, the period of observation has been increased so that in some recent surveys of United States harbors, the minimum period of observation was 1 week, with observations at several stations being continued over a period of 1 to 6 months.

5.26. Tidal current tables and other sources of information.—The navigator should not attempt to predict currents without specific information for the locality in which he is interested. Such information is contained in various forms in many navigational publications.

Tidal current tables, issued annually, list daily predictions of the times and strengths of flood and ebb currents, and of the times of intervening slacks or minima. Due to lack of observational data, coverage is considerably more limited than for the tides. The tidal current tables do include supplemental data by which tidal current predictions can be determined for many places in addition to those for which daily predictions are given. The predictions are made by computers, using current harmonic constants that are obtained by analyzing current observations in the same manner as for tides (art. 5.13). The use of tidal current tables is explained in articles 11.7–11.10.

Sailing directions and **coast pilots** issued by maritime nations include general descriptions of current behavior in various localities throughout the world.

Tidal current charts. A number of important harbors and waterways are covered by sets of tidal current charts showing graphically the hourly current movement.

Tidal Current Diagrams are a series of monthly diagrams used with the tidal current charts. The diagrams directly indicate the chart to use and the speed correction factor to apply to each chart.

The use of tables and charts for tide and current predictions is discussed in chapter XI.

References

Doodson, A. T., and H. D. Warburg. *Admiralty Manual of Tides.* London, H. M. Stationery Office, 1941.

Marmer, H. A. *The Tide.* New York, Appleton, 1926.

Schureman, Paul. *Manual of the Harmonic Analysis and Prediction of Tides.* Rev. ed. U. S. Coast and Geodic Survey Special Publication No. 98, Washington, U. S. Govt. Print. Off., 1940.

Schureman, Paul. *Tide and Current Glossary.* National Ocean Survey. Washington, U. S. Govt. Print. Off., 1975.

U. S. Coast and Geodetic Survey. *Manual of Current Observations.* Special Publication No. 215. Rev. ed. Washington, U. S. Govt. Print. Off., 1950.

U. S. Coast and Geodetic Survey. *Manual of Tide Observations.* Publication 30–1. Washington, U. S. Govt. Print. Off., 1965.

National Ocean Survey. *Tidal Current Charts* and *Tidal Current Diagrams.* Washington, several published periodically.

National Ocean Survey. *Tide Tables* and *Tidal Current Tables.* Washington, U. S. Govt. Print. Off., several volumes of each published annually.

CHAPTER VI

OCEAN CURRENTS

6.1. Introduction.—The movement of water comprising the oceans is one of the principal sources of discrepancy between dead reckoning and actual positions of vessels. Water in essentially horizontal motion is called a **current,** the direction *toward* which it moves being the **set,** and its speed the **drift.** A well-defined current extending over a considerable region of the ocean is called an **ocean current.**

A **periodic current** is one the speed or direction of which changes cyclically at somewhat regular intervals, as a tidal current. A **seasonal current** is one which has large changes in speed or direction due to seasonal winds. A **permanent current** is one which experiences relatively little periodic or seasonal change.

A **coastal current** flows roughly parallel to a coast, outside the surf zone, while a **longshore current** is one parallel to a shore, inside the surf zone, and generated by waves striking the beach at an angle. Any current some distance from the shore may be called an **offshore current,** and one close to the shore an **inshore current.**

A **surface current** is one present at the surface, particularly one that does not extend more than a relatively few feet below the surface. A **subsurface current** is one which is present below the surface only.

There is evidence to indicate that the strongest ocean currents consist of relatively narrow, high-speed streams that follow winding, shifting courses. Often associated with these currents are secondary **countercurrents** flowing adjacent to them but in the opposite direction, and somewhat local, roughly circular, **eddy currents.** A relatively narrow, deep, fast-moving current is sometimes called a **stream current,** and a broad, shallow, slow-moving one a **drift current.**

6.2. Causes of ocean currents.—Although man's knowledge of the processes which produce and maintain ocean currents is far from complete, he does have a general understanding of the principal factors involved. The primary generating forces are wind and the density differences in the water. In addition, such factors as depth of water, underwater topography, shape of the basin in which the current is running, extent and location of land, and deflection by the rotation of the earth all affect the oceanic circulation.

6.3. Wind currents.—The stress of wind blowing across the sea causes the surface layer of water to move. This motion is transmitted to each succeeding layer below the surface, but due to internal friction within the water, the rate of motion decreases with depth. The current is called **Ekman wind current** or simply **wind current.** Although there are many variables, it is generally true that a steady wind for about 12 hours is needed to establish such a current.

A wind-driven current does not flow in the direction of the wind, being deflected by Coriolis force (art. 2.3), due to rotation of the earth. This deflection is toward the *right* in the Northern Hemisphere, and toward the *left* in the Southern Hemisphere. The Coriolis force is greater in higher latitudes, and is more effective in deep water. In general, the difference between wind direction and surface wind-current direction varies from about 15° along shallow coastal areas to a maximum of 45° in the deep oceans. As the motion is transmitted to successive deeper layers, the Coriolis force continues to deflect the current. At several hundred fathoms the current may flow in the opposite direction to the surface current. This shift of current directions with depth combined with the decrease in velocity with depth is called the **Ekman spiral.**

The speed of the current depends upon the speed of the wind, its constancy, the length of time it has blown, and other factors. In general, however, about 2 percent of the wind speed, or a little less, is a good average for deep water where the wind has been blowing steadily for at least 12 hours.

6.4. Currents related to density differences.—As indicated in article 4.9, the density of water varies with salinity, temperature, and pressure. At any given depth, the differences in density are due to differences in temperature and salinity. When suitable information is available, a map showing geographical density distribution at a certain depth could be drawn, with lines connecting points of equal density. These **isopycnic lines,** or lines connecting points at which a given density occurs at the same depth, would be similar to isobars on a weather map (art. 2.27), and would serve an analogous purpose, showing areas of high density and those of low density. In an area of high density, the water surface is lower than in an area of low density, the maximum difference in height being of the order of 1 to 2 feet in 40 miles. Because of this difference, water tends to flow from an area of higher water (low density) to one of lower water (high density), but due to rotation of the earth, it is deflected toward the right in the Northern Hemisphere, and toward the left in the Southern Hemisphere. Thus, a circulation is set up similar to the cyclonic and anticyclonic circulation in the atmosphere. The greater the density gradient (rate of change with distance), the faster the related current.

6.5. Oceanic circulation.—A number of ocean currents flow with great persistence, setting up a circulation that continues with relatively little change throughout the year. Because of the influence of wind in creating current (art. 6.3), there is a relationship between this oceanic circulation and the general circulation of the atmosphere (art. 2.4). The oceanic circulation is shown in figure 6.5, with the names of the major ocean currents. Some differences in opinion exist regarding the names and limits of some of the currents, but those shown are representative. The spacing of the lines is a general indication of speed, but conditions vary somewhat with the season. This is particularly noticeable in the Indian Ocean and along the South China coast, where currents are influenced to a marked degree by the monsoons (art. 2.10).

6.6. Atlantic Ocean currents.—The trade winds (art. 2.6), which blow with great persistence, set up a system of **equatorial currents** which at times extends over as much as 50° of latitude, or even more. There are two westerly flowing currents conforming generally with the areas of trade winds, separated by a weaker, easterly flowing countercurrent.

The **North Equatorial Current** originates to the northward of the Cape Verde Islands and flows almost due west at an average speed of about 0.7 knot.

The **South Equatorial Current** is more extensive. It starts off the west coast of Africa, south of the Gulf of Guinea, and flows in a generally westerly direction at an average speed of about 0.6 knot. However, the speed gradually increases until it may reach a value of 2.5 knots or more off the east coast of South America. As the current approaches Cabo de São Roque, the eastern extremity of South America, it divides, the southern part curving toward the south along the coast of Brazil, and the northern part being deflected by the continent of South America toward the north.

Between the North and South Equatorial Currents a weaker **Equatorial Countercurrent** sets toward the east in the general vicinity of the doldrums (art. 2.5). This is fed by water from the two westerly flowing equatorial currents, particularly the South Equatorial Current. The extent and strength of the Equatorial Countercurrent changes with the seasonal variations of the wind. It reaches a maximum during July

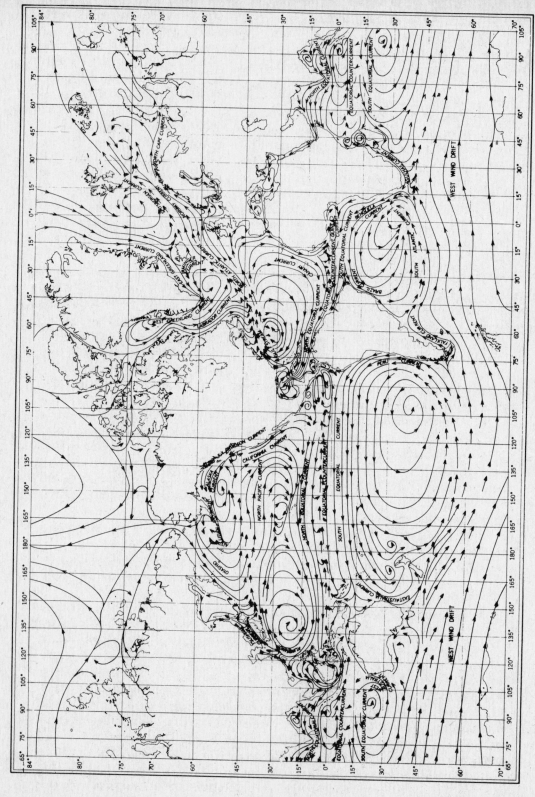

FIGURE 6.5.—Major surface currents of the world (northern hemisphere winter).

and August, when it extends from about 50° west longitude to the Gulf of Guinea. During its minimum, in December and January, it is of very limited extent, the western portion disappearing altogether.

That part of the South Equatorial Current flowing along the northern coast of South America which does not feed the Equatorial Countercurrent unites with the North Equatorial Current at a point west of the Equatorial Countercurrent. A large part of the combined current flows through various passages between the Windward Islands, into the Caribbean Sea. It sets toward the west, and then somewhat north of west, finally arriving off the Yucatan peninsula. From here, some of the water curves toward the right, flowing some distance off the shore of the Gulf of Mexico, and part of it curves more sharply toward the east and flows directly toward the north coast of Cuba. These two parts reunite in the Straits of Florida to form the most remarkable of all ocean currents, the **Gulf Stream.** Off the southeast coast of Florida this current is augmented by a current flowing along the northern coasts of Puerto Rico, Hispaniola, and Cuba. Another current flowing eastward of the Bahamas joins the stream north of these islands.

The Gulf Stream follows generally along the east coast of North America, flowing around Florida, northward and then northeastward toward Cape Hatteras, and then curving toward the east and becoming broader and slower. After passing the Grand Banks, it turns more toward the north and becomes a broad drift current flowing across the North Atlantic. That part in the Straits of Florida is sometimes called the **Florida Current.**

A tremendous volume of water flows northward in the Gulf Stream. It can be distinguished by its deep indigo-blue color, which contrasts sharply with the dull green of the surrounding water. It is accompanied by frequent squalls. When the Gulf Stream encounters the cold water of the Labrador Current, principally in the vicinity of the Grand Banks, there is little mixing of the waters. Instead, the junction is marked by a sharp change in temperature. The line or surface along which this occurs is called the **cold wall.** When the warm Gulf Stream water encounters cold air, evaporation is so rapid that the rising vapor may be visible as frost smoke (art. 2.15). The stream carries large quantities of gulfweed from the Tropics to higher latitudes.

Recent investigations have shown that the current itself is much narrower and faster than previously supposed, and considerably more variable in its position and speed. The maximum current off Florida ranges from about 2 to 4 knots. To the northward the speed is generally less, and decreases further after the current passes Cape Hatteras. As the stream meanders and shifts position, eddies sometimes break off and continue as separate, circular flows until they dissipate. Boats in the Bermuda Race have been known to be within sight of each other and be carried in opposite directions by different parts of the same current. As the current shifts position, its extent does not always coincide with the area of warm, blue water. When the sea is relatively smooth, the edges of the current are marked by ripples.

Information is not yet available to permit prediction of the position and speed of the current at any future time, but it has been found that tidal forces apparently influence the current, which reaches its daily maximum speed about 3 hours after transit of the moon. The current generally is faster at the time of neap tides than at spring tides. When the moon is over the equator, the stream is narrower and faster than at maximum northerly or southerly declination. Variations in the trade winds (art. 2.6) also affect the current.

As the Gulf Stream continues eastward and northeastward beyond the Grand Banks, it gradually widens and decreases speed until it becomes a vast, slow-moving drift current known as the **North Atlantic Current,** in the general vicinity of the prevailing westerlies (art. 2.8). In the eastern part of the Atlantic it divides into the **Northeast Drift Current** and the **Southeast Drift Current.**

The Northeast Drift Current continues in a generally northeasterly direction toward the Norwegian Sea. As it does so, it continues to widen and decrease speed. South of Iceland it branches to form the **Irminger Current** and the **Norway Current.** The Irminger Current curves toward the north and northwest to join the East Greenland Current southwest of Iceland. The Norway Current continues in a northeasterly direction along the coast of Norway. Part of it, the **North Cape Current,** rounds North Cape into the Barents Sea. The other part curves toward the north and becomes known as the **Spitsbergen Current.** Before reaching Svalbard (Spitsbergen), it curves toward the west and joins the cold **East Greenland Current** flowing southward in the Greenland Sea. As this current flows past Iceland, it is further augmented by the Irminger Current.

Off Kap Farvel, at the southern tip of Greenland, the East Greenland Current curves sharply to the northwest following the coastline. As it does so, it becomes known as the **West Greenland Current.** This current continues along the west coast of Greenland, through Davis Strait, and into Baffin Bay. Both East and West Greenland Currents are sometimes known by the single name **Greenland Current.**

In Baffin Bay the Greenland Current follows generally the coast, curving westward off Kap York to form the southerly flowing **Labrador Current.** This cold current flows southward off the coast of Baffin Island, through Davis Strait, along the coast of Labrador and Newfoundland, to the Grand Banks, carrying with it large quantities of ice (ch. X). Here it encounters the warm water of the Gulf Stream, creating the "cold wall." Some of the cold water flows southward along the east coast of North America, inshore of the Gulf Stream, as far as Cape Hatteras. The remainder curves toward the east and flows along the northern edge of the North Atlantic and Northeast Drift Currents, gradually merging with them.

The Southeast Drift Current curves toward the east, southeast, and then south as it is deflected by the coast of Europe. It flows past the Bay of Biscay, toward southeastern Europe and the Canary Islands, where it continues as the **Canary Current.** In the vicinity of the Cape Verde Islands, this current divides, part of it curving toward the west to help form the North Equatorial Current, and part of it curving toward the east to follow the coast of Africa into the Gulf of Guinea, where it is known as the **Guinea Current.** This current is augmented by the Equatorial Countercurrent and, in summer, it is strengthened by monsoon winds. It flows in close proximity to the South Equatorial Current, but in the opposite direction. As it curves toward the south, still following the African coast, it merges with the South Equatorial Current.

The clockwise circulation of the North Atlantic leaves a large central area having no well-defined currents. This area is known as the **Sargasso Sea,** from the large quantities of sargasso or gulfweed encountered there.

That branch of the South Equatorial Current which curves toward the south off the east coast of South America follows the coast as the warm, highly-saline **Brazil Current,** which in some respects resembles the Gulf Stream. Off Uruguay, it encounters the colder, less-salty Falkland Current and the two curve toward the east to form the broad, slow-moving **South Atlantic Current,** in the general vicinity of the prevailing westerlies (art. 2.8). This current flows eastward to a point west of the Cape of Good Hope, where it curves northward to follow the west coast of Africa as the strong **Benguela Current,** augmented somewhat by part of the Agulhas Current flowing around the southern part of Africa from the Indian Ocean. As it continues northward, the current gradually widens and slows. At a point east of St. Helena Island it curves westward to continue as part of the South Equatorial Current, thus completing the counterclockwise circulation of the South Atlantic. The Benguela Current is augmented somewhat by the

West Wind Drift, a current which flows easterly around Antarctica. As the West Wind Drift flows past Cape Horn, that part in the immediate vicinity of the cape is called the **Cape Horn Current.** This current rounds the cape and flows in a northerly and northeasterly direction along the coast of South America as the **Falkland Current.**

6.7. Pacific Ocean currents follow the general pattern of those in the Atlantic. The **North Equatorial Current** flows westward in the general area of the northeast trades, and the **South Equatorial Current** follows a similar path in the region of the southeast trades. Between these two, the weaker **Equatorial Countercurrent** sets toward the east, just north of the equator.

After passing the Mariana Islands, the major part of the North Equatorial Current curves somewhat toward the northwest, past the Philippines and Formosa. Here it is deflected further toward the north, where it becomes known as the **Kuroshio,** and then toward the northeast past the Nansei Shoto and Japan, and on in a more easterly direction. Part of the Kuroshio, called the **Tsushima Current,** flows through Tsushima Strait, between Japan and Korea, and the Sea of Japan, following generally the northwest coast of Japan. North of Japan it curves eastward and then southeastward to rejoin the main part of the Kuroshio. The limits and volume of the Kuroshio are influenced by the monsoons (art. 2.10), being augmented during the season of southwesterly winds, and diminished when the northeasterly winds are prevalent.

The Kuroshio (Japanese for "Black Stream") is so named because of the dark color of its water. It is sometimes called the **Japan Stream.** In many respects it is similar to the Gulf Stream of the Atlantic. Like that current, it carries large quantities of warm tropical water to higher latitudes, and then curves toward the east as a major part of the general clockwise circulation in the Northern Hemisphere. As it does so, it widens and slows. A small part of it curves to the right to form a weak clockwise circulation west of the Hawaiian Islands. The major portion continues on between the Aleutians and the Hawaiian Islands, where it becomes known as the **North Pacific Current.**

As this current approaches the North American continent, most of it is deflected toward the right to form a clockwise circulation between the west coast of North America and the Hawaiian Islands. This part of the current has become so broad that the circulation is generally weak. A small part near the coast, however, joins the southern branch of the Aleutian Current, and flows southeastward as the **California Current.** The average speed of this current is about 0.8 knot. It is strongest near land. Near the southern end of Baja (Lower) California, this current curves sharply to the west and broadens to form the major portion of the North Equatorial Current.

During the winter, a weak countercurrent flows northwestward along the west coast of North America from southern California to Vancouver Island, inshore of the southeasterly flowing California Current. This is called the **Davidson Current.**

Off the west coast of Mexico, south of Baja California, the current flows southeastward, as a continuation of part of the California Current, during the winter. During the summer, the current in this area is northwestward, as a continuation of the Equatorial Countercurrent, before it turns westward to help form the North Equatorial Current.

As in the Atlantic, there is in the Pacific a counterclockwise circulation to the north of the clockwise circulation. Cold water flowing southward through the western part of Bering Strait between Alaska and Siberia is joined by water circulating counterclockwise in the Bering Sea to form the **Oyashio.** As the current leaves the strait, it curves toward the right and flows southwesterly along the coast of Siberia and the Kuril Islands. This current brings quantities of sea ice, but no icebergs. When it encounters the Kuroshio, the Oyashio curves southward and then eastward, the greater portion joining the Kuroshio and North Pacific Current. The northern portion continues eastward to join the curving Aleutian Current.

As this current approaches the west coast of North America, west of Vancouver Island, part of it curves toward the right and is joined by water from the North Pacific Current, to form the California Current. The northern branch of the Aleutian Current curves in a counterclockwise direction to form the **Alaska Current,** which generally follows the coast of Canada and Alaska. When it arrives off the Aleutian Islands, it becomes known as the **Aleutian Current.** Part of it flows along the southern side of these islands to about the 180th meridian, where it curves in a counterclockwise direction and becomes an easterly flowing current, being augmented by the northern part of the Oyashio. The other part of the Aleutian Current flows through various openings between the Aleutian Islands, into the Bering Sea. Here it flows in a general counterclockwise direction, most of it finally joining the southerly flowing Oyashio, and a small part of it flowing northward through the eastern side of the Bering Strait, into the Arctic Ocean.

The South Equatorial Current, extending in width between about 4°N latitude and 10°S, flows westward from South America to the western Pacific. After this current crosses the 180th meridian, the major part curves in a counterclockwise direction, entering the Coral Sea, and then curving more sharply toward the south along the east coast of Australia, where it is known as the **East Australia Current.** In the Tasman Sea, northeast of Tasmania, it is augmented by water from the West Wind Drift, flowing eastward south of Australia. It curves toward the southeast and then the east, gradually merging with the easterly flowing West Wind Drift, a broad, slow-moving current that circles Antarctica.

Near the southern extremity of South America, most of this current flows eastward into the Atlantic, but part of it curves toward the left and flows generally northward along the west coast of South America as the **Peru Current** or **Humboldt Current.** Occasionally a set directly toward land is encountered. At about Cabo Blanco, where the coast falls away to the right, the current curves toward the left, past the Galapagos Islands, where it takes a westerly set and constitutes the major portion of the South Equatorial Current, thus completing the counterclockwise circulation of the South Pacific.

During the northern hemisphere summer, a weak northern branch of the South Equatorial Current, known as the **Rossel Current,** continues on toward the west and northwest along both the southern and northeastern coasts of New Guinea. The southern part flows through Torres Strait, between New Guinea and Australia, into the Arafura Sea. Here, it gradually loses its identity, part of it flowing on toward the west as part of the South Equatorial Current of the Indian Ocean, and part of it following the coast of Australia and finally joining the easterly flowing West Wind Drift. The northern part of the Rossel Current curves in a clockwise direction to help form the Pacific Equatorial Countercurrent. During the northern hemisphere winter, the Rossel Current is replaced by an easterly flowing current from the Indian Ocean.

6.8. Indian Ocean currents follow generally the pattern of the Atlantic and Pacific but with differences caused principally by the monsoons (art. 2.10) and the more limited extent of water in the Northern Hemisphere. During the northern hemisphere winter, the **North Equatorial Current** and **South Equatorial Current** flow toward the west, with the weaker, easterly flowing **Equatorial Countercurrent** flowing between them, as in the Atlantic and Pacific (but somewhat south of the equator). But during the northern hemisphere summer, both the North Equatorial Current and the Equatorial Countercurrent are replaced by the **Monsoon Current,** which flows eastward and southeastward across the Arabian Sea and the Bay of Bengal. Near Sumatra, this current curves in a clockwise direction and flows westward, augmenting the South Equatorial Current and setting up a clockwise circulation in the northern part of the Indian Ocean.

As the South Equatorial Current approaches the coast of Africa, it curves toward the southwest, part of it flowing through the Mozambique Channel between Madagascar and the mainland, and part flowing along the east coast of Madagascar. At the southern end of this island the two join to form the strong **Agulhas Current,** which is analogous to the Gulf Stream.

A small part of the Agulhas Current rounds the southern end of Africa and helps form the Benguela Current. The major portion, however, curves sharply southward and then eastward to join the West Wind Drift. This junction is often marked by a broken and confused sea. During the northern hemisphere winter the northern part of this current curves in a counterclockwise direction to form the **West Australia Current,** which flows northward along the west coast of Australia. As it passes Northwest Cape, it curves northwestward to help form the South Equatorial Current. During the northern hemisphere summer, the West Australia Current is replaced by a weak current flowing around the western part of Australia as an extension of the southern branch of the Rossel Current.

6.9. Polar currents.—The waters of the North Atlantic enter the Arctic Ocean between Norway and Svalbard. The currents flow easterly north of Siberia to the region of the Novosibirskiye Ostrova, where they turn northerly across the North Pole and continue down the Greenland coast to form the **East Greenland Current.** On the American side of the arctic basin, there is a weak, continuous clockwise flow centered in the vicinity of 80°N, 150°W. A current north through Bering Strait along the American coast is balanced by an outward southerly flow along the Siberian coast, which eventually becomes part of the Oyashio. Each of the main islands or island groups in the Arctic, as far as is known, seems to have a clockwise nearshore circulation around it. The Barents Sea, Kara Sea, and Laptev Sea each have a weak counterclockwise circulation. A similar but weaker counterclockwise current system appears to exist in the East Siberian Sea.

In the Antarctic, the circulation is generally from west to east in a broad, slow-moving current extending completely around Antarctica. This is called the **West Wind Drift,** although it is formed partly by the strong westerly wind in this area and partly by density differences. This current is augmented by the Brazil and Falkland Currents in the Atlantic, the East Australia Current in the Pacific, and the Agulhas Current in the Indian Ocean. In return, part of it curves northward to form the Cape Horn, Falkland, and most of the Benguela Currents in the Atlantic, the Peru Current in the Pacific, and west Australia Current in the Indian Ocean.

6.10. Ocean currents and climate.—Many of the ocean currents exert a marked influence upon the climate of the coastal regions along which they flow. Thus, warm water from the Gulf Stream, continuing as the North Atlantic, Northeast Drift, and Irminger Currents, arrives off the southwest coast of Iceland, warming it to the extent that Reykjavík has a higher average winter temperature than New York City, far to the south. Great Britain and Labrador are about the same latitude, but the climate of Great Britain is much milder because of the difference of temperature of currents. The west coast of the United States is cooled in the summer by the California Current, and warmed in the winter by the Davidson Current. As a result of this condition, partly, the range of monthly average temperature is comparatively small.

Currents exercise other influences besides those on temperature. The pressure pattern is affected materially, as air over a cold current contracts as it is cooled, and that over a warm current expands. As air cools above a cold ocean current, fog is likely to form. Frost smoke (art. 2.15) is most prevalent over a warm current which flows into a colder region. Evaporation is greater from warm water than from cold water.

In these and other ways, the climate of the earth is closely associated with the ocean currents, although other factors, such as topography and prevailing winds, are also important.

References

Stream Drift Chart of the World—January. Defense Mapping Agency Hydrographic Center Pilot Charts (various editions).

Stream Drift Chart of the World—July. Defense Mapping Agency Hydrographic Center Pilot Charts (various editions).

Sverdrup, H. U., M. W. Johnson, and R. H. Fleming. *The Oceans, Their Physics, Chemistry and General Biology*. New York, Prentice-Hall, 1942.

CHAPTER VII

OCEAN WAVES

7.1. Introduction.—Undulations of the surface of the water, called **waves**, are perhaps the most widely observed phenomenon at sea, and possibly the least understood by the average seaman. The mariner equipped with a knowledge of the basic facts concerning waves is able to use them to his advantage, and either avoid hazardous conditions or operate with a minimum of danger if such conditions cannot be avoided.

7.2. Causes of waves.—Waves on the surface of the sea are caused principally by wind, but other factors, such as submarine earthquakes, volcanic eruptions, and the tide, also cause waves. If a breeze of less than 2 knots starts to blow across smooth water, small wavelets called **ripples** form almost instantaneously. When the breeze dies, the ripples disappear as suddenly as they formed, the level surface being restored by surface tension of the water. If the wind speed exceeds 2 knots, more stable **gravity waves** gradually form, and progress with the wind.

While the generating wind blows, the resulting waves may be referred to as **sea.** When the wind stops or changes direction, the waves that continue on without relation to local winds are called **swell.**

Unlike wind and current, waves are not deflected appreciably by the rotation of the earth, but move in the direction in which the generating wind blows. When this wind ceases, friction and spreading cause the waves to be reduced in height, or **attenuated,** as they move across the surface. However, the reduction takes place so slowly that swell continues until it reaches some obstruction, such as a shore.

The Fleet Numerical Weather Central, Monterey, California, produces synoptic analyses and predictions of ocean wave heights using a spectral numerical model. The wave information consists of heights and directions for different periods and wavelengths. The model generates and propagates wave energy. Verification has been very good. Information from the model is provided to the U.S. Navy on a routine basis and is a vital input to the Optimum Track Ship Routing program.

7.3. Wave characteristics.—Ocean waves are very nearly in the shape of an inverted **cycloid,** the figure formed by a point inside the rim of a wheel rolling along a level surface. This shape is shown in figure 7.3a. The highest parts of waves are called **crests,** and the intervening lowest parts, **troughs.** Since the crests are steeper and narrower than the troughs, the mean or still water level is a little lower than halfway between the crests and troughs. The vertical distance between trough and crest is called **wave height**, labeled H in figure 7.3a. The horizontal distance between successive crests, measured in the direction of travel, is called **wavelength,** labeled L. The time interval between passage of successive crests at a stationary point is called **wave period (P)**. Wave height, length, and period depend upon a number of factors, such as the wind speed, the length of time it has blown, and its **fetch** (the straight distance it has traveled over the surface). Table 7.3 indicates the relationship between wind

FIGURE 7.3a.—A typical sea wave.

BEAUFORT NUMBER

Fetch	11 P	11 H	11 T	10 P	10 H	10 T	9 P	9 H	9 T	8 P	8 H	8 T	7 P	7 H	7 T	6 P	6 H	6 T	5 P	5 H	5 T	4 P	4 H	4 T	3 P	3 H	3 T
10	5.0	10.0	1.8	4.2	10.0	1.9	4.1	8.0	2.0	3.9	7.3	2.3	3.3	6.0	2.5	3.1	5.0	2.7	2.8	3.5	3.2	2.4	2.6	3.7	2.1	1.8	4.4
20	5.9	16.0	3.0	5.2	14.0	3.2	5.5	12.0	3.5	4.4	10.0	3.9	4.3	8.6	4.2	3.8	7.0	4.7	3.3	4.9	5.4	2.9	3.2	6.2	2.8	2.0	7.1
30	6.3	19.8	4.1	6.0	18.0	4.4	5.9	15.8	4.7	5.4	12.1	5.2	4.6	10.0	5.8	4.2	8.0	6.2	3.7	5.8	7.2	3.3	3.8	8.3	3.0	2.0	9.8
40	6.7	22.5	5.1	6.3	21.0	5.4	6.3	17.7	5.8	5.4	14.0	6.5	4.9	11.2	7.1	4.6	9.0	8.0	4.1	6.3	9.2	3.6	3.9	10.1	3.0	2.0	12.0
50	7.1	25.0	6.1	6.7	23.0	6.4		19.8	6.9	5.6	15.7	7.7	5.2	12.2	8.4	4.8	9.8	9.1	4.4	6.5	11.0	3.8	4.0	12.4	3.2	2.0	14.0
60	7.5	27.5	7.0	7.0	25.0	7.4	6.5	21.0	8.0	6.0	17.0	8.7	5.5	13.2	9.6	5.1	10.3	10.2	4.6	6.8	12.0	4.0	4.0	14.0	3.5	2.0	16.0
70	7.7	29.5	8.0	7.3	26.5	8.3	6.8	22.5	9.0	6.4	18.0	9.9	5.7	13.9	10.5	5.4	11.0	11.9	4.8	7.0	13.5	4.1	4.0	15.8	3.7	2.0	18.0
80	7.9	31.5	8.6	7.7	28.0	9.3	7.1	24.0	10.0	6.6	18.9	11.0	6.0	14.5	12.0	5.6	11.5	13.0	4.9	7.2	15.0	4.2	4.0	17.0	3.8	2.0	20.0
90	8.2	34.0	9.5	7.9	30.0	10.2	7.2	25.0	11.0	6.7	20.0	12.0	6.3	15.5	13.0	5.8	11.4	14.1	5.1	7.3	16.5	4.3	4.0	18.8	3.9	2.0	23.6
100	8.5	35.0	10.3	8.1	32.0	11.0	7.6	26.5	11.9	6.9	20.5	12.8	6.5	15.5	14.0	6.0	11.4	15.1	5.3	7.5	17.5	4.4	4.0	20.0	4.0	2.0	27.1
120	8.8	37.5	11.5	8.4	33.5	12.3	7.9	27.5	13.1	7.3	21.0	14.5	6.7	16.0	15.9	6.2	11.7	17.0	5.4	7.8	20.0	4.7	4.1	22.4	4.2	2.0	31.1
140	9.2	40.0	13.0	8.8	35.5	13.9	8.3	29.0	14.8	7.6	22.0	16.0	7.0	16.2	17.6	6.4	11.9	19.1	5.8	7.9	22.5	4.9	4.2	25.8	4.5	2.0	36.6
160	9.6	42.5	14.5	9.1	37.0	15.1	8.7	30.5	16.4	8.0	23.0	18.0	7.3	16.5	19.5	6.6	12.0	21.1	6.0	8.0	24.3	5.2	4.2	28.8	4.7	2.0	43.2
180	10.0	44.5	16.0	9.5	38.5	16.5	9.0	31.5	18.0	8.3	23.5	19.9	7.5	17.0	21.3	6.8	12.1	23.1	6.2	8.0	27.0	5.4	4.3	30.9	4.9	2.0	50.0
200	10.3	46.0	17.1	9.8	40.0	18.1	9.2	32.5	19.3	8.5	23.5	21.5	7.7	17.5	23.1	7.1	12.1	25.4	6.4	8.0	29.0	5.6	4.3	33.5			
220	10.6	47.5	18.2	10.3	41.5	19.1	9.6	34.0	20.9	8.8	24.0	22.9	8.0	17.9	25.0	7.2	12.3	27.2	6.6	8.0	31.1	5.8	4.4	36.5			
240	10.8	49.0	19.5	10.6	43.0	20.5	9.8	34.5	22.0	9.0	24.5	24.4	8.2	17.9	26.5	7.3	12.4	28.5	6.8	8.0	33.9	5.9	4.4	39.2			
260	11.1	50.5	20.9	10.9	44.0	21.8	10.0	34.5	23.5	9.2	25.0	26.0	8.4	18.0	28.0	7.5	12.6	30.5	6.9	8.0	34.9	6.0	4.4	41.9			
280	11.3	51.5	22.0	10.9	45.0	23.0	10.2	35.0	25.0	9.4	25.0	27.7	8.5	18.0	29.5	7.8	12.8	32.5	7.0	8.0	36.8	6.2	4.4	44.5			
300	11.6	53.0	23.2	11.1	45.0	24.3	10.4	35.0	26.3	9.5	25.0	29.0	8.7	18.0	31.5	8.0	13.1	34.1	7.1	8.0	38.5	6.3	4.4	47.0			
320	11.8	54.0	24.5	11.2	45.5	25.5	10.6	35.5	27.6	9.6	25.0	30.2	8.9	18.0	33.0	8.3	13.3	36.0	7.2	8.0	40.5						
340	12.0	55.0	25.5	11.6	46.0	26.7	10.8	36.0	28.9	9.8	25.0	31.6	9.0	18.1	34.4	8.4	13.4	37.6	7.3	8.0	42.4						
360	12.4	55.5	26.6	11.8	46.5	27.7	10.9	36.5	30.0	9.9	25.0	33.0	9.1	18.1	35.7	8.5	13.4	38.8	7.4	8.0	44.2						
380	12.6	55.5	27.7	11.8	47.0	29.1	11.1	37.0	31.3	10.0	25.5	34.2	9.3	18.4	37.1	8.5	13.5	40.2	7.5	8.0	46.1						
400		56.0	28.9	12.0	47.5	30.2	11.2	37.0	32.5	10.2	26.0	35.6	9.5	18.4	38.8	8.6	13.5	42.2	7.7	8.0	48.0						
420	12.7	56.5	29.6	12.2	47.5	31.5	11.4	37.5	33.7	10.3	26.5	36.9	9.6	18.7	40.0	8.7	13.6	43.5	7.8	8.0	50.0						
440	12.9	57.0	30.9	12.3	48.0	32.5	11.5	37.5	34.8	10.4	27.0	38.1	9.7	18.8	41.3	8.8	13.7	44.7	7.9	8.0	52.0						
460	13.1	57.5	31.8	12.5	48.5	33.5	11.7	37.5	36.0	10.6	27.5	39.5	9.8	19.0	42.8	8.9	13.7	46.2	8.0	8.0	54.0						
480	13.2	57.5	32.7	12.6	49.0	34.5	11.8	37.5	37.0	10.8	27.5	41.0	9.9	19.0	44.0	9.0	13.7	47.8	8.1	8.0	56.0						
500	13.4	58.0	33.9	12.7	49.0	35.5	11.9	38.0	38.3	10.9	27.5	42.1	10.1	19.1	45.5	9.1	13.8	49.2	8.2	8.0	58.0						
550	13.7	59.0	36.5	13.0	50.0	38.2	12.2	38.5	41.0	11.1	27.5	44.9	10.3	19.5	48.5	9.3	13.8	53.0									
600	14.0	60.0	38.7	13.3	50.0	40.3	12.5	39.0	43.6	11.4	27.5	47.7	10.5	19.8	51.8	9.5	13.8	56.3									
650	14.2	60.0	41.0	13.7	50.5	43.5	12.8	39.5	46.4	11.6	27.5	50.3	10.7	19.8	55.0												
700	14.5	60.5	43.5	14.0	50.5	45.4	13.1	40.0	49.0	11.8	27.5	53.2	11.0	19.8	58.5												
750	14.8	61.0	45.8	14.2	51.0	48.0	13.3	40.0	51.0	12.1	27.5	56.2															
800	15.0	61.5	47.8	14.5	51.5	50.6	13.5	40.0	53.8	12.3	27.5	59.2															
850	15.2	62.0	50.0	14.6	52.0	52.5	13.8	40.0	56.2																		
900	15.5	62.5	52.0	14.9	52.0	54.6	14.0	40.0	58.2																		
950	15.7	63.0	54.0	15.1	52.0	57.2																					
1000	16.0	63.0	56.3	15.3	52.0	59.3																					

TABLE 7.3.—Minimum Time (T) in hours that wind must blow to form waves of H significant height (in feet) and P period (in seconds). Fetch in nautical miles.

speed, fetch, length of time the wind blows, wave height, and wave period in deep water.

If the water is deeper than one-half the wavelength (L), this length in feet is theoretically related to period (P) in seconds by the formula

$$L = 5.12P^2.$$

The actual value has been found to be a little less than this for swell, and about two-thirds the length determined by this formula for sea. When the waves leave the generating area and continue as free waves, the wavelength and period continue to increase, while the height decreases. The rate of change gradually decreases.

The speed (S) of a free wave in deep water is nearly independent of its height or steepness. For swell, its relationship in knots to the period (P) in seconds is given by the formula

$$S = 3.03P.$$

The relationship for sea is not known.

The theoretical relationship between speed, wavelength, and period is shown in figure 7.3b. As waves continue on beyond the generating area, the period, wavelength, and speed remains the same. Because the waves of each period have different speeds they tend to sort themselves by periods as they move away from the generating area. The longer period waves move at a greater speed and move ahead. At great enough distances from a storm area the waves will have sorted themselves into packets based on period.

All the waves are attenuated as they propagate but the short period waves attenuate faster so that at a long distance from a storm only the longer waves remain.

The time needed for a wave system to travel some distance is *double* that which would be indicated by the speed of individual waves. This is because the front wave gradually disappears and transfers its energy to succeeding waves. The process is

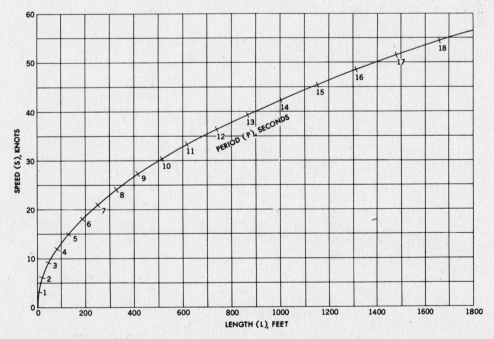

FIGURE 7.3b.—Relationship between speed, length, and period of waves in deep water, based upon the theoretical relationship between period and length.

followed by each front wave in succession, at such a rate that the wave *system* advances at a speed which is just *half* that of *individual* waves. This process can be seen in the bow wave of a vessel. The speed at which the wave system advances is called **group velocity.**

Because of the existence of many independent wave systems at the same time, the sea surface acquires a complex and irregular pattern. Also, since the longer waves outrun the shorter ones, the resulting interference adds to the complexity of the pattern. **The process of interference, illustrated in figure 7.3c, is duplicated many times in the** sea, being the principal reason that successive waves are not of the same height. The irregularity of the surface may be further accentuated by the presence of wave systems crossing at an angle to each other, producing peak-like rises.

In reporting average wave heights, the mariner has a tendency to neglect the lower ones. It has been found that the reported value is about the average for the highest one-third. This is sometimes called the "significant" wave height. The approximate relationship between this height and others, is as follows:

Wave	Relative height
Average	0.64
Significant	1.00
Highest 10 percent	1.29
Highest	1.87

7.4. Path of water particles in a wave.—As shown in figure 7.4, a particle of water on the surface of the ocean follows a somewhat circular orbit as a wave passes, but moves very little in the direction of motion of the wave. The common wave producing this action is called an **oscillatory wave.** As the crest passes, the particle moves forward, giving the water the appearance of moving with the wave. As the trough passes, the motion is in the opposite direction. The radius of the circular orbit decreases with depth, approaching zero at a depth equal to about half the wavelength. In shallower water the orbits become more elliptical, and in very shallow water, as at a beach, the vertical motion disappears almost completely.

Since the speed is greater at the top of the orbit than at the bottom, the particle is not at exactly its original point following passage of a wave, but has moved slightly

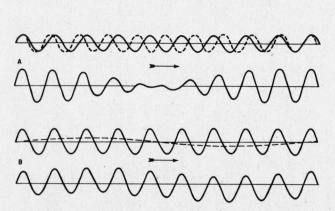

FIGURE 7.3c.—Interference. The upper part of *A* shows two waves of equal height and nearly equal length traveling in the same direction. The lower part of *A* shows the resulting wave pattern. In *B* similar information is shown for short waves and long swell.

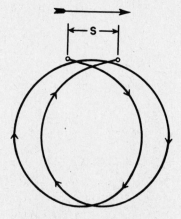

FIGURE 7.4.—Orbital motion and displacement, *s*, of a particle on the surface of deep water during two wave periods.

in the direction of motion of the wave. However, since this advance is small in relation to the vertical displacement, a floating object is raised and lowered by passage of a wave, but moved little from its original position. If this were not so, a slow moving vessel might experience considerable difficulty in making way against a wave train. In figure 7.4 the forward displacement is greatly exaggerated.

7.5. Effects of currents on waves.—A following current increases wavelengths and decreases wave heights. An opposing current has the opposite effect, decreasing the length and increasing the height. A strong opposing current may cause the waves to break. The extent of wave alteration is dependent upon the ratio of the still-water wave speed to the speed of the current.

Moderate ocean currents running at oblique angles to wave directions appear to have little effect, but strong tidal currents perpendicular to a system of waves have been observed to completely destroy them in a short period of time.

7.6. The effect of ice on waves.—When ice crystals form in seawater, internal friction is greatly increased. This results in smoothing of the sea surface. The effect of pack ice is even more pronounced. A vessel following a lead through such ice may be in smooth water even when a gale is blowing and heavy seas are beating against the outer edge of the pack. Hail is also effective in flattening the sea, even in a high wind.

7.7. Waves and shallow water.—When a wave encounters shallow water, the movement of the individual particles of water is restricted by the bottom, resulting in reduced wave speed. In deep water wave speed is a function of period. In shallow water, the wave speed becomes a function of depth. The shallower the water the slower is the wave speed. As the wave speed slows, the period remains the same so the wavelength becomes shorter. Since the energy in the waves remains the same, the shortening of wavelengths results in increased heights. This process is called **shoaling.** If the wave approaches the shoal at an angle, each part is slowed successively as the depth decreases. This causes a change in direction of motion or **refraction,** the wave tending to become parallel to the depth curves. The effect is similar to the refraction of light and other forms of radiant energy.

As each wave slows, the next wave behind it, in deeper water, tends to catch up. As the wavelength decreases, the height generally becomes greater. The lower part of a wave, being nearest the bottom, is slowed more than the top. This may cause the wave to become unstable, the faster-moving top falling or **breaking.** Such a wave is called a **breaker,** and a series of breakers, **surf.** This subject is covered in greater detail in chapter VIII.

Swell passing over a shoal but not breaking undergoes a decrease in wavelength and speed, and an increase in height. Such **ground swell** may cause heavy rolling if it is on the beam and its period is the same as the period of roll of a vessel, even though the sea may appear relatively calm. Figure 7.7 illustrates the approximate alteration of the characteristics of waves as they cross a shoal.

7.8. Energy of waves.—The potential energy of a wave is related to the vertical distance of each particle from its still-water position, and therefore moves with the wave. In contrast, the kinetic energy of a wave is related to the speed of the particles, being distributed evenly along the entire wave.

The amount of kinetic energy in even a moderate wave is tremendous. A 4-foot, 10-second wave striking a coast expends more than 35,000 horsepower per mile of beach. For each 56 miles of coast, the energy expended equals the power generated at Hoover Dam. An increase in temperature of the water in the relatively narrow **surf zone** in which this energy is expended would seem to be indicated, but no pronounced increase has been measured. Apparently, any heat that may be generated is dissipated to the deeper water beyond the surf zone.

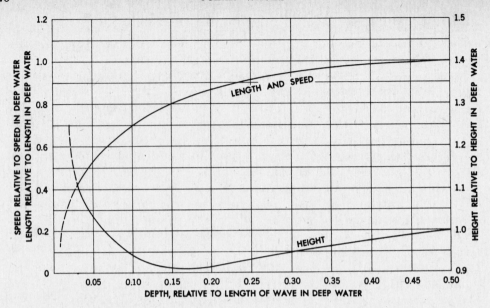

FIGURE 7.7.—Alteration of the characteristics of waves as they cross a shoal.

7.9. Wave measurement aboard ship.—With suitable equipment and adequate training, one can make reasonably reliable measurements of the height, length, period, and speed of waves. However, the mariner's estimates of height and length usually contain relatively large errors. There is a tendency to underestimate the heights of low waves, and overestimate the heights of high ones. There are numerous accounts of waves 75 to 80 feet high, or even higher, although waves more than 55 feet high are very rare. Wavelength is usually underestimated. The motions of the vessel from which measurements are made perhaps contribute to such errors.

Height. Measurement of wave height is particularly difficult. A microbarograph (art. 1.5) can be used if the wave is long enough to permit the vessel to ride up and down with it. If the waves are approaching from dead ahead or dead astern, this requires a wavelength at least twice the length of the vessel. For most accurate results the instrument should be placed at the center of roll and pitch, to minimize the effects of these motions. Wave height can often be estimated with reasonable accuracy by comparing it with freeboard of the vessel. This is less accurate as wave height and vessel motion increase. If a point of observation can be found at which the top of a wave is in line with the horizon when the observer is in the trough, the wave height is equal to height of eye. However, if the vessel is rolling or pitching, this height at the moment of observation may be difficult to determine. The highest wave ever reliably reported was 112 feet observed from the USS *Ramapo* in 1933.

Length. The dimensions of the vessel can be used to determine wavelength. Errors are introduced by perspective and disturbance of the wave pattern by the vessel. These errors are minimized if observations are made from maximum height. Best results are obtained if the sea is from dead ahead or dead astern.

Period. If allowance is made for the motion of the vessel, wave period can be determined by measuring the interval between passages of wave crests past the observer. The correction for the motion of the vessel can be eliminated by timing the passage of successive wave crests past a patch of foam or a floating object at some distance from the vessel. Accuracy of results can be improved by averaging several observations.

Speed can be determined by timing the passage of the wave between measured points along the side of the ship, if corrections are applied for the direction of travel of the wave and the speed of the ship.

The length, period, and speed of waves are interrelated by the relationships indicated in article 7.3. There is no definite mathematical relationship between wave height and length, period, or speed.

7.10. Tsunamis are ocean waves produced by sudden, large-scale motion of a portion of the ocean floor or the shore, as by volcanic eruption, earthquake (sometimes called **seaquake** if it occurs at sea), or landslide. If they are caused by a submarine earthquake, they are usually called **seismic sea waves.** The point directly above the disturbance, at which the waves originate, is called the **epicenter.** Either a tsunami or a storm tide (art. 7.11) that overflows the land is popularly called a **tidal wave**, although it bears no relation to the tide.

If a volcanic eruption occurs below the surface of the sea, the escaping gases cause a quantity of water to be pushed upward in the shape of a dome or mound. The same effect is caused by the sudden rising of a portion of the bottom. As this water settles back, it creates a wave which travels at high speed across the surface of the ocean.

Tsunamis are a series of waves. Near the epicenter, the first wave may be the highest. At greater distances, the highest wave usually occurs later in the series, commonly between the third and the eighth wave. Following the maximum, they again become smaller, but the tsunami may be detectable for several days.

In deep water the wave height of a tsunami is probably never greater than 2 or 3 feet. Since the wavelength is usually considerably more than 100 miles, the wave is not conspicuous at sea. In the Pacific, where most tsunamis occur, the wave period varies between about 15 and 60 *minutes*, and the speed in deep water is more than 400 knots. The approximate speed can be computed by the formula

$$S = 0.6 \sqrt{gd} = 3.4 \sqrt{d},$$

where S is the speed in knots, g is the acceleration due to gravity (32.2 feet per second per second), and d is the depth of water in feet. This formula is applicable to any wave in water having a depth of less than half the wavelength. For most ocean waves it applies only in shallow water, because of the relatively short wavelength.

When a tsunami enters shoal water, it undergoes the same changes as other waves. The formula indicates that speed is proportional to depth of water. Because of the great speed of a tsunami when it is in relatively deep water, the slowing is relatively much greater than that of an ordinary wave crested by wind. Therefore, the increase in height is also much greater. The size of the wave depends upon the nature and intensity of the disturbance. The height and destructiveness of the wave arriving at any place depend upon its distance from the epicenter, topography of the ocean floor, and the coastline. The angle at which the wave arrives, the shape of the coastline, and the topography along the coast and offshore all have their effect. The position of the shore is also a factor, as it may be sheltered by intervening land, or be in a position where waves have a tendency to converge, either because of refraction or reflection, or both.

Tsunamis 50 feet in height or higher have reached the shore, inflicting widespread damage. On April 1, 1946, seismic sea waves originating at an epicenter near the Aleutians spread over the entire Pacific. Scotch Cap Light on Unimak Island, 57 feet above sea level, was completely destroyed. Traveling at an average speed of 490 miles per hour, the waves reached the Hawaiian Islands in 4 hours and 34 minutes, where they arrived as waves 50 feet above the high water level, and flooded a strip of coast more than 1,000 feet wide at some places. They left a death toll of 173, and property damage of $25,000,000. Less destructive waves reached the shores of North and South America, and Australia, 6,700 miles from the epicenter.

After this disaster, a tsunami warning system was set up in the Pacific, even though destructive waves are relatively rare (averaging about one in 20 years in the Hawaiian Islands).

In addition to seismic sea waves, earthquakes below the surface of the sea may produce a longitudinal wave that travels upward toward the surface, at the speed of sound. When a ship encounters such a wave, it is felt as a sudden shock which may be of such severity that the crew thinks the vessel has struck bottom. Because of such reports, some older charts indicated shoal areas at places where the depth is now known to be a thousand fathoms or more.

7.11. Storm tides.—In relatively tideless seas like the Baltic and Mediterranean, winds cause the chief fluctuations in sea level. Elsewhere, the astronomical tide usually masks these variations. However, under exceptional conditions, either severe extra-tropical storms or tropical cyclones can produce changes in sea level that exceed the normal range of tide. Low sea level is of little concern except to shipping, but a rise above ordinary high-water mark, particularly when it is accompanied by high waves, can result in a catastrophe.

Although, like tsunamis, these **storm tides** or **storm surges** are popularly called **tidal waves,** they are not associated with the tide. They consist of a single wave crest and hence have no period or wavelength.

Three effects in a storm induce a rise in sea level. The first is wind stress on the sea surface, which results in a piling-up of water (sometimes called "wind set-up"). The second effect is the convergence of wind-driven currents, which elevates the sea surface along the convergence line. In shallow water, bottom friction and the effects of local topography cause this elevation to persist and may even intensify it. The low atmospheric pressure that accompanies severe storms causes the third effect, which is sometimes referred to as the "inverted barometer." An inch of mercury is equivalent to about 13.6 inches of water (art. 5.15) and the adjustment of the sea surface to the reduced pressure can amount to several feet at equilibrium (art. 3.11).

All three of these causes act independently, and if they happen to occur simultaneously, their effects are additive. In addition, the wave can be intensified or amplified by the effects of local topography. Storm tides may reach heights of 20 feet or more, and it is estimated that they cause three-fourths of the deaths attributed to hurricanes.

7.12. Standing waves and seiches.—Previous articles in this chapter have dealt with **progressive waves** which appear to move regularly with time. When two systems of progressive waves having the same period travel in opposite directions across the same area, a series of **standing waves** may form. These appear to remain stationary.

Another type of standing wave, called a **seiche** (sāsh), sometimes occurs in a confined body of water. It is a long wave, usually having its crest at one end of the confined space, and its trough at the other. Its period may be anything from a few minutes to an hour or more, but somewhat less than the tidal period. Seiches are usually attributed to strong winds or differences in atmospheric pressure.

7.13. Tide waves.—As indicated in chapter V, there are, in general, two regions of high tide separated by two regions of low tide, and these regions move progressively westward around the earth as the moon revolves in its orbit. The high tides are the crests of these **tide waves,** and the low tides are the troughs. The wave is not noticeable at sea, but becomes apparent along the coasts, particularly in funnel-shaped estuaries. In certain river mouths or estuaries of particular configuration, the incoming wave of high water overtakes the preceding low tide, resulting in a high-crested, roaring wave which progresses upstream in one mighty surge called a **bore.**

7.14. Internal waves.—Thus far, the discussion has been confined to waves on the surface of the sea, the boundary between air and water. **Internal waves,** or **boundary waves,** are created below the surface, at the boundaries between water strata of different densities. The density differences between adjacent water strata in the sea are considerably less than that between sea and air. Consequently, internal waves are much more easily formed than surface waves, and they are often much larger. The maximum height of wind waves on the surface is about 60 feet, but internal wave heights as great as 300 feet have been encountered.

Internal waves are detected by a number of observations of the vertical temperature distribution, using recording devices such as the bathythermograph (art. 7.7). They have periods as short as a few minutes, and as long as 12 or 24 hours, these greater periods being associated with the tides.

A slow-moving ship operating in a freshwater layer having a depth approximating the draft of the vessel may produce short-period internal waves. This may occur off rivers emptying into the sea or in polar regions in the vicinity of melting ice. Under suitable conditions, the normal propulsion energy of the ship is expended in generating and maintaining these internal waves and the ship appears to "stick" in the water, becoming sluggish and making little headway. The phenomenon, known as **dead water,** disappears when speed is increased by a few knots.

The full significance of internal waves has not been determined, but it is known that they may cause submarines to rise and fall like a ship at the surface, and they may also affect sound transmission in the sea.

7.15. Waves and ships.—The effects of waves on a ship vary considerably with the type ship, its course and speed, and the condition of the sea. A short vessel has a tendency to ride up one side of a wave and down the other side, while a larger vessel may tend to ride *through* the waves on an even keel. If the waves are of such length that the bow and stern of a vessel are alternately in successive crests and successive troughs, the vessel is subject to heavy sagging and hogging stresses, and under extreme conditions may break in two. A change of heading may reduce the danger. Because of the danger from sagging and hogging, a small vessel is sometimes better able to ride out a storm than a large one.

If successive waves strike the side of a vessel at the same phase of successive rolls, relatively small waves can cause heavy rolling. The effect is similar to that of swinging a child, where the strength of the push is not as important as its timing. The same effect, if applied to the bow or stern in time with the pitch, can cause heavy pitching. A change of either heading or speed can reduce the effect.

A wave having a length twice that of a ship places that ship in danger of falling off into the trough of the sea, particularly if it is a slow-moving vessel. The effect is especially pronounced if the sea is broad on the bow or broad on the quarter. An increase of speed reduces the hazard.

7.16. Use of oil for modifying the effects of breaking waves.— Oil has proved effective in modifying the effects of breaking waves, and has proved useful to vessels at sea, whether making way or stopped, particularly when lowering or hoisting boats. Its effect is greatest in deep water, where a small quantity suffices if the oil can be made to spread to windward. In shallow water where the water is in motion over the bottom, oil is less effective but of some value.

The heaviest oils, notably animal and vegetable oils, are the most effective. Crude petroleum is useful, but its effectiveness can be improved by mixing it with animal and vegetable oils. Gasoline or kerosene are of little value. Oil spreads slowly. In cold weather it may need some thinning with petroleum to hasten the process and produce the desired spread before the vessel is too far away for the effect to be useful.

At sea, best results can be expected if the vessel drifts or runs slowly before the wind, with the oil being discharged on both sides from waste pipes or by other convenient method. If a sea anchor is used, oil can be distributed from a container inserted within it for this purpose. If such a container is not available, an oil bag can be fastened to an endless line rove through a block on the sea anchor. This permits distribution of oil to windward, and provides a means for hauling the bag aboard for refilling. If another vessel is being towed, the oil should be distributed from the towing vessel, forward and on both sides, so that both vessels will be benefited. If a drifting vessel is to be approached, the oil might be distributed from both sides of the drifting vessel or by the approaching vessel, which should distribute it to leeward of the drifting vessel so that that vessel will drift into it. If the vessel being approached is aground, the procedure best suiting the circumstances should be used.

If oil is needed in crossing a bar to enter a harbor, it can be floated in ahead of the vessel if a flood current is running. A considerable amount may be needed. During slack water a hose might be trailed over the bow and oil poured freely through it if no more convenient method is available. With an ebb current oil is of little use, unless it can be distributed from another vessel or in some other manner from the opposite side of the bar.

CHAPTER VIII

BREAKERS AND SURF

8.1. Introduction.—The purpose of this chapter is to acquaint the navigator with the oceanographic factors affecting the safe navigation through the surf zone to the beach.

8.2. Refraction.—As explained in article 7.7, wave speed is slowed in shallow water, causing **refraction** if the waves approach the beach at an angle. Along a perfectly straight beach, with uniform shoaling, the wave fronts tend to become parallel to the shore. Any irregularities in the coastline or bottom contours, however, affect the refraction, causing irregularity. In the case of a ridge perpendicular to the beach, for instance, the shoaling is more rapid, causing greater refraction towards the ridge. The waves tend to align themselves with the bottom contours. Waves on both sides of the ridge have a component of motion toward the ridge. This **convergence** of wave energy toward the ridge causes an increase in wave or breaker height. A submarine canyon or valley perpendicular to the beach, on the other hand, produces **divergence,** with a decrease in wave or breaker height. These effects are illustrated in figure 8.2. Bends in the coastline have a similar effect, convergence occuring at a *point*, and divergence if the coast is *concave* to the sea. Points act as focal areas for wave energy and experience large breakers. Concave bays have small breakers because the energy is spread out as the waves approach the beach.

Under suitable conditions, currents also cause refraction. This is of particular importance at entrances of tidal estuaries. When waves encounter a current running in the opposite direction, they become higher and shorter. This results in a choppy

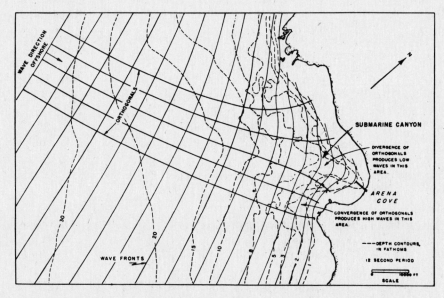

Courtesy of Robert L. Wiegel, Council on Wave Research, University of California.

FIGURE 8.2.—The effect of bottom topography in causing wave convergence and wave divergence.

sea, often with breakers. When waves move in the same direction as current, they decrease in height, and become longer. Refraction occurs when waves encounter a current at an angle.

Refraction diagrams, useful in planning amphibious operations, can be prepared with the aid of nautical charts or aerial photographs. When computer facilities are available, complex computer programs are used to determine refraction diagrams, quickly and accurately.

8.3. Breakers and surf.—In deep water, swell generally moves across the surface as somewhat regular, smooth undulations (ch. VII). When shoal water is reached, the wave period remains the same, but the speed decreases. The amount of decrease is negligible until the depth of water becomes about one-half the wavelength, when the waves begin to "feel" bottom. There is a slight decrease in wave height, followed by a rapid increase, if the waves are traveling perpendicular to a straight coast with a uniformly sloping bottom. As the waves become higher and shorter, they also become steeper, and the crest becomes narrower. When the speed of individual particles at the crest becomes greater than that of the wave, the front face of the wave becomes steeper than the rear face. This process continues at an accelerating rate as the depth of water decreases. At some point the wave may become unstable, toppling forward to form a **breaker.**

There are three general classes of breakers. A **spilling breaker** breaks gradually over a considerable distance. A **plunging breaker** tends to curl over and break with a single crash. A **surging breaker** peaks up, but surges up the beach without spilling or plunging. It is classed as a breaker even though it does not actually break. The type of breaker is determined by the steepness of the beach and the steepness of the wave before it reaches shallow water, as illustrated in figure 8.3.

Longer waves break in deeper water, and have a greater breaker height. The effect of a steeper beach is also to increase breaker height. The height of breakers is less if the waves approach the beach at an acute angle. With a steeper beach slope there is greater tendency of the breakers to plunge or surge. Following the **uprush** of water onto a beach after the breaking of a wave, the seaward **backrush** occurs. The returning water is called **backwash.** It tends to further slow the bottom of a wave, thus increasing its tendency to break. This effect is greater as either the speed or depth of the backwash increases. The still water depth at the point of breaking is approximately 1.3 times the average breaker height.

Surf varies with both position along the beach and time. A change in position often means a change in bottom contour, with the refraction effects discussed in article 8.2. At the same point, the height and period of waves vary considerably from wave to wave. A group of high waves is usually followed by several lower ones. Therefore, passage through surf can usually be made most easily immediately following a series of higher waves.

Since surf conditions are directly related to height of the waves approaching a beach, and the configuration of the bottom, the state of the surf at any time can be predicted if one has the necessary information and knowledge of the principles involved. Height of the sea and swell can be predicted from wind data, and information on bottom configuration can generally be obtained from the nautical chart. In addition, the area of lightest surf along a beach can be predicted if details of the bottom configuration are available.

8.4. Currents in the surf zone.—In and adjacent to the surf zone, currents are generated by waves approaching the bottom contours at an angle, and by irregularities in the bottom.

SPILLING BREAKER

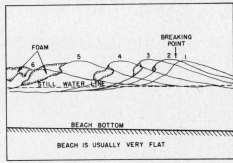

SKETCH SHOWING THE GENERAL CHARACTER
OF SPILLING BREAKERS

PLUNGING BREAKER

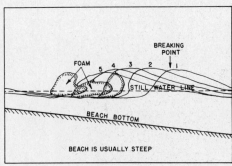

SKETCH SHOWING THE GENERAL CHARACTER
OF PLUNGING BREAKERS

SURGING BREAKER

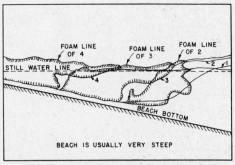

SKETCH SHOWING THE GENERAL CHARACTER
OF SURGING BREAKERS

Courtesy of Robert L. Wiegel, Council on Wave Research, University of California.

FIGURE 8.3.—The three types of breakers.

Waves approaching at an angle produce a **longshore current** parallel to the beach, within the surf zone. Longshore currents are most common along straight beaches. Their speeds increase with increasing breaker height, decreasing wave period, increasing angle of breaker line with the beach, and increasing beach slope. Speed seldom exceeds 1 knot, but sustained speeds as high as 3 knots have been recorded. Longshore currents are usually constant in direction. They increase the danger of landing craft broaching to.

As explained in article 8.2, wave fronts advancing over nonparallel bottom contours are refracted to cause convergence or divergence of the energy of the waves. Energy concentrations, in areas of convergence, form barriers to the returning back-wash, which is deflected *along* the beach to areas of less resistance. Backwash accumulates at weak points, and returns seaward in concentrations, forming **rip currents** through the surf. At these points the large volume of returning water has a retarding effect upon the incoming waves, thus adding to the condition causing the rip current. The waves on one or both sides of the rip, having greater energy and not being retarded by the concentration of backwash, advance faster and farther up the beach. From here, they move *along* the beach as **feeder currents.** At some point of low resistance, the water flows seaward through the surf, forming the **neck** of the rip current. Outside the breaker line the current widens and slackens, forming the **head.** The various parts of a rip current are shown in figure 8.4.

Rip currents may also be caused by irregularities in the beach face. If a beach indentation causes an uprush to advance farther than the average, the backrush is

 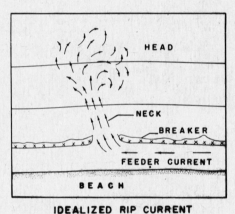

Courtesy of Robert L. Wiegel, Council on Wave Research, University of California.

FIGURE 8.4.—A rip current (left) and a diagram of its parts (right).

delayed and this in turn retards the next incoming **foam line** (the front of a wave as it advances shoreward after breaking) at that point. The foam line on each side of the retarded point continues in its advance, however, and tends to fill in the retarded area, producing a rip current.

8.5. Beach sediments.—In the surf zone, large amounts of sediment are suspended in the water. When the water motion decreases, the sediments are deposited as sand. The water motion can be either waves or currents. Promontories or points are rocky because the large breakers scour the points and small sediments are suspended in the water and carried away. Bays have sandy beaches because of the small wave conditions.

In the winter when storms create larger breakers and surf, the waves erode the beaches and carry the particles offshore where offshore sand bars form; sandy beaches tend to be narrower. In the summer the waves gradually move the sand back to the beaches and the offshore bars decrease; sandy beaches tend to be wider.

Longshore currents move large amounts of sand along the coast. These currents deposit sand on the upcurrent side of a jetty and erode the beach on the downcurrent side.

CHAPTER IX

SOUND IN THE SEA

9.1. Underwater sound and the navigator.—The clarity with which the noises associated with weighing anchor, propelling a ship, and other underwater motions are heard below the waterline and near the skin of a vessel is an indication of the high sound-transmitting qualities of seawater. Water is a better conductor of sound than is air because it absorbs less energy from the sound. There are several ways in which underwater sound can be used in navigation.

The *direction* of travel of sound waves can be measured either by means of **binaural hearing** (hearing with two "ears"), or by equipment which has directional characteristics similar to those of a directional antenna used in radio. Either method can be used for determining the direction from which general noise is coming, but only the latter is used in active sonar equipment for determining direction and distance by reception of an echo from a directional signal, in a manner similar to radar.

Distance can be determined by (1) measuring the elapsed time between transmission of a signal and return of its echo, (2) measuring the elapsed time between transmission of a signal and its receipt at a second station, (3) measuring the time *difference* between reception of a signal transmitted through water and one transmitted through air, (4) measuring the difference in phase between two signals or change of phase of a signal when it returns as an echo, or (5) measuring the angle at which an echo is received from a signal produced at another place. The first method is used in active sonar and echo sounding equipment. The fourth and fifth methods were used in early forms of echo sounders.

9.2. Sources of sound in the ocean.—Underwater sounds intended for navigational use are produced in one of three basic ways: (1) by percussion, as the striking of a bell, gong, or the bottom of the vessel; (2) by oscillator, as the vibration of a diaphragm; (3) by explosion, as by small bomb or depth charge. Certain man-made noises ordinarily produced in water, such as those due to operation of the main engines of a vessel, can be detected by an appropriate listening device.

In addition, many noises are made by animals living in the ocean. Certain shrimp, great numbers of which inhabit some areas, make a snapping noise with their claws. Some fish make a noise by stridulating (scraping). When shellfish are being eaten, a sound is emitted as the shells are broken by the teeth of the fish which are feeding. Grunting noises are made by many kinds of fish, usually by means of their swim bladders. Porpoises produce sounds of a high pitch. Sounds of various frequency and amplitude are produced by other forms of marine life. Where sound-producing marine life is very abundant, it interferes with detection of man-made sounds, requiring a high signal-to-noise ratio. The effect is similar to that of a high atmospheric noise level in radio.

9.3. Speed of sound in seawater.—Three variables govern the speed (S) of sound in a fluid. They are density (ρ), compressibility (β), and the ratio between the specific heats of the fluid at constant pressure and at constant volume (γ). The following formula is sufficiently accurate for most navigational purposes:

$$S = \sqrt{\frac{\gamma}{\rho\beta}}.$$

Compressibility refers to the relative change in volume for a given change in pressure. The compressibility of water is low, and consequently the speed of sound in water is high. The specific heat ratio enters the formula because the energy of a sound impulse is briefly transformed into heat, and then reconverted (with slight loss) into kinetic energy. The ratio rarely exceeds 1.02 in seawater and is commonly taken as unity.

For atmospheric pressure 29.92 inches of mercury, temperature 60°F, and salinity 34.85 parts per thousand, the density of seawater is 64 pounds per cubic foot and the compressibility approximately 0.0000435 per atmosphere (one atmosphere equals 14.696 pounds per square inch). Using these values and 32.174 feet per second per second (the acceleration of gravity at latitude 45°) and 144 square inches per square foot, and taking γ equal to unity, one obtains:

$$S = \sqrt{\frac{1.0 \times 32.174 \times 14.696 \times 144}{64 \times 0.0000435}} = 4945 \text{ ft./sec.}$$

The same formula can be used to determine the speed of sound in air. For atmospheric pressure 29.92 and temperature 60°F, the density of air is 0.0764 pound per cubic foot and, since air is a gas, the compressibility is the reciprocal of the pressure. Taking γ equal to 1.4, one obtains:

$$S = \sqrt{\frac{1.4 \times 32.174 \times 14.696 \times 144}{0.0764 \times 1}} = 1117 \text{ ft./sec.}$$

The speed of sound in water is approximately 4.5 times its speed in air.

An increase in temperature decreases both density and compressibility, resulting in an increase in the speed of sound. In seawater, an increase in pressure or salinity produces a slight increase in density and a larger decrease in compressibility, resulting in a net increase in the speed of sound. Thus, in seawater, an increase in temperature, pressure, or salinity results in greater speed of sound. Of the three, temperature has the greatest influence on the speed of sound in seawater in the upper layers. At depth, pressure, and in coastal areas, changes in salinity, may have the greatest effect.

Normally, the change of these three elements is much more rapid in a vertical direction than in a horizontal direction. The change with depth varies with location. With respect to temperature, much of the ocean is considered to consist of three layers, a **mixed layer** influenced greatly by the temperature of the air above it, a **thermocline** of rapidly decreasing temperature, and a nearly uniform **deep-water layer.** Typical curves showing change of temperature and salinity with depth are shown in figure 9.3a. The increase of pressure with depth is almost uniform, the pressure at 10,000 feet being approximately twice that at 5,000 feet, and 10 times that at 1,000 feet. A typical curve of speed of sound with depth is shown in **figure 9.3b.** In this case there is little or no mixed layer and the temperature decreases rapidly from the surface; therefore, the sound velocity also decreases rapidly. Below the range of temperature decrease, the pressure effect becomes the primary factor and sound velocity starts to increase. Note that the minimum sound velocity is at 2,400 feet. This would be the depth of the deep sound channel.

TEMPERATURE, °F

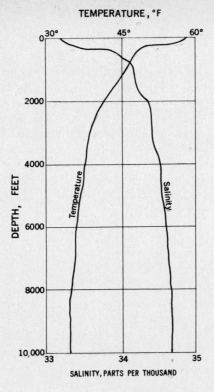

FIGURE 9.3a.—Variation of temperature and salinity with depth at one locality.

SPEED OF SOUND, FT. / SEC.

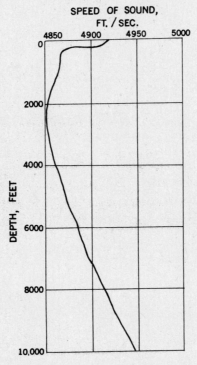

FIGURE 9.3b.—Typical variation of speed of sound with depth in the ocean.

Study of transmission of sound from underwater explosions indicates that near the explosion the speed of sound may be somewhat higher than expected, probably due to increased pressure caused by the disturbance. This effect extends over such a short distance that it is insignificant in ordinary underwater sound transmission.

9.4. Reflection of underwater sound waves.—In water, as in air, sound is reflected by obstructions in the form of solid objects or sharp discontinuities. Thus, sound is reflected from the bottom, the shore, hulls of ships, the surface of the water, etc. It is this reflecting energy that is used in echo sounders to determine depth, and in sonar equipment used for echo ranging.

Reflecting properties of various substances differ markedly. Rock reflects almost all of the sound that strikes its surface, while soft mud absorbs or is penetrated by sound. Thus, in echo sounding, a layer of soft mud over rock may result in two echoes, indicating two depths.

Fish and even tiny sea animals also reflect sound. As a result, echo sounders are widely used among fishermen to locate schools of fish. In deep water it is not unusual for an echo sounder to receive an echo from a depth of about 200 fathoms, although the depth is shallower somewhat at night. This **phantom bottom** or **deep scattering layer,** which is undoubtedly the source of many erroneous shoal sounding reports, is due to large numbers of tiny marine animals, or other marine life.

A sharp discontinuity within the water causes reflection of sound. Thus, an echo sounder may detect the boundary between a layer of freshwater overlying saltwater, a condition which might occur near the mouth of a river.

Sharp, distinct echoes denoting precise depths are difficult to obtain over rough-surfaced bottoms. Therefore, considerable discretion should be exercised in evaluating soundings taken over bottoms possessing a high degree of relief.

9.5. Refraction of underwater sound waves.—The laws of refraction as applied to light and radio waves apply also to sound. Because of differences of velocity of sound in seawater, an advancing sound wave is refracted toward the area of slower sound velocity. If sound is traveling vertically downward, as in echo sounding, the effect of refraction is relatively slight because the layers of water in which velocity differs are approximately horizontal, and when the direction of travel of the sound is normal to the refracting surface or layer, there is no refraction.

If a sound beam is transmitted outward from a source, it will start at a particular sound velocity but the sound velocity will either increase or decrease as the beam moves into water of different temperatures, salinity, or pressure. The beam will refract toward the region where the sound velocity is slower. In a mixed surface layer the temperature is isothermal so the sound velocity increases with depth due to the pressure effect. A sound beam in that layer would be refracted upward to the surface where it would reflect off the surface. Sound beams can be trapped in the mixed layer and create a "surface sound duct." If a beam penetrates below the mixed layer into the thermocline it is in a region where sound velocity decreases rapidly with depth due to the temperature decrease. In the thermocline sound beams are refracted sharply downward. Sonar ranges can be very short in the thermocline layer. This is the region of the "shadow zone."

With typical distribution of sound velocity with depth, as shown in figure 9.3b, sound velocity decreases with depth until a minimum is reached at some level below the surface, and below this it increases. In figure 9.3b minimum velocity occurs at about 2,400 feet. In the Tropics this level of minimum velocity may be as deep as 6,000 feet, and in polar regions it may be at the surface. This level is referred to as the **deep sound channel.** Sound produced at any level tends to be refracted to the level of minimum speed, and to remain there, for as it attempts to leave this level, it is refracted back toward it, as shown in figure 9.5. This, of course, does not refer to sound traveling vertically. If a sound is produced at this level, as by the explosion of a bomb or depth charge, the sound waves start to move outward as expanding spheres, but most of the rays are refracted back toward the minimum speed level. Because of this effect, such a sound may travel great distances with relatively little decrease in intensity. Listening gear placed at this level has detected sounds produced thousands of miles away.

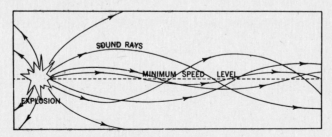

FIGURE 9.5.—Transmission of sound rays along the minimum sound level.

Sound beams that penetrate the deep sound channel without being trapped continue on to regions of increased sound velocity. If the water is deep enough these beams will be refracted upwards towards the surface. If this occurs, the energy converges near the surface at ranges of about 30 nautical miles. This is the **convergence zone.** Convergence zone detection is significant in modern sonar applications.

9.6. Attenuation of sound.—As sound is transmitted from a source, the energy is lost or *attenuated* due to reflection, spreading, and absorption. A sound beam reflected from the bottom or surface loses energy; although the sound energy is concentrated near the source, as the range increases the same energy is spread over a sphere whose radius is the range. The rate of absorption is a function of frequency; high frequency sound is absorbed more than sound of lower frequency.

CHAPTER X

ICE IN THE SEA

10.1. Ice and the navigator.—*Sea ice* has posed a problem to the polar navigator since antiquity. Pytheas of Massalia sighted a strange substance which he described as "neither land nor air nor water" floating upon and covering the northern sea over which the summer sun barely set. Pytheas named this lonely region Thule, hence Ultima Thule (farthest north or land's end). Thus began over 20 centuries of polar exploration.

Ice is of direct concern to the navigator because it restricts and sometimes controls his movements, it affects his dead reckoning by forcing frequent and sometimes inaccurately determined changes of course and speed, it affects his piloting by altering the appearance or obliterating the features of landmarks and by rendering difficult the establishment and maintenance of aids to navigation, it affects his use of electronics by its effect upon propagation of radio waves and the changes it produces both in surface features and radar returns from such features, it affects his celestial navigation by altering the refraction and obscuring his horizon and celestial bodies either directly or by the weather it influences, and it affects his charts by introducing various difficulties to the hydrographic surveyor.

Because of his direct concern with ice, the prospective polar navigator will do well to acquaint himself with its nature and extent in the area he expects to navigate. To this end he should consult the sailing directions for the area, and whatever other literature may be available to him, including reports of previous operations in the same area.

10.2. Formation of ice.—As it cools, water contracts until the temperature of maximum density is reached. Further cooling results in expansion. The maximum density of freshwater occurs at a temperature of $4^\circ.0C$ ($39^\circ.2F$), and freezing takes place at $0°C$ ($32°F$). The addition of salt lowers both the temperature of maximum density and, to a lesser extent, that of freezing. The relationships are shown in figure 10.2. The two lines meet at a salinity of 24.7 parts per thousand, at which maximum density occurs at the freezing temperature of $-1^\circ.3C$ ($29^\circ61F$). At this and greater salinities, the density increases right down to the freezing point. At a salinity of 35 parts per thousand, the approximate average for the oceans, the freezing point is $-1^\circ.88C$ ($28^\circ.6F$).

As the density of surface seawater increases with decreasing temperature, density currents are induced bringing warmer, less dense water to the surface. If the polar seas consisted of water with constant salinity, the entire water column would have to be cooled to the freezing point in this manner before ice would begin to form. This is not the case, however, in the polar regions where the vertical salinity distribution is such that the surface waters are underlaid at shallow depth by waters of higher salinity. In this instance density currents form a shallow mixed layer which subsequently cannot mix with the deep layer of warmer but saltier water. Ice will then begin forming at the water surface when density currents cease and the surface water reaches its freezing point. In shoal water, however, the mixing process can be sufficient to extend the freezing temperature from the surface to the bottom. Ice crystals can, therefore, form at any depth in this case. Because of their decreased density, they tend to rise to the surface unless they form at the bottom and attach themselves there. This ice, called **anchor ice,** may continue to grow as additional ice freezes to that already formed.

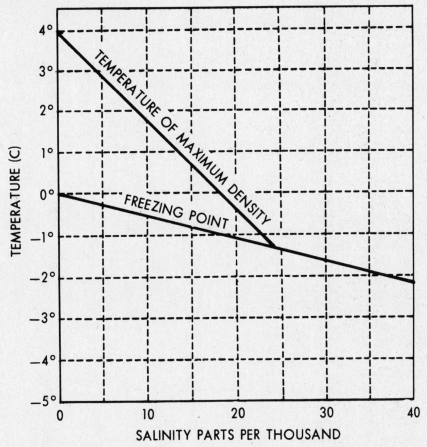

FIGURE 10.2.—Relationship between temperature of maximum density and freezing point for water.

10.3. Ice of land origin is formed on land by the freezing of freshwater or the compacting of snow as layer upon layer adds to the pressure on that beneath.

Under great pressure ice becomes slightly plastic and is forced outward and downward along an inclined surface. If a large area is relatively flat, as on the antarctic plateau, or if the outward flow is obstructed, as on Greenland, an **ice cap** forms and remains winter and summer. The thickness of these ice caps range from nearly 1 kilometer on Greenland to as much as 4.5 kilometers on the Antarctic Continent. Where ravines or mountain passes permit flow of the ice, a **glacier** is formed. This is a mass of snow and ice which continuously flows to lower levels, exhibiting many of the characteristics of rivers of water. The flow may be more than 30 meters per day, but is generally much less. When a glacier reaches a comparatively level area, it spreads out. When a glacier flows into the sea, the buoyant force of the water breaks off pieces from time to time, and these float away as **icebergs.** Icebergs may be described as dome shaped, sloping or pinnacled (fig. 10.3a), tabular (fig. 10.3b), glacier, or weathered.

An iceberg seldom melts uniformly because of lack of uniformity in the ice itself, differences in the temperature above and below the waterline, exposure of one side to the sun, strains, cracks, mechanical erosion, etc. The inclusion of rocks, silt, and other foreign matter further accentuates the differences. As a result, changes in equilibrium take place, which may cause the berg to tilt or capsize. Parts of it may break off or **calve,** forming separate smaller bergs. A relatively large piece of floating ice generally extending 1 to 5 meters above the sea surface and normally about 100 to 300 square meters in area is called a **bergy bit.** A smaller piece of ice but one large enough to inflict serious damage to a vessel is called a **growler** because of the noise it sometimes makes as it bobs up and down in the sea. Growlers extend less than 1 meter above the sea surface and normally occupy an area of about 20 square meters. Bergy bits and growlers are usually pieces calved from icebergs, but they may be formed by the melting of an iceberg. The principal danger from icebergs is their tendency to break or shift position, and possible underwater extensions, called **rams,** which are usually formed due to the more intensive melting or erosion of the unsubmerged portions. Rams may also extend from a vertical **ice cliff,** also known as an **ice front,** which forms the seaward face of a massive ice sheet or floating glacier; or from an **ice wall** which is the ice cliff forming the seaward margin of a glacier which is aground.

As strange as it may seem, icebergs may be helpful to the mariner in some ways. The melt water found on the surface of icebergs is a source of freshwater, and in the past some daring seamen have made their vessels fast to icebergs which, because they are affected more by currents than the wind, have proceeded to tow them out of the ice pack.

10.4. Sea ice forms by the freezing of seawater and accounts for 95 percent of all ice encountered. The first indication of the formation of **new sea ice** (up to 10 centimeters in thickness) is the development of small individual, needlelike crystals of ice, called **spicules,** which become suspended in the top few centimeters of seawater. These spicules, also known as **frazil ice,** give the sea surface an oily appearance. **Grease ice** is formed when the spicules coagulate to form a soupy layer on the surface giving the sea a matte appearance. The next stage in sea ice formation occurs when **shuga,** an accumulation of spongy white ice lumps a few centimeters across, develops from grease ice. Upon further freezing, and depending upon wind exposure, seas, and salinity, shuga and grease ice develop into **nilas,** an elastic crust of high salinity up to 10 centimeters in thickness with a matte surface or into **ice rind,** a brittle, shiny crust of low salinity with a thickness up to approximately 5 centimeters. A layer of 5 centimeters of freshwater ice is brittle but strong enough to support the weight of a heavy man. In contrast, the same thickness of newly formed sea ice will support not more than about 10 percent of this weight, although its strength varies with the temperatures at which it is formed; very cold ice supports a greater weight than warmer ice. As it ages, sea ice becomes harder and more brittle.

New ice may also develop from **slush** which is formed when snow falls into seawater which is near its freezing point, but colder than the melting point of snow. The snow does not melt but floats on the surface, drifting with the wind into beds. If the temperature then drops below the freezing point of the seawater, the slush freezes quickly into a soft ice similar to shuga.

Sea ice is exposed to several forces, including currents, wave motion, tides, wind, and temperature differences. In its early stages, its plasticity permits it to conform readily to virtually any shape required by the forces acting upon it. As it becomes older, thicker, more brittle, and exposed to the influence of wind and wave action, new ice usually separates into circular pieces from 30 centimeters to 3 meters in diameter

FIGURE 10.3.—Pinnacled iceberg.

FIGURE 10.3b.—A tabular iceberg.

and up to approximately 10 centimeters in thickness with raised edges due to individual pieces striking against each other. These circular pieces of ice are called **pancake ice**

FIGURE 10.4a.—Pancake ice, with an iceberg in the background.

(fig. 10.4a) and may break into smaller pieces with strong wave motion. Any single piece of relatively flat sea ice less than 20 meters across is called an **ice cake**. With continued low temperatures individual ice cakes and pancake ice will, depending on wind or wave motion, either freeze together to form a continuous sheet or unite into pieces of ice 20 meters or more across. These larger pieces are then called **ice floes** which may further freeze together to form an ice covered area greater than 10 kilometers across known as an **ice field.** In wind sheltered areas thickening ice usually forms a continuous sheet before it can develop into the characteristic ice cake form. When sea ice reaches a thickness of between 10 to 30 centimeters it is referred to as **grey** and **grey-white ice,** or collectively as **young ice,** and is the transition stage between nilas and **first-year ice.** First-year ice usually attains a thickness of between 30 centimeters and 2 meters in its first winter's growth.

Sea ice may grow to a thickness of 10 to 13 centimeters within 48 hours, after which it acts as an insulator between the ocean and the atmosphere progressively slowing its further growth. However, sea ice may grow to a thickness of between 2 to 3 meters in its first winter. Ice which has survived at least one summer's melt is classified as **old ice.** If it has survived only one summer's melt it is referred to as **second-year ice.** Because it is thicker and less dense than first-year ice, it stands higher out of the water. Old ice which has attained a thickness of 3 meters or more and has survived at least two summers' melt is known as **multiyear ice** and is almost salt free. Old ice may often be recognized by a bluish tone to its surface color in contrast to the greenish tint of first-year ice.

Greater thicknesses in both first and multiyear ice are attained through the deformation of the ice resulting from the movement and interaction of individual floes. Deformation processes occur after the development of new and young ice and are the direct consequence of the effects of winds, tides, and currents. These processes transform a relatively flat sheet of ice into **pressure ice** which has a readily observed roughness in its surface. **Bending,** which is the first stage in the formation of pressure ice, is the upward or downward motion of thin and very plastic ice. **Tenting** occurs when bending produces an upward displacement of ice forming a flat sided arch with a cavity beneath. More frequently, however, **rafting** takes place as one piece of new and young ice overrides another. When pieces of first-year ice are piled haphazardly over one another forming a wall or line of broken ice, referred to as a **ridge,** the process is known as **ridging.** Pressure ice with topography consisting of numerous mounds or hillocks is called **hummocked ice,** each mound being called a **hummock.**

The motion of adjacent floes is seldom equal. The rougher the surface, the greater is the effect of wind, since each piece extending above the surface acts as a sail. Some ice floes are in rotary motion as they tend to trim themselves into the wind. Since ridges extend below as well as above the surface, the deeper ones are influenced more by deep water currents. When a strong wind blows in the same direction for a considerable period, each floe exerts pressure on the next one, and as the distance increases, the pressure becomes tremendous. Ridges on sea ice are generally about 1 meter high and 5 meters deep, but under considerable pressure may attain heights of 30 meters and depths of 150 meters in extreme cases.

The alternate melting and growth of sea ice, combined with the continual motion of various floes that results in separation as well as consolidation, causes widely varying conditions within the ice cover itself. The mean areal density, or **concentration,** of pack ice in any given area is expressed in *oktas* (eighths). Concentrations range from: **open water** (total concentration of all ice does not exceed ⅛), **very open pack** (⅛ to less than ⅜ concentration), **open pack** (⅜ to less than ⅝ concentration), **close pack** (⅝ to less than ⅞ concentration), **very close pack** (⅞ to less than 8/8 concentration), to **compact** or **consolidated pack** (8/8 or complete coverage). The extent to which an ice cover of varying concentrations can be penetrated by a vessel varies from place to place and with changing weather conditions. With a concentration of 1 to 2 oktas in a given area, an unreinforced vessel can generally navigate safely, but the danger of receiving heavy damage is always present. When the concentration increases to between 2 and 4 oktas, the area becomes only occasionally accessible to an unreinforced vessel depending upon the vagaries of wind and current. With concentrations of 4 to 6 oktas, the area becomes accessible only to ice strengthened vessels which on occasion will require icebreaker assistance. Navigation in areas with concentrations of 6 oktas or more should only be attempted by modern icebreakers.

Within the ice cover, openings may develop resulting from a number of deformation processes. Long, jagged cracks may appear first in the ice cover or through a single floe. When these cracks part and reach lengths of a few meters to many kilometers, they are referred to as **fractures.** If they widen further to permit passage of a ship, they are called **leads.** In winter, a thin coating of new ice may cover the water within a lead, but in summer the water usually remains ice-free until a shift in the movement forces the two sides together again. Before this occurs, lateral motion generally occurs between the floes, so that they no longer fit and unless the pressure is extreme, numerous large patches of open water remain. These nonlinear shaped openings enclosed in ice are called **polynyas.** Polynyas may contain small fragments of floating ice and may be covered with miles of new and young ice.

Sea ice which is formed in situ from seawater or by the freezing of pack ice of any age to the shore and which remains attached to the coast, to an ice wall, to an ice front, or between shoals is called **fast ice.** The width of this fast ice varies considerably and may extend for a few meters or several hundred kilometers. In bays and other sheltered areas, fast ice, often augmented by annual snow accumulations and the seaward extension of land ice, may attain a thickness of over 2 meters above the sea surface. When a floating sheet of ice grows to this or a greater thickness and extends over a great horizontal distance, it is called an **ice shelf.** Massive ice shelfs where the ice thickness reaches several hundred meters are found in both the Arctic and Antarctic.

The majority of the icebergs found in the Antarctic do not originate from glaciers as those found in the Arctic, but are calved from the outer edges of broad expanses of shelf ice. Icebergs formed in this manner are called **tabular icebergs,** having a boxlike shape with horizontal dimensions measured in kilometers, and heights above the sea surface approaching 60 meters. The largest antarctic ice shelves are found in the Ross and Weddell Seas. The expression "tabular iceberg" is not applied to bergs which break off from arctic ice shelves; similar formations there are called **ice islands.** These originate when shelf ice, such as that found on the northern coast of Greenland and in the bays of Ellesmere Island, breaks up. As a rule, arctic ice islands are not as large as the tabular icebergs found in the Antarctic. They attain a thickness of up to 55 meters and on the average extend 5 to 7 meters above the sea surface. Both tabular icebergs and ice islands possess a nearly level, but gently rolling surface. Because of their deep draft, they are influenced much more by current than wind. Both the United States and the U.S.S.R. have used arctic ice islands as floating scientific platforms from which polar research has been conducted.

10.5. Thickness of sea ice.—Sea ice has been observed to grow to a thickness of almost 3 meters during its first year. However, the thickness of first-year ice that has not undergone deformation does not generally exceed 2 meters. In coastal areas where the melting rate is less than the freezing rate, the thickness may increase during succeeding winters, being augmented by compacted and frozen snow, until a maximum thickness of about 3.5 to 4.5 meters may eventually be reached. Old sea ice may also attain a thickness of over 4 meters in this manner, or when summer melt water from its surface or from snow cover runs off into the sea and refreezes under the ice where the seawater temperature is below the freezing point of the fresher melt water.

The growth of sea ice is dependent upon a number of meteorological and oceanographic parameters. Such parameters include air temperature, initial ice thickness, snow depth and density, wind speed, seawater salinity and density, and the specific heats of sea ice and seawater. Investigations, however, have shown that the most influential parameters affecting sea ice growth are air temperature, wind speed, snow depth and density, and initial ice thickness. Many complex equations have been formulated to predict ice growth using these five parameters. However, except for the first two, these parameters are not routinely observed for remote polar locations.

In the early 1940's a Russian geographical scientist, N. N. Zubov, formulated an ice growth equation as a function of air temperature alone and based on his empirical observations of ice formation along portions of the northern Russian arctic coast. Air temperatures are translated into accumulated frost degree days from which theoretical ice thicknesses are calculated using the equation:

$$T_j = \frac{-50 + \sqrt{2500 + 32 \sum_{i=1}^{j} DDi}}{2},$$

where T_j is the ice thickness in centimeters on day j and DDi is the frost degree day accumulation in degrees Celsius on day i.

A **frost degree day** is defined as a day with a mean temperature of 1° below an arbitrary base. The base most commonly used is the freezing point of freshwater (0°C). If, for example, the mean temperature on a given day is 5° below freezing, then five frost degree days are collected for that day. These frost degree days are then added to those collected the very next day to obtain an accumulated value, which is then added to the number of degree days collected the following day. This process is repeated daily throughout the ice growing season. Temperatures usually fluctuate above and below freezing for several days before remaining below freezing. Therefore, frost degree day accumulations are initiated on the first day of the period when temperatures remain below freezing. The relationship between frost degree day accumulations and theoretical ice growth curves at Point Barrow, Alaska is shown in figure 10.5a. Similar curves for other arctic stations are contained in publications available from the U. S. Naval Oceanographic Office. Figure 10.5b graphically depicts the relationship between accumulated frost degree days (°C) and ice thickness in centimeters.

During the winter the ice usually becomes covered with snow which insulates the ice beneath and tends to slow down its rate of growth. This thickness of snow cover varies considerably from region to region as a result of differing climatic conditions. Its depth may also vary widely within very short distances in response to variable winds and ice topography. While this snow cover persists, almost 90 percent of the incoming radiation is reflected back to space. Eventually, however, the snow begins to melt as the air temperature rises above 0° C in early summer and the resulting freshwater forms puddles on the surface. These puddles absorb about 90 percent of the incoming radiation and rapidly enlarge as they melt the surrounding snow or ice. Eventually the puddles penetrate to the bottom surface of the floes and are known as **thawholes**. This slow process is characteristic of ice in the Arctic Ocean and seas where movement is restricted by the coastline or islands. Where ice is free to drift into warmer waters (e.g., the Antarctic, East Greenland, and the Labrador Sea) decay is accelerated in response to wave erosion as well as warmer air and sea temperatures.

10.6. Salinity of sea ice.—Sea ice forms first as salt-free crystals near the surface of the sea. As the process continues, these crystals are joined together and, as they do so, small quantities of brine are trapped within the ice. On the average, new ice 15 centimeters thick contains 5 to 10 parts of salt per thousand. With lower temperatures, freezing takes place faster. With faster freezing, a greater amount of salt is trapped in the ice.

Depending upon the temperature, the trapped brine may either freeze or remain liquid, but because its density is greater than that of the pure ice, it tends to settle down through the pure ice. As it does so, the ice gradually freshens, becoming clearer, stronger, and more brittle. At an age of 1 year, sea ice is sufficiently fresh that its melt water, if found in **puddles** of sufficient size, and not contaminated by spray from

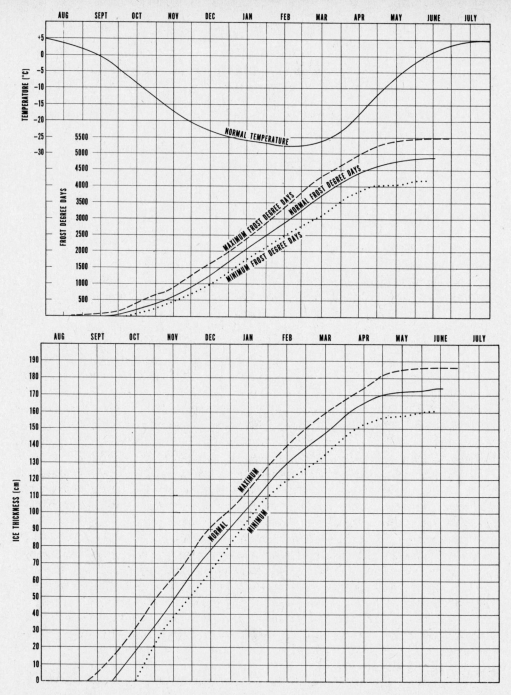

FIGURE 10.5a.—Relationship between accumulated frost degree days and theoretical ice thickness at Point Barrow, Alaska.

the sea, can be used to replenish the freshwater supply of a ship. However, ponds of sufficient size to water ships are seldom found except in ice of great age, and then much of the melt water is from snow which has accumulated on the surface of the ice. When sea ice reaches an age of about 2 years, virtually all of the salt has been eliminated. Icebergs contain no salt, and uncontaminated melt water obtained from them is fresh.

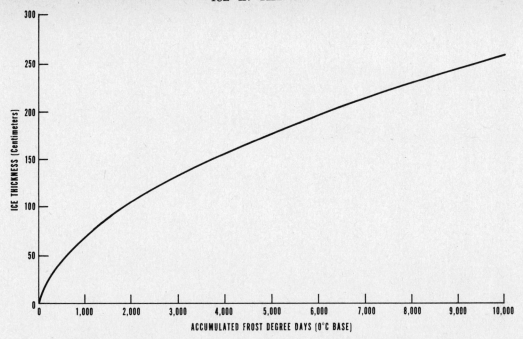

FIGURE 10.5b.—Relationship between accumulated frost degree days (°C) and ice thickness (cm).

The settling out of the brine gives sea ice a honeycomb structure which greatly hastens its disintegration when the temperature rises above freezing. In this state, when it is called **rotten ice,** much more surface is exposed to warm air and water, and the rate of melting is increased. In a day's time, a floe of apparently solid ice several inches thick may disappear completely.

10.7. Density of ice.—The density of freshwater ice at its freezing point is 0.917. Newly formed sea ice, due to its salt content, is more dense, 0.925 being a representative value. The density decreases as the ice freshens (art. 10.6). By the time it has shed most of its salt, sea ice is less dense than freshwater ice, because ice formed in the sea contains more air bubbles. Ice having no salt but containing air to the extent of 8 percent by volume (an approximately maximum value for sea ice) has a density of 0.845.

The density of land ice varies over even wider limits. That formed by freezing of freshwater has a density of 0.917, as stated above. Much of the land ice, however, is formed by compacting of snow. This results in the entrapping of relatively large quantities of air. **Névé,** a snow which has become coarse grained and compact through temperature change, forming the transition stage to glacier ice, may have an air content of as much as 50 percent by volume. By the time the ice of a glacier reaches the sea, its density approaches that of freshwater ice. A sample taken from an iceberg on the Grand Banks had a density of 0.899.

When ice floats, part of it is above water and part is below the surface. The percentage of the mass below the surface can be found by dividing the average density of the ice by the density of the water in which it floats. Thus, if an iceberg of density 0.920 floats in water of density 1.028 (corresponding to a salinity of 35 parts per thousand and a temperature of −1°C, or 30°F), 89.5 percent of its mass will be below the surface. That is, about nine-tenths of the mass will be below the surface, and only about one-tenth will be above the surface.

The height to draft ratio for a blocky or tabular iceberg probably varies fairly closely about 1:5. This average ratio was computed for icebergs south of Newfoundland by considering density values and a few actual measurements, and by seismic means at a number of locations along the edge of the Ross Ice Shelf near Little America Station. It was also substantiated by density measurements taken in a nearby hole drilled through the 256-meter thick ice shelf. The height to draft ratios of icebergs become significant when determining their drift (art. 10.9).

10.8. Drift of ice.—Although surface currents have some effect upon the drift of pack ice, the principal factor is wind. Due to Coriolis force (art. 2.3), ice does not drift in the direction of the wind, but varies from approximately 18° to as much as 90° from this direction, depending upon the force of the surface wind and the ice thickness. In the Northern Hemisphere, this drift is to the *right* of the direction toward which the wind blows, and in the Southern Hemisphere it is toward the *left*. Although early investigators computed average angles of approximately 28° or 29° for the drift of close multiyear pack ice, large drift angles were usually observed with low rather than high wind speeds. The relationship between surface wind speed, ice thickness, and drift angle, shown in figure 10.8, was derived theoretically for the drift of consolidated pack under equilibrium (a balance of forces acting on the ice) conditions, and shows that the drift angle increases with increasing ice thickness and decreasing surface wind speed. A slight increase also occurs with higher latitude.

Since the cross-isobar deflection of the surface wind over the oceans is approximately 20°, the deflection of the ice varies from approximately along the isobars to as much as 70° to the right of the isobars, with low pressure on the left and high pressure on the right in the Northern Hemisphere. The positions of the low and high pressure areas are, of course, reversed in the Southern Hemisphere. The drift angles that are given in figure 10.8a may be used for all ice concentrations and polar latitudes.

The rate of drift, compiled from observations of ice drift along the northern Russian coast bordering the Chukchi Sea, is presented in table 10.8. Rates are given as a percentage of the surface wind speed and depend upon the roughness of the surface and the concentration of the ice. Percentages vary from approximately one quarter of 1 percent to almost 8 percent of the surface wind speed measured approximately 6 meters above the ice surface. Low concentrations of heavily ridged or hummocked floes drift faster than high concentrations of lightly ridged or hummocked floes with the same wind speed. From table 10.8 it can be seen that sea ice of 6 or 7 okta concentrations and six tenths hummocking or close multiyear ice will drift at approximately 2 percent of the surface wind speed. Additionally, the response factors of 1 and 4 okta ice concentrations respectively, are approximately three times and twice the magnitude of the response factor for 7 okta ice concentrations with the same extent of surface roughness. Although a maximum ice drift to surface wind speed ratio of approximately 8 percent is indicated by table 10.8, isolated ice floes have been observed to drift as fast as 10 percent to 12 percent of strong surface winds.

The rates with which sea ice drifts have been quantified through empirical observation. The drift angle, however, has been determined theoretically for 8 okta ice concentrations. This relationship presently is extended to the drift of all ice concentrations due to the lack of basic knowledge of the dynamic forces that act upon and result in redistributions of sea ice in the polar regions.

10.9. Iceberg drift.—Icebergs extend a considerable distance below the surface and have relatively small "sail areas" compared to their subsurface mass. Therefore, the near-surface current is thought to be primarily responsible for drift; however, observations have shown that wind can be the dominant force that governs iceberg drift at a particular location or time. Also, the current and wind may contribute nearly equally to the resultant drift.

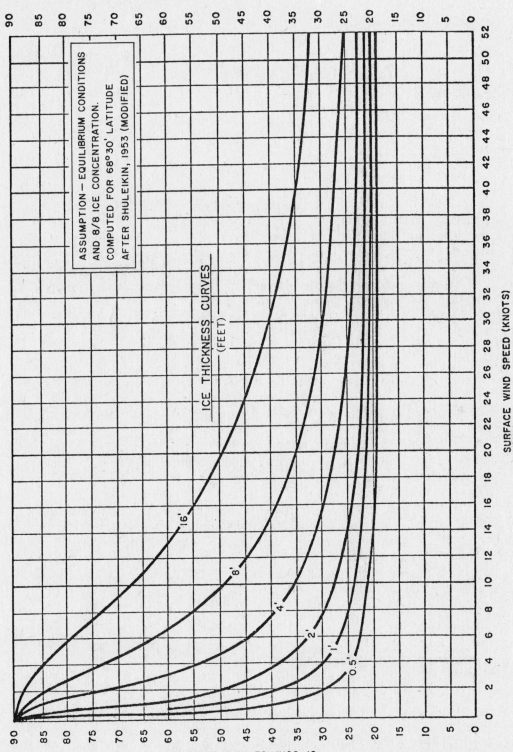

ASSUMPTION — EQUILIBRIUM CONDITIONS
AND 8/8 ICE CONCENTRATION.
COMPUTED FOR 68°30' LATITUDE
AFTER SHULEIKIN, 1953 (MODIFIED)

ICE THICKNESS CURVES
(FEET)

SURFACE WIND SPEED (KNOTS)

DRIFT ANGLE (DEGREE) TO RIGHT (N. HEMISPHERE) OR LEFT (S. HEMISPHERE)
OF SURFACE WIND DIRECTION

FIGURE 10.8.—Ice drift direction for varying wind speed and ice thickness.

ICE CONCENTRATION (OKTAS)

EXTENT OF RIDGING & HUMMOCKING (TENTHS)		1	2	3	4	5	6	7
	1	0.88	0.75	0.63	0.50	0.39	0.33	0.26
	2	1.75	1.50	1.25	1.05	0.85	0.65	0.53
	3	2.64	2.33	2.05	1.75	1.44	1.13	0.85
	4	3.53	3.13	2.69	2.30	1.86	1.45	1.08
	5	4.40	3.90	3.40	2.90	2.40	1.90	1.40
	6	5.28	4.65	4.06	3.50	2.94	2.35	1.73
	7	6.66	5.48	4.79	4.10	3.41	2.73	2.04
	8	7.03	6.25	5.55	4.75	3.95	3.18	2.36
	9	7.93	7.05	6.10	5.20	4.35	3.50	2.61

TABLE 10.8.—Rate of the wind drift of sea ice (given as a percent of the surface wind speed) for varying ice concentration and surface roughness.

Two other major forces which act on a drifting iceberg are the Coriolis force and, to a lesser extent, the pressure gradient force which is caused by gravity owing to a tilt of the sea surface and is important only for iceberg drift in a major current. Near-surface currents are generated by a variety of factors such as horizontal pressure gradients owing to density variations in the water, rotation of the earth, gravitational attraction of the moon, and slope of the sea surface. Wind not only acts directly on an iceberg, but also indirectly by generating waves and a surface current in about the same direction as the wind. Because of inertia, an iceberg may continue to move from the influence of wind for some time after the wind stops or changes direction.

The relative influence of currents and winds on the drift of an iceberg varies according to the direction and magnitude of the forces acting on its sail area and subsurface cross-sectional area. The resultant force therefore involves the proportions of the iceberg above and below the sea surface in relation to the velocity and depth of the current, and the velocity and duration of the wind. Studies tend to show that, generally, where strong currents prevail, the current is dominant. In regions of weak currents, however, winds that blow for a number of hours in a steady direction materially affect the drift of icebergs. Generally, it can be stated that currents tend to have a greater effect on deep-drafted icebergs, while winds tend to have a greater effect on shallow-drafted icebergs.

As icebergs waste through melting, erosion, and calving, observations indicate the height to draft ratio may approach 1:1 during their last stage of decay when they are referred to as valley, winged, horned, or spired icebergs. The height to draft ratios found for icebergs in their various stages are presented in table 10.9a. Since wind tends to have a greater effect on shallow than deep-drafted icebergs, the wind can be expected to exert increasing influence on iceberg drift as wastage increases.

Iceberg type	Height to draft ratio
Blocky or tabular	1:5
Rounded or domed	1:4
Picturesque or Greenland (sloping)	1:3
Pinnacled or ridged.	1:2
Horned, winged, valley, or spired (weathered)	1:1

TABLE 10.9a.—Height to draft ratios for various types of icebergs.

Simple equations which precisely define iceberg drift cannot be formulated at present because of the uncertainty in the water and air drag coefficients associated with iceberg motion. Values for these parameters not only vary from iceberg to iceberg, but they probably change for the same iceberg over its period of wastage.

Present investigations utilize an analytical approach facilitated by computer calculations in which the air and water drag coefficients are varied within reasonable limits. Combinations of these drag values are then used in several increasingly complex water models that try to duplicate observed iceberg trajectories. The results indicate that with a wind generated current, Coriolis force, and a uniform wind, but without a gradient current, small and medium icebergs will drift with the percentages of the wind as given in table 10.9b. The drift will be to the right in the Northern Hemisphere and to the left in the Southern Hemisphere.

Wind Speed (knots)	Ice Speed/Wind Speed (percent)		Drift Angle (degrees)	
	Small Berg	Med. Berg	Small Berg	Med. Berg
10	3.6	2.2	12	69
20	3.8	3.1	14	55
30	4.1	3.4	17	36
40	4.4	3.5	19	33
50	4.5	3.6	23	32
60	4.9	3.7	24	31

TABLE 10.9b.—Drift of iceberg as percentage of wind speed.

When gradient currents are introduced, trajectories vary considerably depending on the magnitude of the wind and current and whether they are in the same or opposite direction. When a 1-knot current and wind are in the same direction, drift is to the right of both wind and current with drift angles increasing linearly from approximately 5° at 10 knots to 22° at 60 knots. When the wind and a 1-knot current are in opposite directions, drift is to the left of the current with the angle increasing from approximately 3° at 10 knots, to 20° at 30 knots, and to 73° at 60 knots. As a limiting case for increasing wind speeds, drift may be approximately normal (to the right) to the wind direction. This indicates that the wind generated current is clearly dominating the drift. In general, the various models used demonstrated that a combination of the wind and current was responsible for the drift of icebergs.

10.10. Extent of ice in the sea.—When an area of sea ice, no matter what form it takes or how it is disposed, is described, it is referred to as **pack ice.** In both polar regions the pack ice is a very dynamic feature with wide deviations in its areal extent dependent upon changing oceanographic and meteorological phenomena. In winter the arctic pack extends over the entire Arctic Ocean and for a varying distance outward from it; the limits receding considerably during the warmer summer months. Each year a large portion of the ice from the Arctic Ocean moves outward between Greenland and Spitsbergen, into the North Atlantic, and is replaced by new ice. Relatively little of the arctic pack ice is more than 10 years old. An example of the variance possible in the outer limit of the arctic ice pack is shown in figure 10.10a where the average positions of the maximum and minimum extents of sea ice are plotted.

Ice covers a large portion of the antarctic waters and is probably the greatest single factor contributing to the isolation of the Antarctic Continent. During the austral winter (June through September), ice completely surrounds the continent, forming an almost impassable barrier that extends northward on the average to about 54°S in the Atlantic and to about 62°S in the Pacific. Disintegration of the pack ice during the austral summer months of December through March allows the limits of the ice edge to recede considerably opening some coastal areas of the Antarctic to navigation. The mean maximum and mean minimum positions of the antarctic ice limit are shown in figure 10.10b.

Historical information on sea conditions for specific localities and time periods can be found in publications of the U. S. Naval Oceanographic Office and the Defense Mapping Agency Hydrographic Center. Such publications include sailing directions, forecasting guides, and ice atlases.

10.11. Ice in the North Atlantic.—Sea level glaciers exist on a number of landmasses bordering the northern seas, including Alaska, Greenland, Svalbard (Spitsbergen), Zemlya Frantsa-Iosifa (Franz Josef Land), Noyaya Zemlya, and Severnaya Zemlya (Nicholas II Land). Except in Greenland, the rate of calving is relatively slow, and the few icebergs produced melt near their points of formation. Many of those produced along the coasts of Greenland, however, are eventually carried into the shipping lanes of the North Atlantic, where they constitute a major menace to ships.

Generally the majority of icebergs produced along the east coast of Greenland remain near their source of origin. However, a small number of bergy bits, growlers, and small icebergs are transported from this region by the East Greenland Current

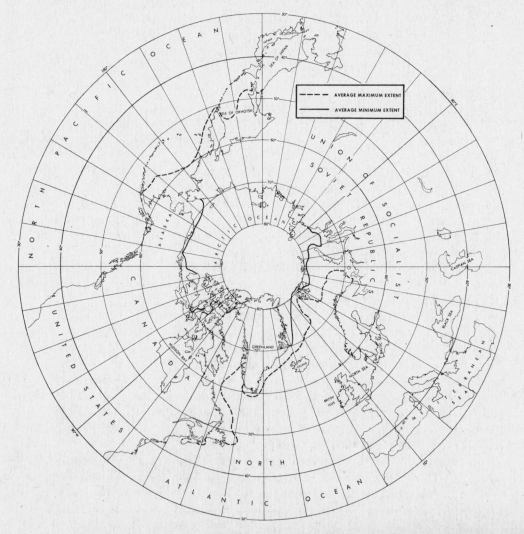

FIGURE 10.10a.—Average maximum and minimum extent of arctic sea ice.

around Kap Farvel at the southern tip of Greenland and then northward by the West Greenland Current into Davis Strait to the vicinity of 67°N. Relatively few of these icebergs menace shipping but some are carried to the south and southeast of Kap Farvel by a counterclockwise current gyre centered near 57°N and 43°W.

The main source of the icebergs encountered in the North Atlantic is the west coast of Greenland between 67°N and 76°N where approximately 7,500 icebergs are formed each year. In this area there are about 100 low lying coastal glaciers, 20 of them being the principal producers of icebergs. Of these 20 major glaciers, 2 located in Disko Bugt between 69°N and 70°N are estimated to contribute 28 percent of all icebergs appearing in Baffin Bay and the Labrador Sea. The West Greenland Current carries icebergs from this area northward and then westward until they encounter the south flowing Labrador Current. West Greenland icebergs generally spend their first winter locked in the Baffin Bay pack ice; however, a large number can also be found within the sea ice extending along the entire Labrador coast by late winter. During the next spring and summer, when they are freed by the break up of the pack ice, they are transported further southward by the Labrador Current. The general drift patterns of icebergs that are prevalent in the eastern portion of the North American Arctic are shown in figure 10.11a. Observations over a 69-year period show that an average of 365 icebergs per year reach latitudes south of 48°N, with approximately 10 percent of this

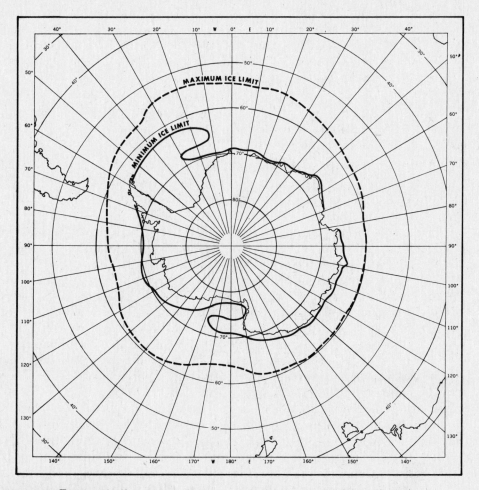

FIGURE 10.10b.—Average maximum and minimum extent of antarctic sea ice.

total carried south of the Grand Banks (43°N) before they melt. Icebergs may be encountered during any part of the year, but in the Grand Banks area they are most numerous during spring. The maximum monthly average of iceberg sightings occur during April, May, and June, with May having the highest average of 124.

The variation from average conditions is considerable. More than 1,587 icebergs have been sighted south of latitude 48°N in a single year (1972), while in 1966 not a single iceberg was encountered in this area. In the years of 1940 and 1958 only one iceberg was observed south of 48°N. Although this variation has not been fully explained, it is apparently related to wind conditions, the distribution of pack ice in Davis Strait, and to the amount of pack ice off Labrador. It has been suggested that the distribution of the Davis Strait-Labrador Sea pack ice influences the effectiveness of this ice in holding back the icebergs. According to this theory, when pack ice is heavy along the Labrador coast, the icebergs are forced well offshore, where warmer water causes them to melt before they reach the North Atlantic shipping lanes; but when the pack ice is not sufficient for this, the icebergs drift closer to shore, where there is colder water which prolongs their existence.

Average iceberg and pack ice limits in this area during April, May, and June are shown in figures 10.11b, 10.11c, and 10.11d. Icebergs have been observed in the vicinity of Bermuda, the Azores, and within 400 to 500 kilometers of Great Britain.

Pack ice may also be found in the North Atlantic, some having been brought south by the Labrador Current and some coming through Cabot Strait after having formed in the Gulf of St. Lawrence.

10.12. The International Ice Patrol was established in 1914 by the *International Convention for the Safety of Life at Sea*, held in 1913 as a result of the sinking of the SS *Titanic* in 1912. On its maiden voyage this vessel struck an iceberg and sank with the loss of 1,513 lives. In accordance with the agreement reached at the convention, this patrol is conducted by the U. S. Coast Guard, which is responsible for the observations and dissemination of information concerning ice conditions in the North Atlantic. Information on ice conditions for the Gulf of St. Lawrence and the coastal waters of Newfoundland and Labrador, including the Strait of Belle Isle to west of Belle Isle itself, is provided by the Canadian Ministry of Transport between the months of December through late June. Ice data for these areas are obtained from the Ice Operations Officer located at Dartmouth, Nova Scotia via Sidney or Halifax marine radio.

During each ice season, aerial reconnaissance surveys are made in the vicinity of the Grand Banks of Newfoundland to determine the southeastern, southern, and southwestern limit of the seaward extent of icebergs. During the war years of 1916–18 and 1941–45 the patrol was suspended. Aircraft were added to the patrol force following World War II, and today perform the majority of the work. Reports of ice sightings are also requested and collected from ships transiting the Grand Banks area. When reporting ice, vessels are requested to detail the type of ice (icebergs or sea ice) sighted, its position, concentration and thickness (for sea ice), and size and shape (for icebergs).

In addition to ice reports, masters who do not issue routine weather reports, are urged to make sea surface temperature and weather reports to the Ice Patrol every 6 hours within latitudes 40° to 50°N and longitudes 42° to 60°W.

Operations of the Ice Patrol are directed from the U. S. Coast Guard Base, Governors Island, New York. Regularly scheduled bulletins are issued by the Ice Patrol twice daily during the ice season by radio and landline communications from Boston, Massachusetts. When icebergs are sighted outside the known limits of ice, special broadcasts are issued from St. Johns, Newfoundland, between those regularly scheduled. Iceberg positions in the ice bulletins are updated for drift at 12-hour intervals. A radio-facsimile chart is also broadcast once a day throughout the ice season. The Ice Patrol,

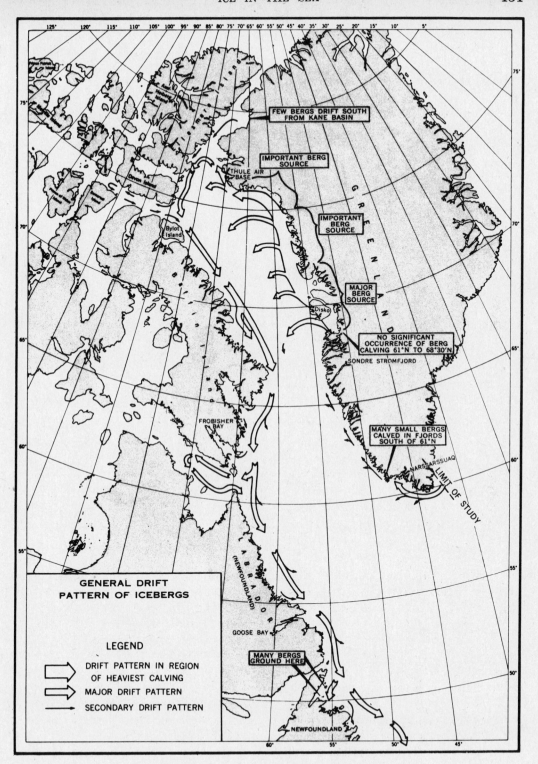

FIGURE 10.11a.—General drift pattern of icebergs.

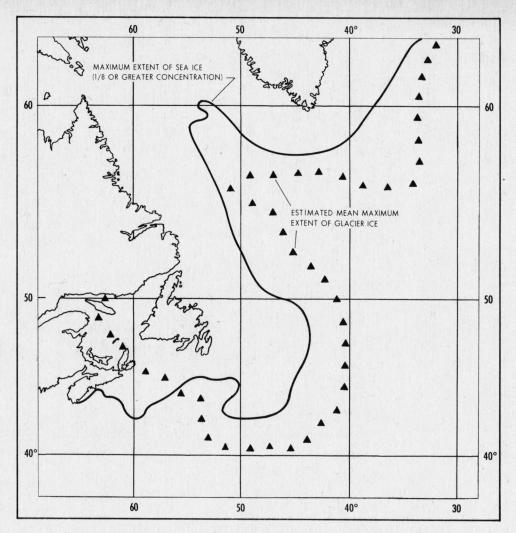

FIGURE 10.11b.—Average iceberg and pack ice limits during the month of April.

in addition to patrolling possible iceberg areas, conducts oceanographic surveys, maintains up-to-date records of the currents in its area of operation to aid in predicting the drift of icebergs, studies sea ice conditions in general, and offers assistance, if the need arises, to ships within the limits of its operation.

10.13. Ice detection.—Safe navigation in the polar seas depends on a number of factors, not the least of which is accurate knowledge of the location and amount of sea ice that lies between the mariner and his destination. Sophisticated electronic equipment such as radar, sonar, and the visible, infrared, and microwave radiation remote sensors on board earth orbiting satellites have joined forces with the polar traveler's own eyesight and, in some cases, hearing to aid him in detecting ice in the sea.

As a ship proceeds into higher latitudes, the first ice it encounters is likely to be in the form of icebergs, because such large pieces require a longer time to disintegrate. Icebergs can easily be avoided if detected soon enough. The distance at which an iceberg can be seen visually depends upon meteorological visibility, height of the iceberg, source and condition of lighting, and the observer. On a clear day with excellent visibility, a large iceberg, due to its brilliant luster, might be sighted at a distance of almost 35 kilometers. With a low-lying haze around the horizon, this distance may be

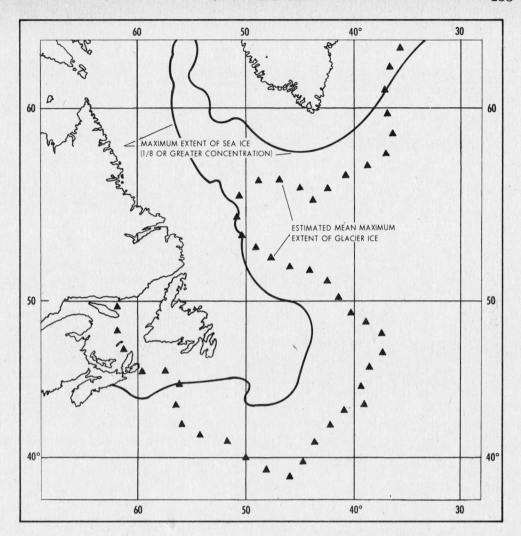

FIGURE 10.11c.—Average iceberg and pack ice limits during the month of May.

reduced by one-half. In light fog or drizzle this distance is further reduced from 1,850 meters to 5.5 kilometers.

In a dense fog an iceberg may not be perceptible at a distance of over 100 meters or until it is close aboard where it will appear in the form of a luminous, white object if the sun is shining; or as a dark, somber mass with a narrow streak of blackness at the waterline if the sun is not shining. If the layer of fog is not too thick, an iceberg may be sighted from aloft sooner than from a point lower in the vessel, but this fact should not be considered justification for omitting a bow lookout. The diffusion of light in a fog will produce a blink, or area of whiteness, above and at the sides of an iceberg which will appear to increase the apparent size of its mass.

On dark, clear nights icebergs may be seen at a distance of from 1,850 meters to 4 kilometers, appearing either as white or black objects with an occasional light spot where a wave breaks against it. Under such conditions of visibility growlers are a greater menace to vessels, and the vessel's speed should be reduced and a sharp lookout maintained.

The moon may either help or hinder, depending upon its phase and position relative to ship and iceberg. A full moon in the direction of the iceberg interferes with

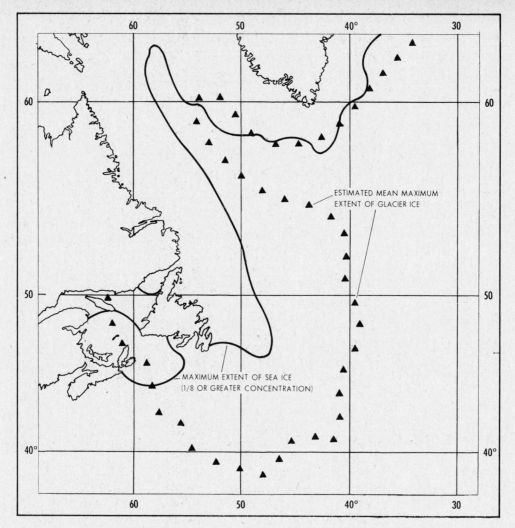

FIGURE 10.11d.—Average iceberg and pack ice limits during the month of June.

its detection, while light from one in the opposite direction may produce a blink which renders the iceberg visible for a greater distance, possibly as much as 5.5 kilometers. A clouded sky at night, through which the moonlight is intermittent, also renders ice detection difficult. A night sky with heavy passing clouds may also dim or obscure any object which has been sighted, and fleecy cumulus and cumulonimbus clouds often may give the appearance of blink from icebergs.

If an iceberg is in the process of disintegration, its presence may be detected by the cracking sound as a piece breaks off, or by the thunderous roar as a large piece falls into the water. The appearance of smaller pieces of ice in the water often indicates the presence of an iceberg nearby. In calm weather such pieces may form a curved line with the parent iceberg on the concave side. Some of the pieces broken from an iceberg are themselves large enough to be a menace to ships.

As the ship moves closer towards areas known to contain sea ice, one of the most reliable signs that pack ice is being approached is the absence of swell or wave motion in a fresh breeze or a sudden flattening of the sea, especially from leeward. The observation of icebergs in itself is not a good indication that pack ice will be encountered soon, since icebergs may be found at great distances from pack ice. If the sea ice is approached from windward, it is usually compacted and the edge will be sharply defined. However, if it is approached from leeward, the ice is likely to be loose and somewhat scattered, often in long narrow arms.

Another reliable sign of the approach of pack ice, not yet in sight, is the appearance of a pattern, or **sky map,** on the horizon or on the underside of distant, extensive cloud areas, created by the varying amounts of light reflected from different materials on the sea or earth's surface. A bright white glare, or **snow blink,** will be observed above a snow covered surface. When the reflection on the underside of clouds is caused by an accumulation of distant ice, the glare is a little less bright and is referred to as an **ice blink.** A relatively dark pattern is reflected on the underside of clouds when it is over land that is not snow covered. This is known as a **land sky.** The darkest pattern will occur when the clouds are above an open water area, and is called a **water sky.** A mariner experienced in recognizing the sky maps detailed above will find them useful in avoiding ice or searching out openings which may permit his vessel to make progress while proceeding through an ice field.

Another indication of the presence of sea ice is the formation of thick bands of fog over the ice edge as moisture condenses from warm air as it passes over the colder ice. An abrupt change in air or sea temperature or seawater salinity is not a reliable sign of the approach of icebergs or pack ice. However, a drop in the seawater temperature to 1°1C may be an indication that a ship is within 90 kilometers of pack ice.

The presence of certain species of animals and birds can also indicate that pack ice is in close proximity. The sighting of walruses, seals, or polar bears in the Arctic should warn the mariner that pack ice is close at hand. In the Antarctic, the usual precursors of sea ice are penguins, terns, fulmars, petrels, and skuas. The mariner will do well to observe the habits of all species encountered, for the information gained will be useful on subsequent journeys.

When visibility becomes limited, radar can prove to be an invaluable tool for the polar mariner. Although many icebergs will be observed visually on clear days before there is a return on the radarscope, radar under bad weather conditions will detect the average iceberg at a range of about 15 to 18 kilometers. The intensity of the return is a function of the nature of the iceberg's exposed surface (slope, surface roughness); however, it is unusual to find an iceberg which will not produce a detectable echo.

Large, vertical-sided tabular icebergs of the antarctic and arctic ice islands are usually detected by radar at ranges of 28 to 55 kilometers, with ranges of 68.5 kilometers having been reported.

Whereas a large iceberg is almost always detected by radar in time to be avoided, a growler large enough to be a serious menace to a vessel may be lost in the sea return and escape detection. If an iceberg or growler is detected by radar, tracking is sometimes necessary to distinguish it from a rock, islet, or another ship.

Radar can be of great assistance to one experienced in interpreting the radarscope. Smooth sea ice, like smooth water, returns little or no echo, but small floes of rough, hummocky sea ice capable of inflicting damage to a ship can be detected in a smooth sea at a range of about 4 to 6 kilometers. The return may be similar to sea return, but the same echoes appear at each sweep. A lead in smooth ice is clearly visible on a radarscope, even though a thin coating of new ice may have formed in the opening. A light covering of snow obliterating many of the features to the eye has little effect upon a radar return. The ranges at which ice can be detected by radar are somewhat dependent upon refraction, which is sometimes quite abnormal in polar regions. Adequate training and experience are essential if full benefit is to be realized from radar.

Echoes from the ship's whistle or horn will sometimes indicate the presence of icebergs. Such echoes can give an indication of direction. If the time interval between the

sound and its echo is measured, the distance in meters can be determined by multiplying the number of seconds by 168. However, echoes are not a reliable indication
because only those pieces of ice with large vertical areas facing the ship return enough
echo to be heard. Also, echoes might be received from land or a fog bank.

At relatively short ranges, sonar is sometimes helpful in locating ice. The initial
detection of icebergs may be made at a distance of about 5 kilometers or more, but
usually considerably less. Growlers may be detected at a distance of 900 meters to 2.5
kilometers, and even smaller pieces may be detected in time to avoid them. Since
one-half to seven-eighths of the mass of an iceberg may lie below the surface, the
underwater portion presents a better target than the portion above water.

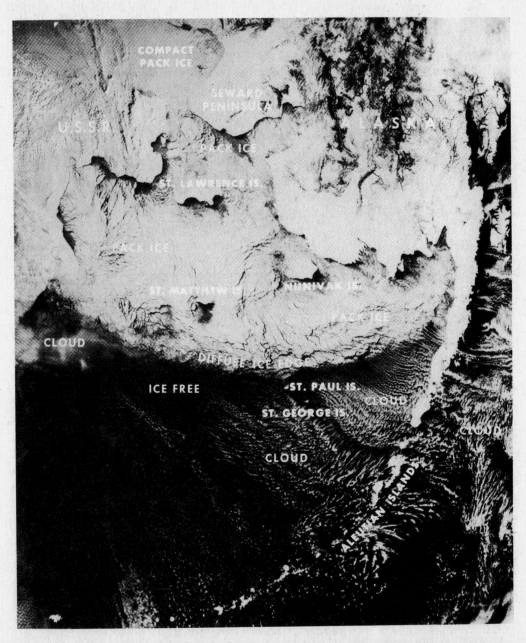

FIGURE 10.13a.—Example of satellite imagery with a resolution of 0.9 kilometer.

Ice in the polar regions is best detected and observed from the air either from aircraft or by satellite remote sensing surveillance systems. Fixed-winged aircraft have been utilized extensively for obtaining detailed aerial ice reconnaissance information since the early 1930's and will no doubt continue to provide this invaluable service for many years to come. Some ships, particularly icebreakers, proceeding into high latitudes carry helicopters, which are invaluable in locating ice and determining the relative navigability of different portions of the ice pack. If these helicopters, their support vessels, or aircraft flying aerial reconnaissance can be contacted by radio, much useful information will be obtained from them. Ice reports from personnel at arctic and antarctic coastal shore stations can also prove valuable to the polar mariner.

The enormous ice reconnaissance capabilities of meteorological satellites were confirmed within hours of the launch by the National Aeronautics and Space Administration (NASA) of the first experimental meteorological satellite, TIROS I, on April 1, 1960. Although this satellite was placed in an equatorial orbit, it was able to detect ice covered waters in the Gulf of St. Lawrence-Newfoundland region. With the advent of the polar-orbiting meteorological satellites during the mid and late 1960's,

FIGURE 10.13b.—Example of satellite imagery with a resolution of 80 meters.

the U. S. Navy initiated an operational satellite ice reconnaissance program which could, depending upon solar illumination, observe on a daily basis ice and its movement in any region of the globe. With the further addition of improved sensors such as high resolution infrared and visible scanning radiometers (SR); very high resolution radiometers (VHRR), also in the visible and infrared spectrum; and microwave systems; detailed global satellite ice data were made available under all weather and lighting conditions with resolutions in some cases below 100 meters. Examples of satellite imagery of ice covered waters are shown in figures 10.13a and 10.13b.

Utilizing portable Automatic Picture Transmission (APT) equipment, which can easily be installed aboard ships or aircraft, visible and infrared radiation data transmitted from operational satellites can be collected during a satellite's passage overhead. In this manner ice data from the satellite's scanning radiometers can be received by APT stations anywhere in the polar regions during the time they are in the line of sight of the satellite. Portable APT equipment is generally small and inexpensive, usually consisting of a receiver with a camera pack, a scanning radiometer adapter, a tape recorder for later data playback, and an omnidirectional antenna for ship and aircraft use. A printed display is available with the addition of a mini-computer that also enhances the image and a small printer to display the digitized data. General information relating to operational satellites, various APT systems, types and modes of satellite data available, and transmission frequencies and times can be obtained from the National Environmental Satellite Service, National Oceanic and Atmospheric Administration, Washington, D.C.

10.14. Operations in ice.—Operations in the polar regions necessarily require considerable advanced planning and many more precautionary measures than those taken prior to a typical open ocean voyage. The crew, large or small, of a polar-bound vessel should be thoroughly indoctrinated in the fundamentals of polar operations, utilizing the best information sources available. The subjects covered should include training in shiphandling in ice, polar navigation, effects of low temperatures on materials and equipment, damage control procedures, communications problems inherent in polar regions, polar meteorology, sea ice terminology, ice observing and reporting procedures (including classification and codes) and polar survival. Training materials should consist of reports on previous arctic and antarctic voyages, sailing directions, ice atlases, training films on polar operations, and U. S. Navy service manuals detailing the recommended procedures to follow during high latitude missions. Information relating to sources of information can be obtained from the Director, Naval Oceanography and Meteorology Command, Bay St. Louis, Mississippi, and from the Office of Polar Programs, National Science Foundation, Washington, D.C.

The preparation of a vessel for polar operations is of extreme importance and the considerable experience gained from previous operations should be drawn upon to bring the ship to optimum operating condition. At the very least, operations conducted in ice infested waters require that the vessel's hull and propulsion system undergo certain modifications.

The bow and waterline of the forward part of the vessel should be heavily reinforced. Similar reinforcement should also be considered for the propulsion spaces of the vessel. Cast iron propellers and those made of a bronze alloy do not possess the strength necessary to operate safely in ice. Therefore, it is strongly recommended that propellers made of these materials be replaced by those fabricated from steel. Other desirable features are the absence of vertical sides, deep placement of the propellers, a blunt bow, metal guards to protect propellers from ice damage, and lifeboats for 150 percent of personnel aboard. The complete list of desirable features depends upon the area of operations, types of ice to be encountered, length of stay in the vicinity of ice, anticipated assistance by icebreakers, and possibly other factors. Strength requirements and the minimum thicknesses deemed necessary for the vessel's frames and additional plating to be used as reinforcement, as well as other procedures needed to outfit a vessel for ice operations, can be obtained from the American Bureau of Shipping. For a more definitive and complete guide to the ice strengthening of ships, the mariner may desire to consult the procedures outlined in *Rules for Ice Strengthening of Ships*, from the Board of Navigation, Helsinki, Finland.

Equipment necessary to meet the basic needs of the crew and to insure the successful and safe completion of the polar voyage should not be overlooked. A minimum list of essential items should consist of polar clothing and footware, food, vitamins, medical supplies, fuel, storage batteries, antifreeze, explosives, detonators, fuses, meteorological supplies, and survival kits containing sleeping bags, trail rations, firearms, ammunition, fishing gear, emergency medical supplies, and a repair kit.

Whatever the nature of the vessel, it will be subjected to various hazards which may cause damage. Its safety depends largely upon the thoroughness of advance preparations, the alertness and skill of its crew, and their ability to make repairs if damage is incurred. Spare propellers, rudder assemblies, and patch materials, together with the equipment necessary to effect emergency repairs of structural damage should be carried. Examples of repair materials needed include quick setting cement, oakum, canvas, timbers, planks, pieces of iron of varying shapes, welding equipment, clamps, and an assortment of nuts, bolts, washers, screws, and nails.

Ice and snow accumulation on portions of the vessel poses a definite safety hazard. Therefore, mallets, hammers, and scrapers to aid in the removal of heavy accumulations of ice, together with a supply of snow shovels and stiff brooms for snow removal should be provided.

Navigation in polar waters is, even under optimum conditions, difficult and, during poor conditions, almost impossible. Environmental conditions encountered in the high latitudes such as fog, storms, compass anomalies, atmospheric effects, and, of course, ice, hinder polar operations. Also, deficiencies in the reliability and detail of hydrographic and geographical information presented on polar navigation charts coupled with a distinct lack of reliable bathymetry, current, and tidal data add to the problems of polar navigation. Much work is being carried out in the polar regions to improve the geodetic control, triangulation, and quality of hydrographic and topographic information necessary for accurate polar charts; however, until this massive task is completed, the only resource open to the polar navigator, especially during periods of poor environmental conditions, is to rely upon the basic principles of navigation and adapt them to unconventional methods when abnormal situations arise.

Upon the approach to pack ice, a careful decision is needed to determine the best action. Often it is possible to go around the ice, rather than through it. Unless the pack is quite loose, this action usually gains rather than loses time. When skirting an ice field or an iceberg, do so to windward, if a choice is available, to avoid projecting tongues of ice or individual pieces that have been blown away from the main body of ice.

When it becomes necessary to enter pack ice, a thorough examination of the distribution and extent of the ice conditions should be made beforehand from the highest possible location. Aircraft (particularly helicopters) and direct satellite readouts are of great value in determining the nature of the ice to be encountered. The most important features to be noted include the location of open water such as leads and polynyas which may be manifested by water sky, icebergs, and the presence or absence of both ice under pressure and rotten ice. Some protection may be offered the propeller and rudder assemblies by trimming the vessel down by the stern slightly (at no time more than 60 or 90 centimeters) prior to entering the ice; however, this precaution usually impairs the maneuvering characteristics of most vessels not specifically built for icebreaking.

Selection of the point of entry into the pack should be undertaken with great care; and if the ice boundary consists of closely packed ice or ice under pressure, it is advisable to skirt the edge until a more desirable point of entry is located. Seek areas with low ice concentrations, areas of rotten ice or those containing navigable leads, and if possible enter from leeward on a course perpendicular to the ice edge. It is also advisable to take into consideration the direction and force of the wind, and the set and drift of the prevailing currents when determining the point of entry and the course followed thereafter. Due to wind induced wave action, ice floes close to the periphery of the ice pack will take on a bouncing motion which can be quite hazardous to the hull of thin-skinned vessels. In addition, keep in mind that pack ice will drift slightly to the right of the lee of the true wind in the Northern Hemisphere and to the left of the lee in the Southern Hemisphere (art. 10.8), and that leads opened by the force of the wind will appear perpendicular to the wind direction. If a suitable entry point cannot be located due to less than favorable conditions, one should be patient. Unfavorable conditions generally improve over a short period of time by a change in the wind, tide, or sea state.

Having entered the pack, *always work with the ice, not against it, and keep moving,* but do not rush the work of negotiating the pack. Patience may pay big dividends. Respect the ice but do not fear it. Proceed at slow speed at first, staying in open water or in areas of weak ice if possible. The vessel's speed may be safely increased after it has been ascertained how well it handles under the varying ice conditions encountered. Remember that it is always better to make good progress in the *general* direction desired than to fight large thick floes in the *exact* direction to be made good. However, avoid the temptation to proceed far to one side of the intended track; it is almost always better to back out and seek a more penetrable area. During those situations when it becomes necessary to back, always do so with extreme caution.

Ice conditions may change rapidly while a vessel is working in pack ice, necessitating some quick maneuvering. It must never be forgotten that conventional vessels, even though ice strengthened, are not built for ice navigation. The vessel should be conned to first attempt to place it in leads or polynyas, giving due consideration to wind conditions. The age, thickness, and size of ice which can be broken depends upon the type, size, strength, and shaft horsepower of the vessel employed. If contact with an ice floe is unavoidable, never strike it a glancing blow. This maneuver may cause the ship to veer off in a direction which will swing the stern into the ice. If possible seek weak spots in the floe and hit it head-on at slow speed. Unless the ice is rotten or very young, do not attempt to break through the floe, but rather make an attempt to swing it aside as speed is slowly increased. Keep clear of corners and projecting points of ice, but do so without making sharp turns which may throw the stern against the ice, resulting in a damaged propeller, propeller shaft, or rudder. The use of full rudder, in non-emergency situations, is not recommended because it may swing either the stern or mid-section of the vessel into the ice. Keep a sharp watch on the propellers and rudder, fending off with long ice poles pieces of ice which might damage these vital parts. Stop the propellers only if ice cannot be avoided.

Offshore winds may open relatively ice free navigable coastal leads, but such leads should not be entered without benefit of icebreaker escort. If it becomes necessary to enter coastal leads, narrow straits, or bays, an alert watch should be maintained since a shift in the wind may force drifting ice down upon the vessel. An increase in wind on the windward side of a prominent point, grounded iceberg, or land ice tongue extending into the sea will similarly endanger a vessel. It will always be wiser to seek out leads toward the windward side of the main body of the ice pack. In the event that the vessel is under imminent danger of being trapped close to shore by pack ice, immediately attempt to orient the vessel's bow seaward. This will help to take advantage of the little manuevering room available in the open water areas found between ice floes. Work carefully through these areas, easing the ice floes aside while maintaining a close watch on the general movement of the ice pack.

If the vessel is completely halted by pack ice, it is best to keep the rudder amidships and the propellers turning at slow speed. The wash of the propellers may help to clear ice away from the stern, making it possible to back down safely. When the vessel is stuck fast as is the case when the bow is forced up onto a massive ice floe, an attempt first should be made to free the vessel by going full speed astern. If this manuever proves ineffective, it may be possible to get the vessel's stern to move slightly, thereby causing the bow to shift, by shifting the rudder from one side to the other while going full speed ahead. Another attempt at going astern should then free the vessel. The vessel may also be freed by either transferring water from the ballast tanks causing the vessel to list, or by alternately flooding and emptying the fore and aft tanks. Men wielding crowbars may also be able to split the ice at the pressure points. If all these methods fail, the utilization of deadmen (2- to 4-meter lengths of timber buried in holes out in the ice and to which a vessel is moored) and ice anchors (a stockless, single-fluked hook embedded in the ice) may be helpful. With a deadman or ice anchors attached to the ice astern, the vessel may be warped off the ice by winching while the engines are going full astern. If all the foregoing methods fail, explosives placed in holes cut nearly to the bottom of the ice approximately 10 to 12 meters off the beam of the vessel and detonated while the engines are working full astern should succeed in freeing the vessel. A vessel may also be sawed out of the ice if the ambient air temperature is above the freezing point of seawater.

When a vessel becomes so closely surrounded by ice that all steering control is lost and it is unable to move, it is **beset.** It may then be carried by the drifting pack into shallow water or areas containing thicker ice or icebergs with their accompanying dangerous underwater projections. If ice forcibly presses itself against the hull, the vessel is said to be **nipped,** whether or not damage is sustained. When this occurs, the gradually increasing pressure may be capable of holing the vessel's bottom. When a vessel is beset or nipped, freedom may be achieved through the careful manuevering procedures, the physical efforts of the crew, or by the use of explosives similar to those previously detailed. Under severe conditions the mariner's best ally may be patience since there will be many times when nothing can be done to improve the vessel's plight until there is a change in meteorological conditions. It is a time to preserve fuel and perform any needed repairs to the vessel and its engines. Damage to the vessel while it is beset is usually attributable to collisions or pressure exerted between the vessel's hull, propellers, or rudder assembly and the sharp corners of ice floes. These collisions can be minimized greatly by attempting to align the vessel in such a manner as to insure that the pressure from the surrounding pack ice is distributed as evenly as possible over the hull. This is best accomplished when medium or large ice floes encircle the vessel.

In the vicinity of icebergs, either in or outside of the pack ice, a sharp lookout should be kept and all icebergs given a wide berth. The commanding officers and masters of all vessels, irrespective of their size, should treat all icebergs with due respect. The best locations for lookouts are generally in a crow's nest rigged in the foremast or housed in a shelter built specifically for a bow lookout in the eyes of a vessel. Telephone communications between these sites and the navigation bridge on larger vessels will prove invaluable. It is dangerous to approach close to an iceberg of any size because of the possibility of encountering underwater extensions, and because icebergs that are disintegrating may suddenly capsize or readjust their masses to new positions of equilibrium. In periods of low visibility the utmost caution is needed at all times. Vessel speed should be reduced and the watch prepared for quick maneuvering. Radar becomes an effective tool in this case, but does not negate the need for trained lookouts.

Since icebergs may have from eight to nine-tenths of their masses below the water surface, their drift (art. 10.9) is generally influenced more by currents than winds, particularly under light wind conditions. The drift of pack ice, on the other hand, is usually dependent upon the wind. Under these conditions, icebergs within the pack may be found moving at a different rate and in a different direction from that of the pack ice. In regions of strong currents, icebergs should always be given a wide berth because they may often travel upwind at great speeds under the influence of contrary currents, wreaking heavy pack in their paths and endangering those vessels that are unable to work clear. In these situations, open water will generally be found leeward of the iceberg, with piled up pack ice to windward. Where currents are weak and a strong wind predominates, similar conditions will be observed as the wind driven ice pack overtakes an iceberg and piles up to windward with an open water area lying to leeward.

Under ice submarine operations require knowledge of prevailing and expected sea ice conditions to ensure maximum operational efficiency and safety. The most important ice features are the frequency and extent of downward projections (bummocks and ice keels) from the underside of the **ice canopy** (pack ice and enclosed water areas from the point of view of the submariner), the distribution of thin ice areas through which submarines can attempt to surface, and the probable location of the outer pack edge where submarines can remain surfaced during emergencies to rendezvous with surface ship or helicopter units.

Bummocks are the subsurface counterpart of hummocks, and **ice keels** are similarly related to ridges. When the physical nature of these ice features is considered, it is apparent that ice keels may have considerable horizontal extent whereas individual bummocks can be expected to have little horizontal extent. In shallow water lanes to the Arctic Basin such as the Bering Strait and the adjoining portions of the Bering Sea and Chukchi Sea, deep bummocks and ice keels may leave little vertical leeway for submarine passage. Widely separated bummocks may be circumnavigated but make for a hazardous passage. Extensive ice areas with numerous bummocks or ice keels which cross the lane, however, may effectively block passage to the Arctic Basin.

Bummocks and ice keels extend downward approximately five times their vertical extent above the ice surface. Therefore, observed ridges of approximately 10 meters may extend as much as 50 meters below sea level. Owing to the direct relation of the frequency and vertical extent between these surface features and their subsurface counterparts, aircraft reconnaissance of ice conditions over a planned submarine cruise track should be conducted before under ice operations are commenced.

Skylights are defined as thin places (usually less than 1 meter thick) in the ice canopy and appear from below as relatively light translucent patches in dark surroundings. The undersurface of a skylight is usually flat; not having been subjected to great pressure although the ice canopy may have a concentration of nearly 8 oktas. Skylights are called large if big enough for a submarine to attempt to surface through them; that is, have a linear extent of at least 120 meters. Skylights smaller than 120 meters are called small. An ice canopy along a submarine's track that contains a number of large skylights or other features such as leads and polynyas which permit a submarine to surface more frequently than 10 times per 56 kilometers is called **friendly ice.** An ice canopy containing no large skylights or other features which permit a submarine to surface is called **hostile ice.**

For a more comprehensive guide to operations in ice, it is recommended that the mariner refer to *Polar Operations*, by Captain Edwin A. MacDonald, USN (Ret.), published by the United States Naval Institute, Annapolis, Maryland.

10.15. Great Lakes ice.—Large vessels have been navigating the Great Lakes since the early 1760's. This large expanse of navigable water has since become one of the world's busiest waterways. Due to the northern geographical location of the Great Lakes Basin and its susceptibility to arctic outbreaks of polar air during winter, the formation of ice plays a major role, albeit a disruptive one, in this region's economically vital marine industry. Because of the relatively large size of the five Great Lakes, the ice cover which forms on them is affected by the wind and currents to a greater degree than that on smaller lakes. The Great Lakes northern location results in a long ice growth season which in combination with the effect of wind and current imparts to their ice covers some of the characteristics and behavior of an arctic ice pack. For these reasons, this article is being included in this chapter on ice in the sea.

Since the five Great Lakes extend over a distance of approximately 800 kilometers in a north-south direction, each lake is influenced by varying degrees of meteorological parameters. These parameters, in combination with the fact that each lake also possesses differing hydrometeorological characteristics, materially affects the extent and distribution of their respective ice covers. The largest, deepest, and most northern of the five Great Lakes is Lake Superior. Ice not under pressure, especially along this lake's northern shores, can attain a thickness of between 70 to 100 centimeters which is equivalent to medium first-year ice. Winds and currents acting upon the ice have been known to cause ridging with heights approaching 10 meters. The great depth of Lake Superior, however, provides it with a large heat storage capacity which hinders the growth of ice somewhat, particularly during the period of initial ice formation. During a normal winter, it can be expected that 60 percent of the surface area of Lake Superior will become covered by ice. This value increases to 95 percent during a severe winter and decreases to 40 percent during a mild winter. Under average conditions, ice which presents an obstacle to navigation on Lake Superior appears during the last week of December along both the north and south shores with the maximum extent of ice cover occurring between March 30 and April 10.

Lake Michigan extends in a north-south direction for approximately 480 kilometers and possesses the third largest mean depth of the Great Lakes. Its north-south alignment causes ice to accumulate in the northern portions of the lake initially and then grow in a southerly direction as the winter progresses. Ice thickness ranges from an average of 10 to 20 centimeters (grey to grey-white ice) in the southern portion to 50 to 80 centimeters (thin to medium first-year ice) in the northern portion. The ice cover becomes a hazard to navigation in the northern sector of Lake Michigan during the last week of December with the average date of maximum extent of ice cover ranging from March 10 in the Chicago area to March 28 at the northern end of Green Bay. During a severe winter, 80 percent of Lake Michigan's surface area will be covered by ice. This value reduces to 40 percent for a normal winter and to only 10 percent during a mild winter.

Ice formation on Lake Huron generally commences along the northeastern and western shorelines during the last week in December of each year. The deep north central basin of the lake does not, except during severe winters, generally acquire even a partial ice cover. The ice on Lake Huron will on the average consist predominately of thin first-year ice (30- 70 centimeters), with some medium first-year ice (70–120 centimeters) forming during a severe winter. The time interval between dates of the average maximum ice cover extent for this lake ranges from March 11 in the extreme southern portion to March 28 in the northern sector. The percent of lake surface area that will become ice covered is 60 percent for a normal winter, 40 percent for a mild winter, and 80 percent for a severe winter.

Lake Erie is the shallowest of the Great Lakes with a mean depth of just under 20 meters. Because of its shallowness, this lake is greatly influenced by seasonal temperature changes and will accumulate a considerable ice cover over a short period of time. Ice will begin to form first in the very shallow western portion of the lake during mid-December. During its growth period, the lake ice is acted upon by the prevailing winds and currents which concentrate it at the northeastern end of the lake. Generally, Lake Erie's ice cover is made up of a combination of grey-white ice (15–30 centimeters) and thin first-year ice (30–70 centimeters). The average dates on which the maximum extent of ice cover is attained on Lake Erie varies from March 5 in the western sector to March 15 for the northeastern portion of the lake. Since it reacts rapidly to the change in the seasons, Lake Erie will attain an ice cover that blankets 95 to 100 percent of its surface area during a normal winter. During a mild winter, the ice will occupy an area covering 50 percent of the surface area.

Lake Ontario has the smallest surface area of the five Great Lakes, but the second greatest mean depth, second only to Lake Superior. Like Lake Superior, its large mean depth gives Lake Ontario a large heat storage capacity which, in combination with its small surface area, causes the lake to respond slowly to changing meteorological conditions. This, in turn, produces the smallest amount of ice cover found on the Great Lakes. Ice will begin forming during mid-December in the northeastern section of the lake, and wind and current conditions similar to those found on Lake Erie will confine the majority of the ice cover to that section of the lake. The majority of the ice formed will consist of thin first-year ice (30–70 centimeters) with a small concentration of grey-white ice (15–30 centimeters). The date at which the ice cover on Lake Ontario reaches its maximum extent varies on the average from March 10 to March 20. During a mild winter only 8 percent of the lake surface area will be covered by ice. This value increases to 15 percent for a normal winter and to 25 percent for a severe winter.

The maximum ice cover distribution attained by each of the Great Lakes for a normal, mild, and severe winter are shown in figures 10.15a, 10.15b, and 10.15c. It should be noted that although the average maximum ice cover distributions for all five lakes appear on a single chart, they occur during the average time periods detailed above for each lake.

Information concerning analyses, forecasts, and climatology of Great Lakes ice can be obtained from the Great Lakes Environmental Research Laboratory or the National Weather Service Forecast Office, both located in Ann Arbor, Michigan.

10.16. Ice observing, reporting, and forecasting.—Advance knowledge of ice conditions to be encountered and knowledge of how these conditions will change over specified time periods are invaluable in both the planning and operational phases of a voyage undertaken in polar regions. Typical ice support services offered to the polar navigator generally include analyses of current ice conditions; short-range (24 to 48 hour), weekly (5 to 7 day), and long range (15 to 30 day) ice forecasts; seasonal long-range (60 to 90 day) ice outlooks which are ordinarily updated by 15 to 30 day forecasts as the season progresses; and world ship weather routing service through ice infested waters. Generally an ice analysis or forecast will depict the current or expected configuration and location of the pack ice edge, ice concentrations within the pack itself, and the locations of features such as leads, polynyas, fast ice, and areas of open water.

The single most important input into an ice forecast of any duration is an accurate, current, ice analysis based on the latest ice observations available. As stated previously, ice in the sea can be observed from vessels, fixed winged aircraft, helicopters, and by

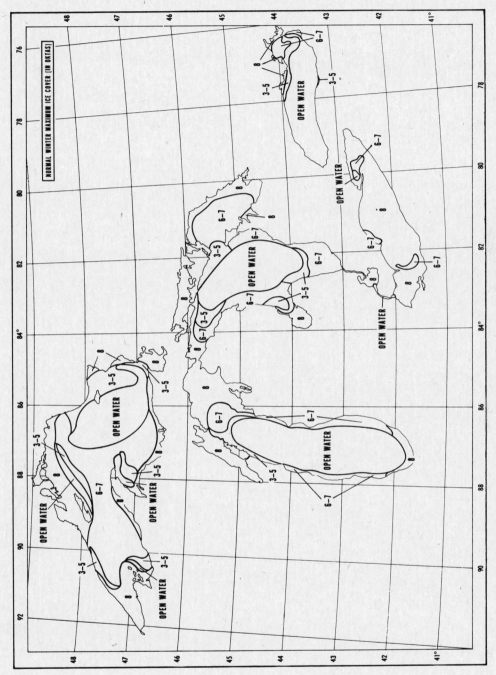

FIGURE 10.15a.—Great Lakes maximum ice cover during a normal winter.

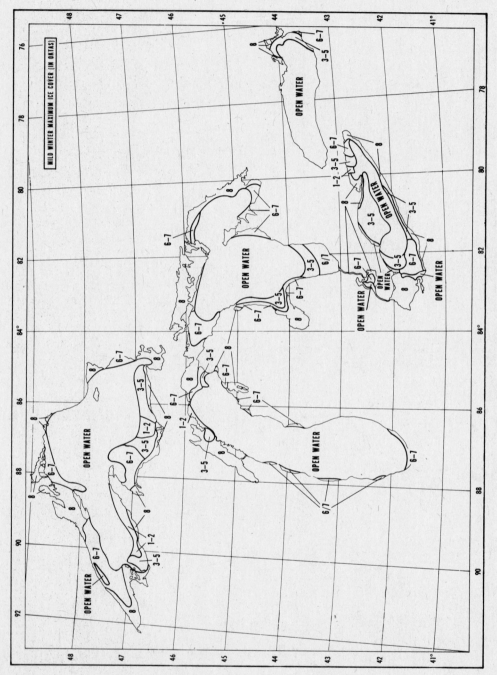

FIGURE 10.15b.—Great Lakes maximum ice cover during a mild winter.

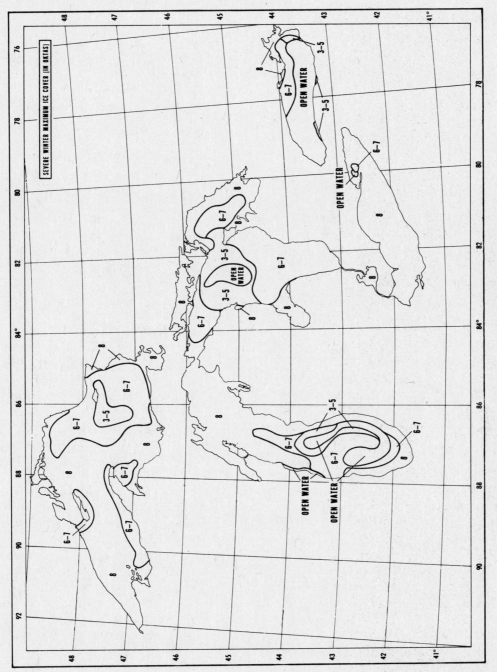

FIGURE 10.15c.—Great Lakes maximum ice cover during a severe winter.

earth orbiting satellites, as well as by personnel at arctic and antarctic coastal shore stations. By means of modern communications networks, the reports of these observations are relayed to the offices of both federal government agencies and private commercial companies.

Federal agencies providing both ice observing and forecasting services on an operational basis include the Department of Defense, the Department of Transportation, and the Department of Commerce. The Department of Defense, principally through its Department of the Navy, provides ice support services to all U. S. and allied military units. Additional support may be provided to commercial concerns in some instances with the prior approval of the Chief of Naval Operations. Specific information relating to ice observation; forecasting freeze-up, ice growth, movement, concentration, and break-up; ship weather routing services; as well as the methods used to disseminate ice information can be obtained from Director, Naval Oceanography and Meteorology Command, Bay St. Louis, Mississippi. The Department of Transportation, through the U. S. Coast Guard, provides ice breaker support for polar operations, and the International Ice Patrol provides the ice support services detailed in article 10.12. Ice forecasts for Alaskan waters are provided to commercial interests by the National Weather Service Forecast Offices in Anchorage and Fairbanks, Alaska. Inquiries concerning assistance available from the Coast Guard while conducting programs in ice should be sent to Commandant, U. S. Coast Guard Headquarters, Washington, D.C.

A network of polar orbiting meteorological satellites and a worldwide communications system provide the National Oceanic and Atmospheric Administration (NOAA) with great quantities of detailed ice information from all areas of the world. Inquiries concerning the products and services available to the polar navigator should be directed to either the National Weather Service or the National Environmental Satellite Service, NOAA, Department of Commerce, Washington, D.C. Listings of consulting meteorologists and oceanographers, and firms employing such personnel, that provide ice support services to individuals or commercial organizations can be usually located in the professional directories of the leading meteorology and oceanography journals.

Mariners operating in and around sea ice can do much to contribute to the overall effectiveness of many of the services provided them. To assist these programs it is essential that all vessels and aircraft operating in ice areas submit as detailed an ice report as possible to interested agencies. Several codes are now available for reporting ice conditions. The code normally used by those trained to make meteorological observations, but not specifically sea ice observations, consists of a five-character group appended to the World Meteorological Organization (WMO) weather reporting codes FM 21–V, FM 22–V, and FM 23–V. These codes are completely described in volume I of World Meteorological Organization, *Manual on Codes*, WMO No. 306, available from the Secretariat of the World Meteorological Organization, Geneva, Switzerland. A more complete and detailed reporting code (ICEOB) has been in use in the United States since 1972. This code and the procedures for its use are contained in the *Ice Observers Manual*, U. S. Naval Oceanographic Office Technical Note 3700–49–76 (unpublished manuscript).

These codes make use of special nomenclature which has been precisely defined by the World Meteorological Organization in several languages. *Sea-Ice Nomenclature*, (WMO No. 259.TP.145) contains the nomenclature along with photography of most ice features. This publication is very useful for those who plan to submit ice condition reports.

In addition, *Guide to Meteorological Instruments and Observing Practices* (WMO No. 8.TP.31) contains a chapter on ice development, dynamic processes, and observing procedures and should also prove extremely useful.

The mariner who regularly sends as complete an ice report as possible contributes substantially to an increase in the knowledge of synoptic ice conditions and therefore to the accuracy and timeliness of subsequent ice analyses and forecasts.

CHAPTER XI

TIDE AND CURRENT PREDICTIONS

11.1. Tidal effects.—The daily rise and fall of the **tide**, with its attendant flood and ebb of **tidal current,** is familiar to every mariner. He is aware, also, that at **high water** and **low water** the depth of water is momentarily constant, a condition called **stand.** Similarly, there is a moment of **slack water** as a tidal current reverses direction. As a general rule, the *change* in height or the current speed is at first very slow, increasing to a maximum about midway between the two extremes, and then decreasing again. If plotted against time, the height of tide or speed of a tidal current takes the general form of a sine curve. Sample curves, and more complete information about causes, types, and features of tides and tidal currents, are given in chapter V. The present chapter is concerned primarily with the application of tides and currents to piloting, and predicting the tidal conditions that might be encountered at any given time.

Although tides and tidal currents are caused by the same phenomena, the time relationship between them varies considerably from place to place. For instance, if an estuary has a wide entrance and does not extend far inland, the time of maximum speed of current occurs at about the mid time between high water and low water. However, if an extensive tidal basin is connected to the sea by a small opening, the maximum current may occur at about the time of high water or low water outside the basin, when the difference in height is maximum.

The *height of tide* should not be confused with *depth of water*. For reckoning tides a reference level is selected. Soundings shown on the largest scale charts are the vertical distances from this level to the bottom. At any time the actual depth is this charted depth *plus* the height of tide. In most places the reference level is some form of low water. But all low waters at a place are not the same height, and the selected reference level is seldom the *lowest* tide that occurs at the place. When lower tides occur, these are indicated by a negative sign. Thus, at a spot where the charted depth is 15 feet, the actual depth is 15 feet plus height of tide. When the tide is three feet, the depth is $15+3=18$ feet. When it is $(-)$ 1 foot, the depth is $15-1=14$ feet. It is well to remember that *the actual depth can be less than the charted depth.* In an area where there is a considerable **range of tide** (the difference between high water and low water), the height of tide might be an important consideration in using soundings to assist in determining position, or whether the vessel is in safe water.

One should remember that heights given in the tide tables are *predictions*, and that when conditions vary considerably from those used in making the predictions, the heights shown may be considerably in error. Heights lower than predicted are particularly to be anticipated when the atmospheric pressure is higher than normal, or when there is a persistent strong offshore wind. Along coasts where there is a large inequality between the two high or two low tides during a tidal day the height predictions are less reliable than elsewhere.

The current encountered in pilot waters is due primarily to tidal action, but other causes are sometimes present. The tidal current tables give the best prediction of total current, regardless of cause. The predictions for a river may be considerably in error following heavy rains or a drought. The effect of current is to alter the course and speed made good over the bottom. Due to the configuration of land (or shoal areas) and water, the set and drift may vary considerably over different parts of a harbor. Since this is generally an area in which small errors in position of a vessel are of considerable importance to its safety, a knowledge of predicted currents can be critical, particularly if the visibility is reduced by fog, snow, etc. If the vessel is proceeding at reduced speed, the effect of current with respect to distance traveled is greater than normal. Strong currents are particularly to be anticipated in narrow passages connecting larger bodies of water. Currents of more than five knots are encountered from time to time in the Golden Gate at San Francisco. Currents of more than 13 knots sometimes occur at Seymour Narrows, British Columbia.

In straight portions of rivers and channels the strongest currents usually occur in the middle, but in curved portions the swiftest currents (and deepest water) usually occur near the outer edge of the curve. Countercurrents and eddies may occur on either side of the main current of a river or narrow passage, especially near obstructions and in bights.

In general, the range of tide and the speed of tidal current are at a minimum upon the open ocean or along straight coasts. The greatest tidal effects are usually encountered in rivers, bays, harbors, inlets, bights, etc. A vessel proceeding along a coast can be expected to encounter stronger sets toward or away from the shore while passing an indentation than when the coast is straight.

11.2. Predictions of tides and currents to be expected at various places are published annually by the National Ocean Survey. These are supplemented by eleven sets of tidal current charts (art. 11.11), each set consisting of charts for each hour of the tidal cycle. On these charts the set of the current at various places in the area is shown by arrows, and the drift by numbers. Since these are *average* conditions, they indicate in a general way the tidal conditions on any day and during *any* year. They are designed to be used with tidal current diagrams (art. 11.11) or the tidal current tables (except those for New York Harbor, and Narragansett Bay, which are used with the tide tables). These charts are available for Boston Harbor, Narragansett Bay to Nantucket Sound, Narragansett Bay, Long Island Sound and Block Island Sound, New York Harbor, Delaware Bay and River, upper Chesapeake Bay, Charleston Harbor, San Francisco Bay, Puget Sound (northern part), and Puget Sound (southern part). Current arrows are sometimes shown on nautical charts. These represent average conditions and should not be considered reliable predictions of the conditions to be encountered at any given time. When a strong current sets over an irregular bottom, or meets an opposing current, ripples may occur on the surface. These are called **tide rips.** Areas where they occur frequently are shown on charts.

Usually, the mariner obtains tidal information from tide and tidal current tables. However, if these are not available, or if they do not include information at a desired place, the mariner may be able to obtain locally the **mean high water lunitidal interval** or the **high water full and change.** The approximate *time* of high water can be found by adding either interval to the time of transit (either upper or lower) of the moon. Low water occurs approximately ¼ tidal day (about $6^h 12^m$) before and after the time of high water. The actual interval varies somewhat from day to day, but approximate results can be obtained in this manner. Similar information for tidal currents (**lunicurrent interval**) is seldom available.

11.3. Tide tables for various parts of the world are published in four volumes by the National Ocean Survey. Each volume is arranged as follows:

Table 1 contains a complete list of the predicted times and heights of the tide for each day of the year at a number of places designated as **reference stations.**

Table 2 gives differences and ratios which can be used to modify the tidal information for the reference stations to make it applicable to a relatively large number of **subordinate stations.**

Table 3 provides information for use in finding the approximate height of the tide at any time between high water and low water.

Table 4 is a sunrise-sunset table at five-day intervals for various latitudes from 76°N to 60°S (40°S in one volume).

Table 5 provides an adjustment to convert the local mean time of table 4 to zone or standard time.

Table 6 (two volumes only) gives the zone time of moonrise and moonset for each day of the year at certain selected places.

Certain astonomical data are contained on the inside back cover of each volume.

Extracts from tables 1, 2, and 3 for the East Coast of North and South America are given in appendix C.

11.4. Tide predictions for reference stations.—The first page of appendix C is the table 1 daily predictions for New York (The Battery) for the first quarter of 1975. As indicated at the bottom of the page, times are for Eastern Standard Time (+5 zone, time meridian 75°W). Daylight saving time is not used. Times are given on the 24-hour basis. The tidal reference level for this station is mean low water.

For each day, the date and day of week are given, and the time and height of each high and low water are given in chronological order. Although high and low waters are not labeled as such, they can be distinguished by the relative heights given immediately to the right of the times. Since *two* high tides and *two* low tides occur each tidal day, the type of tide at this place is *semidiurnal*. The *tidal* day being longer than the *civil* day (because of the revolution of the moon eastward around the earth), any given tide occurs *later* from day to day. Thus, on Saturday, March 29, 1975, the first tide that occurs is the lower low water (−1.2 feet at 0334). The following high water (lower high water) is 4.9 feet above the reference level (a 6.1 foot rise from the preceding low water), and occurs at 0942. This is followed by the higher low water (−0.9 feet) at 1547, and then the higher high water of 5.5 feet at 2206. The cycle is repeated on the following day with variations in height, and later times.

Because of later times of corresponding tides from day to day, certain days have only one high water or only one low water. Thus, on January 17 high tides occur at 1120 and 2357. The next following high tides are at 1154 on January 18 and 0029 on January 19. Thus, only one high tide occurs on January 18, the previous one being shortly before midnight on the seventeenth, and the next one occurring early in the morning of the nineteenth, as shown.

11.5. Tide predictions for subordinate stations.—The second page of appendix C is a page of table 2 of the tide tables. For each subordinate station listed, the following information is given:

Number. The stations are listed in geographical order and given consecutive numbers. At the end of each volume an alphabetical listing is given, and for each entry the consecutive number is shown, to assist in finding the entry in table 2.

Place. The list of places includes both subordinate and reference stations, the latter being given in bold type.

Position. The approximate latitude and longitude are given to assist in locating the station. The latitude is north or south, and the longitude east or west, depending upon the letters (N, S, E, W) next *above* the entry. These may not be the same as those at the *top* of the column.

Differences. The differences are to be applied to the predictions for the reference station shown in bold capitals next *above* the entry on the page. Time and height differences are given separately for high and low waters. Where differences are omitted, they are either unreliable or unknown.

The time difference is the number of hours and minutes to be applied to the time at the reference station to find the time of the corresponding tide at the subordinate station. This interval is added if preceded by a plus sign (+), and subtracted if preceded by a minus sign (−). The results obtained by the application of the time differences will be in the zone time of the time meridian shown directly above the difference for the subordinate station. Special conditions occurring at a few stations are indicated by footnotes on the applicable pages. In some instances, the corresponding tide falls on a different date at reference and subordinate stations.

Height differences are shown in a variety of ways. For most entries separate height differences in feet are given for high water and low water. These are applied to the height given for the reference station. In many cases a *ratio* is given for either high water or low water, or both. The height at the reference station is multiplied by this ratio to find the height at the subordinate station. For a few stations, *both* a ratio and difference are given. In this case the height at the reference station is first multiplied by the ratio, and the difference is then applied. An example is given in each volume of tide tables. Special conditions are indicated in the table or by footnote. Thus, a footnote on the second page of appendix C indicates that "Values for the Hudson River above George Washington Bridge are based upon averages for the six months May to October, when the fresh-water discharge is a minimum."

Ranges. Various ranges are given, as indicated in the tables. In each case this is the difference in height between high water and low water for the tides indicated.

Example.—List chronologically the times and heights of all tides at Yonkers. (No. 1531) on January 2, 1975.

Solution.—

Date	January 2, 1975
Subordinate station	Yonkers
Reference station	New York
High water time difference	(+) 1^{h}09^m
Low water time difference	(+) 1^{h}10^m
High water height difference	(−) 0.8 ft.
Low water height difference	0.0 ft.

	New York			Yonkers	
HW	2321 (1st)	4.6 ft.		0030	3.8 ft.
LW	0516	(−) 0.6 ft.		0626	(−) 0.6 ft.
HW	1138	4.9 ft.		1247	4.1 ft.
LW	1749	(−) 0.9 ft.		1859	(−) 0.9 ft.

11.6. Finding height of tide at any time.—Table 3 of the tide tables provides means for determining the approximate height of tide at any time. It is based upon the assumption that a plot of height versus time is a sine curve. Instructions for use of the table are given in a footnote below the table, which is reproduced in appendix C.

Example 1.—Find the height of tide at Yonkers (No. 1531) at 1000 on January 2, 1975.

Solution.—The given time is between the low water at 0626 and the high water at 1247 (example of art. 1205). Therefore, the tide is rising. The duration of rise is 1247−0626=6^{h}21^m. The range of tide is 4.1−(−0.6)=4.7 feet. The given time is 2^{h}47^m *before* high water, the nearest tide. Enter the upper part of the table with duration of rise 6^{h}20^m (the nearest tabulated value to 6^{h}21^m), and follow the line horizontally to 2^{h}45^m (the nearest tabulated value to 2^{h}47^m). Follow this column vertically downward to the entry 1.8 feet in the line for a range of tide of 4.5 feet (the nearest tabulated value to 4.7 feet). This is the correction to be applied to the nearest tide. Since the nearest tide is high water, subtract 1.8 from 4.1 feet. The answer, 2.3 feet, is the height of tide at the given time.

Answer.—Ht. of tide at 1,000, 2.3 ft. A suitable form (fig. 11.6a) is used to facilitate the solution.

Interpolation in this table is not considered justified.

TIDE AND CURRENT TABLES
SRNC-USNA-NC&M-3161/31(1-71)

<u>NAVIGATION DEPARTMENT</u>   DIVISION OF NAVAL COMMAND AND MANAGEMENT

<u>COMPLETE TIDE TABLE</u>

Date: Jan. 2, 1975

Substation	Yonkers
Reference Station	New York
HW Time Difference	(+) 1^h 09^m
LW Time Difference	(+) 1^h 10^m
Difference in height of HW	(−) 0.8ft.
Difference in height of LW	0.0ft.

<u>Reference Station</u>			<u>Substation</u>	
HW	2231	4.6ft.	0030	3.8ft.
LW	0516	(−) 0.6ft.	0626	(−) 0.6ft.
HW	1138	4.9ft.	1247	4.1ft.
LW	1749	(−) 0.9ft.	1859	(−) 0.9ft.
HW				
LW				

HEIGHT OF TIDE AT ANY TIME

Locality: Yonkers   Time: 1000   Date: Jan. 2, 1975

Duration of Rise or Fall:	6^h 21^m
Time from Nearest Tide:	2^h 47^m
Range of Tide:	4.7ft.
Height of Nearest Tide:	4.1ft.
Corr. from Table 3:	1.8ft.
Height of Tide at: 1000	2.3ft.

FIGURE 11.6a.—U.S. Naval Academy tide form.

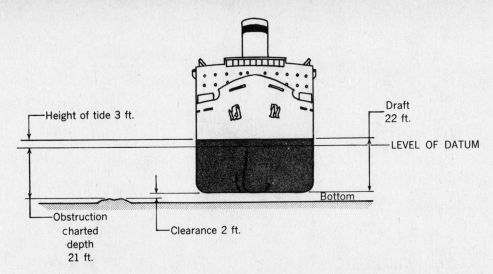

FIGURE 11.6b.—Height of tide required to pass clear of charted obstruction.

It may be desired to know at what time a given depth of water will occur. In this case, the problem is solved in reverse.

Example 2.—The captain of a vessel drawing 22 feet wishes to pass over a temporary obstruction near Days Point, Weehawken (No. 1521), having a charted depth of 21 feet, passage to be made during the morning of January 31, 1975. Refer to figure 11.6b.

Required.—The earliest time after 0800 that this passage can be made, allowing a safety margin of two feet.

Solution.—The least acceptable depth of water is 24 feet, which is three feet more than the charted depth. Therefore, the height of tide must be three feet or more. At the New York reference station a low tide of (−)0.9 foot occurs at 0459, followed by a high tide of 4.9 feet at 1120. At Days Point the corresponding low tide is (−)0.9 foot at 0522, and the high tide is 4.6 feet at 1144. The duration of rise is 6^{h}22^m, and the range of tide is 5.5 feet. The least acceptable tide is 3.0 feet, or 1.6 feet less than high tide. Enter the *lower* part of table 3 with range 5.5 feet and follow the horizontal line until 1.6 feet is reached. Follow this column vertically *upward* until the value of 2^{h}19^m is reached on the line for a duration of 6^{h}20^m (the nearest tabulated value to 6^{h}22^m). The minimum depth will occur about 2^{h}19^m *before* high water or at about 0925.

Answer.—A depth of 24 feet occurs at 0925.

If the range of tide is more than 20 feet, *half* the range (*one third* if the range is greater than 40 feet) is used to enter table 3, and the correction to height is *doubled* (*trebled* if one third is used).

A diagram for a graphical solution is given in figure 11.6c. Eye interpolation can be used if desired. The steps in this solution are as follows:

1. Enter the upper graph with the duration of rise or fall. This is represented by a horizontal line.

2. Find the intersection of this line and the curve representing the interval from the nearest *low* water (point A).

3. From A, follow a vertical line to the sine curve of the lower diagram (point B).

4. From B, follow horizontally to the vertical line representing the range of tide (point C).

5. Using C, read the correction from the series of curves.

6. Add (algebraically) the correction of step 5 to the *low* water height, to find the height at the given time.

Interval in hours from LOW water (curves)

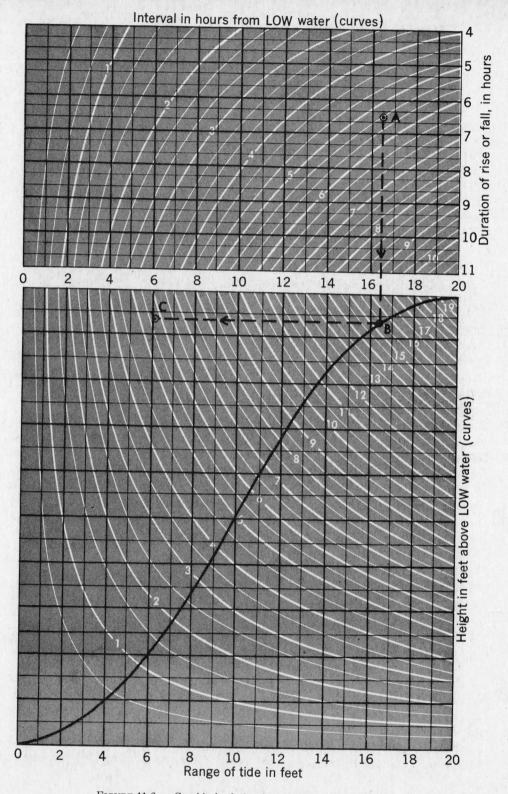

FIGURE 11.6c.—Graphical solution for height of tide at any time.

The problem illustrated in figure 11.6c is similar to that of example 1 given above. The duration of rise is 6^h25^m, and the interval from *low* water is 5^h23^m. The range of tide is 6.1 feet. The correction (by interpolation) is 5.7 feet. If the height of the preceding low tide is (−)0.2 foot, the height of tide at the given time is (−)0.2+5.7=5.5 feet. To solve example 2 by the graph, enter the lower graph and find the intersection of the vertical line representing 5.5 feet and the curve representing 3.9 feet (the minimum acceptable height above low water). From this point follow horizontally to the sine curve, and then vertically to the horizontal line in the upper figure representing the duration of rise of 6^h22^m. From the curve, determine the interval 4^h00^m. The earliest time is about 4^h00^m *after* low water, or at about 0922.

11.7. Tidal current tables are somewhat similar to tide tables, but the coverage is less extensive, being given in two volumes. Each volume is arranged as follows:

Table 1 contains a complete list of predicted times of maximum currents and slack, with the velocity (speed) of the maximum currents, for a number of reference stations.

Table 2 gives differences, ratios, and other information related to a relatively large number of subordinate stations.

Table 3 provides information for use in finding the speed of the current at any time between tabulated entries in tables 1 and 2.

Table 4 gives the number of minutes the current does not exceed stated amounts, for various maximum speeds.

Table 5 (Atlantic Coast of North America only) gives information on rotary tidal currents.

Each volume contains additional useful information related to currents. Extracts from the tables for the Atlantic Coast of North America are given in appendix D.

11.8. Tidal current predictions for reference stations.—The extracts of appendix D are for The Narrows, New York Harbor. Times are given on the 24-hour basis, for meridian 75°W. *Daylight saving time is not used.*

For each day, the date and day of week are given, with complete current information. Since the cycle is repeated twice each tidal day, currents at this place are semidiurnal. On most days there are four slack waters and four maximum currents, two of them floods (F) and two of them ebbs (E). However, since the tidal day is longer than the civil day, the corresponding condition occurs later from day to day, and on certain days there are only three slack waters or three maximum currents. At some places, the current on some days runs maximum flood twice, but ebb only once, a minimum flood occurring in place of the second ebb. The tables show this information.

As indicated by appendix D, the sequence of currents at The Narrows on Monday, February 3, 1975, is as follows:

0000 Flood current, 5^m after maximum velocity (speed).

0305 Slack, ebb begins.

0621 Maximum ebb of 2.0 knots, setting 160°.

1005 Slack, flood begins.

1222 Maximum flood of 1.5 knots, setting 340°.

1516 Slack, ebb begins.

1839 Maximum ebb of 1.9 knots, setting 160°.

2216 Slack, flood begins.

2400 Flood current, 56^m before maximum velocity (speed).

Only one maximum flood occurs on this day, the previous one having occurred 5 minutes before the day began, and the following one predicted for 56 minutes after the day ends.

11.9. Tidal current predictions for subordinate stations.—For each subordinate station listed in table 2 of the tidal current tables, the following information is given:

Number. The stations are listed in geographical order and given consecutive numbers, as in the tide tables (art. 11.5). At the end of this volume an alphabetical listing is given, and for each entry the consecutive number is shown, to assist in finding the entry in table 2.

Place. The list of places includes both subordinate and reference stations, the latter being given in bold type.

Position. The approximate latitude and longitude are given to assist in locating the station. The latitude is north or south and the longitude east or west as indicated by the letters (N, S, E, W) next *above* the entry. The current given is for the center of the channel unless another location is indicated by the station name.

Time difference. Two time differences are tabulated. One is the number of hours and minutes to be applied to the tabulated times of slack water at the reference station to find the times of slack waters at the subordinate station. The other time difference is applied to the times of maximum current at the reference station to find the times of the corresponding maximum current at the subordinate station. The intervals, which are added or subtracted in accordance with their signs, include any difference in time between the two stations, so that the answer is correct for the standard time of the subordinate station. Limited application and special conditions are indicated by footnotes.

Velocity (speed) ratios. Speed of the current at the subordinate station is found by multiplying the speed at the reference station by the tabulated ratio. Separate ratios may be given for flood and ebb currents. Special conditions are indicated by footnotes.

As indicated in appendix D, the currents at The Battery (No. 2375) can be found by *adding* 1^h30^m for slack water and 1^h35^m for maximum current to the times for The Narrows, and multiplying flood currents by 0.9 and ebb currents by 1.2. Applying these to the values for Monday, February 3, 1975, the sequence is as follows:

0000 Flood current, 1^h30^m before maximum velocity (speed).
0130 Maximum flood of 1.8 knots, setting 015°.
0435 Slack, ebb begins.
0756 Maximum ebb of 2.4 knots, setting 195°.
1135 Slack, flood begins.
1357 Maximum flood of 1.4 knots, setting 015°.
1646 Slack, ebb begins.
2014 Maximum ebb of 2.3 knots setting 195°.
2346 Slack, flood begins.
2400 Flood current, 14^m after slack.

11.10. Finding speed of tidal current at any time.—Table 3 of the tidal current table provides means for determining the approximate velocity (speed) at any time. Instructions for its use are given below the table, which is reproduced in appendix D.

Example 1.—Find the speed of the current at The Battery at 1500 on February 3, 1975.

Solution.—The given time is between the maximum flood of 1.4 knots at 1357 and the slack at 1646 (art. 11.9). The interval between slack and maximum current (1646 −1357) is 2^h49^m. The interval between slack and the desired time (1646−1500) is 1^h46^m. Enter the table (A) with 2^h40^m at the top, and 1^h40^m at the left side (the nearest tabulated values to 2^h49^m and 1^h46^m, respectively), and find the factor 0.8 in the body of the table. The approximate speed at 1500 is $0.8\times1.4=1.1$ knots, and it is flooding.

Answer.—Speed 1.1 kn. A suitable form (fig. 11.10) is used to facilitate the solution.

It may be desired to determine the period during which the current is less (or greater) than a given amount. Table 4 of the tidal current tables can be used to determine the period during which the speed does not exceed 0.5 knot. For greater

NAVIGATION DEPARTMENT   DIVISION OF NAVAL COMMAND AND MANAGEMENT

COMPLETE CURRENT TABLE

Locality: The Battery   Date: Feb. 3, 1975

Reference Station: The Narrows

Time Difference:	Slack Water:	(+) 1^h 30^m
	Maximum Current:	(+) 1^h 35^m
Velocity Ratio:	Maximum Flood:	0.9
	Maximum Ebb:	1.2

Flood Direction: 015°
Ebb Direction: 195°

Reference Station: The Narrows _____ Locality: The Battery _____

			0000	F
2355	2.0F		0130	1.8F
0305	0		0435	0
0621	2.0E		0756	2.4E
1005	0		1135	0
1222	1.5F		1357	1.4F
1516	0		1646	0
1839	1.9E		2014	2.3E
2216	0		2346	0
2400	F		2400	F

VELOCITY OF CURRENT AT ANY TIME

Int. between slack and desired time:	1^h 46^m	
Int. between slack and maximum current:	2^h 49^m	(Ebb) (Flood)
Maximum current:	1.4kn	
Factor, Table 3	0.8	
Velocity:	1.1kn	
Direction:	015°	

DURATION OF SLACK

Times of maximum current:	0756	1357
Maximum current:	2.4kn	1.4kn
Desired maximum:	0.3	0.3
Period – Table 4:	35^m	46^m
Sum of periods:		81^m
Average period:		40^m
Time of slack:		1135
Duration of slack: From: 1115 To: 1155		

FIGURE 11.10.—U.S. Naval Academy tidal current form.

speeds, and for more accurate results under some conditions, table 3 of the tidal current tables can be used, solving by reversing the process used in example 1.

Example 2.—During what period on the evening of February 3, 1975, does the ebb current equal or exceed 1.0 knot at The Battery?

Solution.—The maximum ebb of 2.3 knots occurs at 2014. This is preceded by a slack at 1646, and followed by the next slack at 2346. The interval between the earlier slack and the maximum ebb is 3^{h}28^m, and the interval between the ebb and following slack is 3^{h}32^m. The desired factor is $\frac{1.0}{2.3} = 0.4$. Enter table A with 3^{h}20^m (the nearest tabulated value to 3^h28^m) at the top, and follow down the column to 0.4 (midway between 0.3 and 0.5). At the left margin the interval between slack and the desired time is found to be 0^h50^m (midway between 0^h40^m and 1^h00^m). Therefore, the current becomes 1.0 knot at $1646+0^h50^m=1736$. Next, enter table A with 3^h40^m (the nearest tabulated value to 3^h32^m) at the top, and follow down the column to 0.4. Follow this line to the left margin, where the interval between slack and desired time is found to be 1^h00^m. Therefore, the current is 1.0 knot or greater until $2346-1^h00^m=2246$. If the two intervals between maximum current and slack were nearest the same 20^m interval, table A would have to be entered only once.

Answer.—The speed equals or exceeds 1.0 knot between 1736 and 2246.

The predicted times of slack water given in the tidal current tables indicate the instant of zero velocity. There is a period each side of slack water, however, during which the current is so weak that for practical purposes it may be considered as negligible. Table 4 of the tidal current tables gives, for various maximum currents, the approximate period of time during which weak currents not exceeding 0.1 to 0.5 knot will be encountered. This duration includes the last of the flood or ebb and the beginning of the following flood or ebb, that is, half of the duration will be before and half after the time of slack water.

When there is a difference between the velocities of the maximum flood and ebb preceding and following the slack for which the duration is desired, it will be sufficiently accurate for practical purposes to find a separate duration for each maximum velocity and take the average of the two as the duration of the weak current.

Of the two subtables of table 4, table A should be used for all places *except* those listed for table B; table B should be used for all places listed, and all stations in table 2 which are referred to them.

Example 3.—Find the period from just before until just after the slack at The Battery at 1135 on February 3, 1975, that the current does not exceed 0.3 kn.

Solution.—Refer to table 4. Table A of table 4 of the tidal current tables is entered with the maximum current before the slack to find the period during which the current does not exceed 0.3 kn. Since there is a difference between the velocities of the maximum ebb and flood preceding and following the slack for which the duration is desired, table A is re-entered with the maximum current after the slack to find the period during which the current does not exceed 0.3 kn. The average of the two values so found is taken as the duration of the weak current. The form shown in figure 11.10 is used to facilitate the solution.

Answer.—Duration 40 min. (from 1115 to 1155).

11.11. Tidal current charts present a comprehensive view of the hourly speed and direction of the current in 11 bodies of water (art. 11.2). They also provide a means for determining the speed and direction of the current at various localities throughout these bodies of water. The arrows show the direction of the current; the figures give the speed in knots at the time of spring tides, that is, during the time of new or full moon when the currents are stronger than average. When the current is given as weak, the speed is less than 0.1 knot. The decimal point locates the position of the station.

The charts depict the flow of the tidal current under normal weather conditions. Strong winds and freshets, however, bring about nontidal currents which may modify considerably the speed and direction shown on the charts.

The speed of the tidal current varies from day to day principally in accordance with the phase, distance, and declination of the moon. Therefore, to obtain the speed for any particular day and hour, the *spring speeds* shown on the charts must be modified by correction factors. A correction table given in the charts can be used for this purpose.

The **tidal current diagrams** are a series of 12 monthly diagrams to be used with the tidal current charts. There is one diagram for each month of the year. A new set of diagrams must be used each year. The diagrams are computer constructed lines that locate each chart throughout all hours of every month. The diagrams indicate directly the chart and the speed correction factor to use at any desired time.

11.12. Current diagrams.—A current diagram is a graph showing the speed of the current along a channel at different stages of the tidal current cycle. The current tables include such diagrams for Vineyard and Nantucket Sounds (one diagram); East River, New York; New York Harbor; Delaware Bay and River (one diagram); and Chesapeake Bay. The diagram for New York Harbor is reproduced in appendix D.

On this diagram each vertical line represents a given instant identified in terms of the number of hours before or after slack at The Narrows. Each horizontal line represents a distance from Ambrose Channel Entrance, measured along the usually traveled route. The names along the left margin are placed at the correct distances from Ambrose Channel Entrance. The current is for the center of the channel opposite these points. The intersection of any vertical line with any horizontal line represents a given moment in the current cycle at a given place in the channel. If this intersection is in a shaded area, the current is flooding; if in an unshaded area, it is ebbing. The speed in knots can be found by interpolation (if necessary) between the numbers given in the body of the diagram. The given values are *averages*. To find the value at any given time, multiply the speed found from the diagram by the ratio of *maximum speed of the current involved* to the *maximum shown on the diagram*, both values being taken for The Narrows. If the diurnal inequality is large, the accuracy can be improved by altering the width of the shaded area to fit conditions. The diagram covers 1½ current cycles, so that the right-hand third is a duplication of the left-hand third.

If the current for a single station is desired, table 1 or 2 should be used. The current diagrams are intended for use in either of two ways: First, to determine a favorable time for passage through the channel. Second, to find the average current to be expected during any passage through the channel. For both of these uses a number of "speed lines" are provided. When the appropriate line is transferred to the correct part of the diagram, the current to be encountered during passage is indicated along the line.

Example.—During the morning of January 3, 1975, a ship is to leave Pier 83 at W. 42nd St., and proceed down the bay at ten knots.

Required.—(1) Time to get underway to take maximum advantage of a favorable current, allowing 15 minutes to reach mid channel.

(2) Average speed over the bottom during passage down the bay.

Solution.—(1) Transfer the line (slope) for ten knots southbound to the diagram, locating it so that it is centered on the unshaded ebb current section between W. 42nd St. and Ambrose Channel Entrance. This line crosses a horizontal line through W. 42nd St. about one-half of the distance between the vertical lines representing three and two hours, respectively, after ebb begins at The Narrows. The setting is not critical. Any time within about half an hour of the correct time will result in about the same current. Between the points involved, the entire speed line is in the ebb current area.

(2) Table 1 indicates that on the morning of January 3 ebb begins at The Narrows at 0132. Two hours twenty-eight minutes after ebb begins, the time is 0400. Therefore, the ship should reach mid channel at 0400. It should get underway 15 minutes earlier, at 0345.

(3) To find the average current, determine the current at intervals (as every two miles), add, and divide by the number of entries.

Distance	Current
18	1.2
16	1.4
14	1.9
12	1.5
10	2.0
8	1.9
6	1.3
4	1.2
2	1.4
0	1.2
sum	15.0

The sum of 15.0 is for ten entries. The average is therefore $15.0 \div 10 = 1.5$ knots.

(4) This value of current is correct only if the ebb current is an average one. From table 1 the maximum ebb involved is 2.2 knots. From the diagram the maximum value at The Narrows is 2.0 knots. Therefore, the average current found in step (3) should be increased by the ratio $2.2 \div 2.0 = 1.1$. The average for the run is therefore $1.5 \times 1.1 = 1.6$ knots. Speed over the botton is $10 + 1.6 = 11.6$ knots.

Answers.—(1) T 0345, (2) S 11.6 kn.

In the example, an ebb current is carried throughout the run. If the transferred speed line had been partly in a flood current area, all ebb currents (those increasing the ship's speed) should be given a positive sign $(+)$, and all flood currents a negative sign $(-)$. A separate ratio should be determined for each current (flood or ebb), and applied to the entries for that current. In Chesapeake Bay it is not unusual for an outbound vessel to encounter three or even four separate currents during passage down the bay. Under the latter condition, it is good practice to multiply *each* current taken from the diagram by the ratio for the current involved.

If the time of starting the passage is fixed, and the current during passage is desired, the starting time is identified in terms of the reference tidal cycle. The speed line is then drawn through the intersection of this vertical time line and the horizontal line through the place. The average current is then determined in the same manner as when the speed line is located as described above.

Problems

11.2. The mean high water lunitidal interval at a certain port is $2^h 17^m$.

Required.—The approximate times of each high and low water on a day when the moon transits the local meridian at 1146.

Answers.—HW at 0139 and 1403, LW at 0751 and 2015.

11.4. List chronologically the times and heights of all tides at New York (The Battery) on February 11, 1975.

Answer.—

Time	Tide	Height
0222	LW	$(-)$ 0.4 ft.
0829	HW	4.6 ft.
1449	LW	$(-)$ 0.6 ft.
2053	HW	4.2 ft.

11.5. List chronologically the times and heights of all tides at Castle Point, Hoboken, N.J. (No. 1519) on March 18, 1975.

Answer.—

Time	Tide	Height
0533	LW	0.2 ft.
1141	HW	3.5 ft.
1724	LW	0.3 ft.
0003	HW	4.1 ft.

11.6a. Find the height of tide at Union Stock Yards, New York (No. 1523) at 0600 on February 6, 1975.

*Answer.—*Ht. of tide at 0600, 3.8 ft.

11.6b. The captain of a vessel drawing 24 feet wishes to pass over a temporary obstruction near Bayonne, N.J. (No. 1505) having a charted depth of 23 feet, passage to be made during the afternoon of March 5, 1975.

*Required.—*The earliest and latest times that the passage can be made, allowing a safety margin of two feet.

*Answers.—*Earliest time 1316, latest time 1531.

11.8. Determine the sequence of currents at The Narrows on January 15, 1975.

Answer.—

0000 Ebb current, 42^m after slack.

0231 Maximum ebb of 1.9 knots.

0557 Slack, flood begins.

0822 Maximum flood of 1.7 knots.

1137 Slack, ebb begins.

1455 Maximum ebb of 2.1 knots.

1836 Slack, flood begins.

2051 Maximum flood of 1.5 knots.

2400 Flood current, 2^m before slack.

11.9. Determine the sequence of currents at Ambrose Channel Entrance (No. 2310) on January 12, 1975.

Answer.—

0000 Ebb current, 42^m after maximum velocity (speed).

0241 Slack, flood begins.

0533 Maximum flood of 2.0 knots, setting 310°.

0828 Slack, ebb begins.

1155 Maximum ebb of 2.5 knots.

1527 Slack, flood begins.

1801 Maximum flood of 1.5 knots, setting 310°.

2040 Slack, ebb begins.

2400 Ebb current, 3^m before maximum velocity (speed).

11.10a. Find the speed of the current at Bear Mountain Bridge (No. 2445) at 0900 on February 19, 1975.

*Answer.—*Speed 0.8 kn.

11.10b. At about what time during the afternoon of February 3, 1975, does the flood current northwest of The Battery (No. 2375) reach a speed of 1.0 knot?

*Answer.—*T 1245.

11.12. A vessel arrives at Ambrose Channel Entrance two hours after flood begins at The Narrows on the morning of February 16, 1975.

Required.—(1) The speed through the water required to take fullest advantage of the flood tide in steaming to Chelsea Docks.

(2) The average current to be expected.

(3) Estimated time of arrival off Chelsea Docks.

Answers.—(1) S 9 kn., (2) S 1.4 kn., (3) ETA 1035.

APPENDIX A

EXPLANATION OF TABLES

Table 1. Conversion Table for Millibars and Inches and Millimeters of Mercury.—The reading of a barometer in inches or millimeters of mercury corresponding to a given reading in millibars can be found directly in this table.

Table 2. Correction of Barometer Reading for Height Above Sea Level.—For barometer readings taken at different heights to be of maximum value convert them to a standard height, usually sea level. To convert, enter this table with the outside temperature and height of the barometer above sea level. The correction taken from this table applies to any type barometer and is always added to the observed readings unless the barometer is below sea level.

Table 3. Correction of Barometer Reading for Gravity.—*This correction doesn't apply to aneroid barometers.* The height of a mercury column is affected by the force of gravity. Gravity changes with latitude. Enter the table with your latitude, take out the correction, and apply in accordance with the sign given to convert to standard latitude, 45° 32′ 40″.

Table 4. Correction of Barometer Reading for Temperature.—*This correction doesn't apply to aneroid barometers.* Because of the difference in expansion of the mercury column and the brass scale, a correction should be applied when the temperature is different than the calibrated temperature. To find the correction, enter this table with the temperature in degrees Fahrenheit and the barometer reading. Apply the correction in accordance with the sign given.

Table 5. Direction and Speed of True Wind.—This table provides a means of converting apparent winds to true wind. To use the table, divide the apparent wind, in knots, by the vessel's speed, also in knots. Enter the table with this value and the difference between the heading and the apparent wind direction. The values taken from the table are (1) the difference between the heading and the true wind direction, and (2) the speed of the true wind in units of ship's speed. The true wind is on the same side as the apparent wind, and from a point farther aft. To convert wind speed in units of ship's speed to speed in knots, multiply by the ship's speed in knots. If speed of the true wind and relative direction of the apparent wind are known, enter the column for direction of the apparent wind, and find the speed of the true wind in units of ship's speed. The number to the left is the relative direction of the true wind. The number on the same line in the side columns is the speed of the apparent wind in units of ship's speed. Two solutions are possible if speed of the true wind is less than the ship's speed.

Table 6. Conversion Table for Thermometer Scales.—Enter this table with temperature in degrees Fahrenheit, Celsius, or Kelvin and take out the corresponding readings on the other two temperature scales. *Note:* freezing point of pure water at standard sea level—32° F., 0° C., 273.15° K.; boiling point of pure water at standard sea level—212° F., 100° C., 373.15° K.

184

Table 7. Relative Humidity.—To determine relative humidity enter this table with the dry-bulb (air) temperature (F) and the difference between the dry-bulb and wet-bulb temperatures (F). The value is the approximate relative humidity. If the temperatures are the same, relative humidity is 100 percent.

Table 8. Dew Point.—To determine the dew point, enter this table with the dry-bulb (air) temperature (F) and the difference between the dry-bulb and wet-bulb temperatures (F). The value taken is the dew point in degrees Fahrenheit. If the values are the same, the air is at or below the dew point.

TABLE 1

Conversion Table for Millibars, Inches of Mercury, and Millimeters of Mercury

Millibars	Inches	Millimeters	Millibars	Inches	Millimeters	Millibars	Inches	Millimeters
900	26. 58	675. 1	960	28. 35	720. 1	1020	30. 12	765. 1
901	26. 61	675. 8	961	28. 38	720. 8	1021	30. 15	765. 8
902	26. 64	676. 6	962	28. 41	721. 6	1022	30. 18	766. 6
903	26. 67	677. 3	963	28. 44	722. 3	1023	30. 21	767. 3
904	26. 70	678. 1	964	28. 47	723. 1	1024	30. 24	768. 1
905	26. 72	678. 8	965	28. 50	723. 8	1025	30. 27	768. 8
906	26. 75	679. 6	966	28. 53	724. 6	1026	30. 30	769. 6
907	26. 78	680. 3	967	28. 56	725. 3	1027	30. 33	770. 3
908	26. 81	681. 1	968	28. 58	726. 1	1028	30. 36	771. 1
909	26. 84	681. 8	969	28. 61	726. 8	1029	30. 39	771. 8
910	26. 87	682. 6	970	28. 64	727. 6	1030	30. 42	772. 6
911	26. 90	683. 3	971	28. 67	728. 3	1031	30. 45	773. 3
912	26. 93	684. 1	972	28. 70	729. 1	1032	30. 47	774. 1
913	26. 96	684. 8	973	28. 73	729. 8	1033	30. 50	774. 8
914	26. 99	685. 6	974	28. 76	730. 6	1034	30. 53	775. 6
915	27. 02	686. 3	975	28. 79	731. 3	1035	30. 56	776. 3
916	27. 05	687. 1	976	28. 82	732. 1	1036	30. 59	777. 1
917	27. 08	687. 8	977	28. 85	732. 8	1037	30. 62	777. 8
918	27. 11	688. 6	978	28. 88	733. 6	1038	30. 65	778. 6
919	27. 14	689. 3	979	28. 91	734. 3	1039	30. 68	779. 3
920	27. 17	690. 1	980	28. 94	735. 1	1040	30. 71	780. 1
921	27. 20	690. 8	981	28. 97	735. 8	1041	30. 74	780. 8
922	27. 23	691. 6	982	29. 00	736. 6	1042	30. 77	781. 6
923	27. 26	692. 3	983	29. 03	737. 3	1043	30. 80	782. 3
924	27. 29	693. 1	984	29. 06	738. 1	1044	30. 83	783. 1
925	27. 32	693. 8	985	29. 09	738. 8	1045	30. 86	783. 8
926	27. 34	694. 6	986	29. 12	739. 6	1046	30. 89	784. 6
927	27. 37	695. 3	987	29. 15	740. 3	1047	30. 92	785. 3
928	27. 40	696. 1	988	29. 18	741. 1	1048	30. 95	786. 1
929	27. 43	696. 8	989	29. 21	741. 8	1049	30. 98	786. 8
930	27. 46	697. 6	990	29. 23	742. 6	1050	31. 01	787. 6
931	27. 49	698. 3	991	29. 26	743. 3	1051	31. 04	788. 3
932	27. 52	699. 1	992	29. 29	744. 1	1052	31. 07	789. 1
933	27. 55	699. 8	993	29. 32	744. 8	1053	31. 10	789. 8
934	27. 58	700. 6	994	29. 35	745. 6	1054	31. 12	790. 6
935	27. 61	701. 3	995	29. 38	746. 3	1055	31. 15	791. 3
936	27. 64	702. 1	996	29. 41	747. 1	1056	31. 18	792. 1
937	27. 67	702. 8	997	29. 44	747. 8	1057	31. 21	792. 8
938	27. 70	703. 6	998	29. 47	748. 6	1058	31. 24	793. 6
939	27. 73	704. 3	999	29. 50	749. 3	1059	31. 27	794. 3
940	27. 76	705. 1	1000	29. 53	750. 1	1060	31. 30	795. 1
941	27. 79	705. 8	1001	29. 56	750. 8	1061	31. 33	795. 8
942	27. 82	706. 6	1002	29. 59	751. 6	1062	31. 36	796. 6
943	27. 85	707. 3	1003	29. 62	752. 3	1063	31. 39	797. 3
944	27. 88	708. 1	1004	29. 65	753. 1	1064	31. 42	798. 1
945	27. 91	708. 8	1005	29. 68	753. 8	1065	31. 45	798. 8
946	27. 94	709. 6	1006	29. 71	754. 6	1066	31. 48	799. 6
947	27. 96	710. 3	1007	29. 74	755. 3	1067	31. 51	800. 3
948	27. 99	711. 1	1008	29. 77	756. 1	1068	31. 54	801. 1
949	28. 02	711. 8	1009	29. 80	756. 8	1069	31. 57	801. 8
950	28. 05	712. 6	1010	29. 83	757. 6	1070	31. 60	802. 6
951	28. 08	713. 3	1011	29. 85	758. 3	1071	31. 63	803. 3
952	28. 11	714. 1	1012	29. 88	759. 1	1072	31. 66	804. 1
953	28. 14	714. 8	1013	29. 91	759. 8	1073	31. 69	804. 8
954	28. 17	715. 6	1014	29. 94	760. 6	1074	31. 72	805. 6
955	28. 20	716. 3	1015	29. 97	761. 3	1075	31. 74	806. 3
956	28. 23	717. 1	1016	30. 00	762. 1	1076	31. 77	807. 1
957	28. 26	717. 8	1017	30. 03	762. 8	1077	31. 80	807. 8
958	28. 29	718. 6	1018	30. 06	763. 6	1078	31. 83	808. 6
959	28. 32	719. 3	1019	30. 09	764. 3	1079	31. 86	809. 3
960	28. 35	720. 1	1020	30. 12	765. 1	1080	31. 89	810. 1

TABLE 2

Correction of Barometer Reading for Height Above Sea Level

All barometers.   All values positive.

Height in feet	Outside temperature in degrees Fahrenheit													Height in feet
	−20°	−10°	0°	10°	20°	30°	40°	50°	60°	70°	80°	90°	100°	
	Inches	*Inches*	*Inches*	*Inches*	*Inches*	*Inches*	*Inches*	*Inches*	*Inches*	*Inches*	*Inches*	*Inches*	*Inches*	
5	0.01	0.01	0.01	0.01	0.01	0.01	0.01	0.01	0.01	0.01	0.01	0.01	0.01	5
10	0.01	0.01	0.01	0.01	0.01	0.01	0.01	0.01	0.01	0.01	0.01	0.01	0.01	10
15	0.02	0.02	0.02	0.02	0.02	0.02	0.02	0.02	0.02	0.02	0.02	0.02	0.02	15
20	0.03	0.02	0.02	0.02	0.02	0.02	0.02	0.02	0.02	0.02	0.02	0.02	0.02	20
25	0.03	0.03	0.03	0.03	0.03	0.03	0.03	0.03	0.03	0.03	0.03	0.03	0.03	25
30	0.04	0.04	0.04	0.04	0.04	0.04	0.03	0.03	0.03	0.03	0.03	0.03	0.03	30
35	0.04	0.04	0.04	0.04	0.04	0.04	0.04	0.04	0.04	0.04	0.04	0.04	0.04	35
40	0.05	0.05	0.05	0.05	0.05	0.05	0.04	0.04	0.04	0.04	0.04	0.04	0.04	40
45	0.06	0.06	0.05	0.05	0.05	0.05	0.05	0.05	0.05	0.05	0.05	0.05	0.05	45
50	0.06	0.06	0.06	0.06	0.06	0.06	0.06	0.06	0.05	0.05	0.05	0.05	0.05	50
55	0.07	0.07	0.07	0.07	0.06	0.06	0.06	0.06	0.06	0.06	0.06	0.06	0.06	55
60	0.08	0.07	0.07	0.07	0.07	0.07	0.07	0.07	0.06	0.06	0.06	0.06	0.06	60
65	0.08	0.08	0.08	0.08	0.08	0.07	0.07	0.07	0.07	0.07	0.07	0.07	0.07	65
70	0.09	0.09	0.09	0.08	0.08	0.08	0.08	0.08	0.08	0.07	0.07	0.07	0.07	70
75	0.10	0.09	0.09	0.09	0.09	0.09	0.08	0.08	0.08	0.08	0.08	0.08	0.08	75
80	0.10	0.10	0.10	0.10	0.09	0.09	0.09	0.09	0.09	0.09	0.08	0.08	0.08	80
85	0.11	0.11	0.10	0.10	0.10	0.10	0.10	0.09	0.09	0.09	0.09	0.09	0.09	85
90	0.11	0.11	0.11	0.11	0.11	0.10	0.10	0.10	0.10	0.10	0.09	0.09	0.09	90
95	0.12	0.12	0.12	0.11	0.11	0.11	0.11	0.10	0.10	0.10	0.10	0.10	0.10	95
100	0.13	0.12	0.12	0.12	0.12	0.11	0.11	0.11	0.11	0.11	0.10	0.10	0.10	100
105	0.13	0.13	0.13	0.13	0.12	0.12	0.12	0.12	0.11	0.11	0.11	0.11	0.11	105
110	0.14	0.14	0.13	0.13	0.13	0.13	0.12	0.12	0.12	0.12	0.11	0.11	0.11	110
115	0.15	0.14	0.14	0.14	0.13	0.13	0.13	0.13	0.12	0.12	0.12	0.12	0.12	115
120	0.15	0.15	0.15	0.14	0.14	0.14	0.13	0.13	0.13	0.13	0.12	0.12	0.12	120
125	0.16	0.16	0.15	0.15	0.15	0.14	0.14	0.14	0.13	0.13	0.13	0.13	0.12	125

TABLE 3

Correction of Barometer Reading for Gravity

Mercurial barometers only.

Latitude	Correction	Latitude	Correction	Latitude	Correction	Latitude	Correction
°	*Inches*	°	*Inches*	°	*Inches*	°	*Inches*
0	−0.08	25	−0.05	50	+0.01	75	+0.07
5	−0.08	30	−0.04	55	+0.03	80	+0.07
10	−0.08	35	−0.03	60	+0.04	85	+0.08
15	−0.07	40	−0.02	65	+0.05	90	+0.08
20	−0.06	45	0.00	70	+0.06		

TABLE 4

Correction of Barometer Reading for Temperature
Mercurial barometers only.

Temp. F	Height of barometer in inches								Temp. F
	27.5	28.0	28.5	29.0	29.5	30.0	30.5	31.0	
°	*Inches*	*Inches*	*Inches*	*Inches*	*Inches*	*Inches*	*Inches*	*Inches*	°
−20	+0.12	+0.12	+0.13	+0.13	+0.13	+0.13	+0.14	+0.14	−20
18	0.12	0.12	0.12	0.12	0.13	0.13	0.13	0.13	18
16	0.11	0.11	0.12	0.12	0.12	0.12	0.12	0.13	16
14	0.11	0.11	0.11	0.11	0.11	0.12	0.12	0.12	14
12	0.10	0.10	0.11	0.11	0.11	0.11	0.11	0.11	12
−10	+0.10	+0.10	+0.10	+0.10	+0.10	+0.11	+0.11	+0.11	−10
8	0.09	0.09	0.10	0.10	0.10	0.10	0.10	0.10	8
6	0.09	0.09	0.09	0.09	0.09	0.09	0.10	0.10	6
4	0.08	0.08	0.08	0.09	0.09	0.09	0.09	0.09	4
−2	0.08	0.08	0.08	0.08	0.08	0.08	0.09	0.09	−2
0	+0.07	+0.07	+0.07	+0.08	+0.08	+0.08	+0.08	+0.08	0
+2	0.07	0.07	0.07	0.07	0.07	0.07	0.07	0.08	+2
4	0.06	0.06	0.06	0.07	0.07	0.07	0.07	0.07	4
6	0.06	0.06	0.06	0.06	0.06	0.06	0.06	0.06	6
8	0.05	0.05	0.05	0.05	0.06	0.06	0.06	0.06	8
+10	+0.05	+0.05	+0.05	+0.05	+0.05	+0.05	+0.05	+0.05	+10
12	0.04	0.04	0.04	0.04	0.04	0.05	0.05	0.05	12
14	0.04	0.04	0.04	0.04	0.04	0.04	0.04	0.04	14
16	0.03	0.03	0.03	0.03	0.03	0.03	0.03	0.04	16
18	0.03	0.03	0.03	0.03	0.03	0.03	0.03	0.03	18
+20	+0.02	+0.02	+0.02	+0.02	+0.02	+0.02	+0.02	+0.02	+20
22	0.02	0.02	0.02	0.02	0.02	0.02	0.02	0.02	22
24	0.01	0.01	0.01	0.01	0.01	0.01	0.01	0.01	24
26	+0.01	+0.01	+0.01	+0.01	+0.01	+0.01	+0.01	+0.01	26
28	0.00	0.00	0.00	0.00	0.00	0.00	0.00	0.00	28
+30	0.00	0.00	0.00	0.00	0.00	0.00	0.00	0.00	+30
32	−0.01	−0.01	−0.01	−0.01	−0.01	−0.01	−0.01	−0.01	32
34	0.01	0.01	0.01	0.01	0.01	0.01	0.01	0.02	34
36	0.02	0.02	0.02	0.02	0.02	0.02	0.02	0.02	36
38	0.02	0.02	0.02	0.02	0.02	0.03	0.03	0.03	38
+40	−0.03	−0.03	−0.03	−0.03	−0.03	−0.03	−0.03	−0.03	+40
42	0.03	0.03	0.03	0.04	0.04	0.04	0.04	0.04	42
44	0.04	0.04	0.04	0.04	0.04	0.04	0.04	0.04	44
46	0.04	0.04	0.04	0.05	0.05	0.05	0.05	0.05	46
48	0.05	0.05	0.05	0.05	0.05	0.05	0.05	0.05	48
+50	−0.05	−0.05	−0.06	−0.06	−0.06	−0.06	−0.06	−0.06	+50
52	0.06	0.06	0.06	0.06	0.06	0.06	0.06	0.07	52
54	0.06	0.06	0.07	0.07	0.07	0.07	0.07	0.07	54
56	0.07	0.07	0.07	0.07	0.07	0.07	0.08	0.08	56
58	0.07	0.07	0.08	0.08	0.08	0.08	0.08	0.08	58
+60	−0.08	−0.08	−0.08	−0.08	−0.08	−0.09	−0.09	−0.09	+60
62	0.08	0.08	0.09	0.09	0.09	0.09	0.09	0.09	62
64	0.09	0.09	0.09	0.09	0.09	0.10	0.10	0.10	64
66	0.09	0.09	0.10	0.10	0.10	0.10	0.10	0.10	66
68	0.10	0.10	0.10	0.10	0.11	0.11	0.11	0.11	68
+70	−0.10	−0.10	−0.11	−0.11	−0.11	−0.11	−0.11	−0.12	+70
72	0.11	0.11	0.11	0.11	0.12	0.12	0.12	0.12	72
74	0.11	0.11	0.12	0.12	0.12	0.12	0.13	0.13	74
76	0.12	0.12	0.12	0.12	0.13	0.13	0.13	0.13	76
78	0.12	0.12	0.13	0.13	0.13	0.13	0.14	0.14	78
+80	−0.13	−0.13	−0.13	−0.13	−0.14	−0.14	−0.14	−0.14	+80
82	0.13	0.14	0.14	0.14	0.14	0.14	0.15	0.15	82
84	0.14	0.14	0.14	0.15	0.15	0.15	0.15	0.16	84
86	0.14	0.15	0.15	0.15	0.15	0.16	0.16	0.16	86
88	0.15	0.15	0.15	0.16	0.16	0.16	0.16	0.17	88
+90	−0.15	−0.16	−0.16	−0.16	−0.16	−0.17	−0.17	−0.17	+90
92	0.16	0.16	0.16	0.17	0.17	0.17	0.17	0.18	92
94	0.16	0.17	0.17	0.17	0.17	0.18	0.18	0.18	94
96	0.17	0.17	0.17	0.18	0.18	0.18	0.19	0.19	96
98	0.17	0.18	0.18	0.18	0.18	0.19	0.19	0.19	98
100	0.18	0.18	0.18	0.19	0.19	0.19	0.20	0.20	100

TABLE 5
Direction and Speed of True Wind in Units of Ship's Speed

Apparent wind speed	Difference between the heading and apparent wind direction										Apparent wind speed
	0°		10°		20°		30°		40°		
0.0	180	1.00	180	1.00	180	1.00	180	1.00	180	1.00	0.0
0.1	180	0.90	179	0.90	178	0.91	177	0.91	176	0.93	0.1
0.2	180	0.80	178	0.80	175	0.81	173	0.83	171	0.86	0.2
0.3	180	0.70	176	0.71	172	0.73	169	0.76	166	0.79	0.3
0.4	180	0.60	173	0.61	168	0.64	163	0.68	160	0.74	0.4
0.5	180	0.50	170	0.51	162	0.56	156	0.62	152	0.70	0.5
0.6	180	0.40	166	0.42	155	0.48	148	0.57	144	0.66	0.6
0.7	180	0.30	159	0.33	145	0.42	138	0.53	136	0.65	0.7
0.8	180	0.20	147	0.25	132	0.37	127	0.50	127	0.64	0.8
0.9	180	0.10	126	0.19	117	0.34	116	0.50	118	0.66	0.9
1.0	calm	0.00	95	0.17	100	0.35	105	0.52	110	0.68	1.0
1.1	0	0.10	66	0.21	85	0.38	95	0.55	103	0.72	1.1
1.2	0	0.20	49	0.28	73	0.43	86	0.60	96	0.78	1.2
1.3	0	0.30	39	0.36	64	0.50	79	0.66	90	0.84	1.3
1.4	0	0.40	33	0.45	57	0.57	73	0.73	85	0.90	1.4
1.5	0	0.50	29	0.54	51	0.66	68	0.81	81	0.98	1.5
1.6	0	0.60	26	0.64	47	0.74	64	0.89	78	1.05	1.6
1.7	0	0.70	24	0.74	44	0.83	61	0.97	75	1.13	1.7
1.8	0	0.80	22	0.83	42	0.93	58	1.06	72	1.22	1.8
1.9	0	0.90	21	0.93	40	1.02	56	1.15	70	1.30	1.9
2.0	0	1.00	20	1.03	38	1.11	54	1.24	68	1.39	2.0
2.5	0	1.50	17	1.52	32	1.60	47	1.71	60	1.85	2.5
3.0	0	2.00	15	2.02	29	2.09	43	2.19	56	2.32	3.0
3.5	0	2.50	14	2.52	28	2.58	41	2.68	53	2.81	3.5
4.0	0	3.00	13	3.02	26	3.08	39	3.17	51	3.30	4.0
4.5	0	3.50	13	3.52	25	3.58	38	3.67	50	3.79	4.5
5.0	0	4.00	12	4.02	25	4.08	37	4.16	49	4.28	5.0
6.0	0	5.00	12	5.02	24	5.07	36	5.16	47	5.27	6.0
7.0	0	6.00	12	6.02	23	6.07	35	6.15	46	6.27	7.0
8.0	0	7.00	11	7.02	23	7.07	34	7.15	45	7.26	8.0
9.0	0	8.00	11	8.02	22	8.07	34	8.15	44	8.26	9.0
10.0	0	9.00	11	9.02	22	9.06	33	9.15	44	9.26	10.0

Apparent wind speed	50°		60°		70°		80°		90°		Apparent wind speed
0.0	180	1.00	180	1.00	180	1.00	180	1.00	180	1.00	0.0
0.1	175	0.94	175	0.95	174	0.97	174	0.99	174	1.00	0.1
0.2	170	0.88	169	0.92	169	0.95	168	0.99	169	1.02	0.2
0.3	164	0.84	163	0.89	163	0.94	163	0.99	163	1.04	0.3
0.4	158	0.80	157	0.87	156	0.94	157	1.01	158	1.08	0.4
0.5	151	0.78	150	0.87	150	0.95	152	1.04	153	1.12	0.5
0.6	143	0.77	143	0.87	145	0.97	147	1.07	149	1.17	0.6
0.7	136	0.77	137	0.89	139	1.01	142	1.12	145	1.22	0.7
0.8	128	0.78	131	0.92	134	1.05	138	1.17	141	1.28	0.8
0.9	121	0.81	125	0.95	129	1.09	134	1.22	138	1.35	0.9
1.0	115	0.85	120	1.00	125	1.15	130	1.29	135	1.41	1.0
1.1	109	0.89	115	1.05	121	1.21	127	1.35	132	1.49	1.1
1.2	104	0.95	111	1.11	118	1.27	124	1.42	130	1.56	1.2
1.3	99	1.01	107	1.18	114	1.34	121	1.50	128	1.64	1.3
1.4	95	1.08	104	1.25	112	1.42	119	1.57	126	1.72	1.4
1.5	92	1.15	101	1.32	109	1.49	117	1.65	124	1.80	1.5
1.6	89	1.23	98	1.40	107	1.57	115	1.73	122	1.89	1.6
1.7	86	1.31	96	1.48	105	1.65	113	1.82	120	1.97	1.7
1.8	84	1.39	94	1.56	103	1.73	111	1.90	119	2.06	1.8
1.9	81	1.47	92	1.65	101	1.82	110	1.99	118	2.15	1.9
2.0	79	1.56	90	1.73	100	1.91	108	2.07	117	2.24	2.0
2.5	72	2.01	83	2.18	94	2.35	103	2.53	112	2.69	2.5
3.0	68	2.48	79	2.65	89	2.82	99	2.99	108	3.16	3.0
3.5	65	2.96	76	3.12	87	3.29	96	3.47	106	3.64	3.5
4.0	63	3.44	74	3.61	84	3.78	94	3.95	104	4.12	4.0
4.5	61	3.93	72	4.09	83	4.26	93	4.44	103	4.61	4.5
5.0	60	4.42	71	4.58	81	4.75	92	4.93	101	5.10	5.0
6.0	58	5.41	69	5.57	79	5.74	90	5.91	99	6.08	6.0
7.0	57	6.40	68	6.56	78	6.72	88	6.90	98	7.07	7.0
8.0	56	7.40	67	7.55	77	7.72	87	7.89	97	8.06	8.0
9.0	55	8.39	66	8.54	76	8.71	86	8.88	96	9.06	9.0
10.0	55	9.39	65	9.54	76	9.70	86	9.88	96	10.01	10.0

TABLE 5 (cont'd.)
Direction and Speed of True Wind in Units of Ship's Speed

Apparent wind speed	Difference between the heading and apparent wind direction										Apparent wind speed
	90°		100°		110°		120°		130°		
0.0	180	1.00	180	1.00	180	1.00	180	1.00	180	1.00	0.0
0.1	174	1.00	174	1.02	175	1.04	175	1.05	176	1.07	0.1
0.2	169	1.02	169	1.05	170	1.08	171	1.11	172	1.14	0.2
0.3	163	1.04	164	1.09	166	1.14	167	1.18	169	1.21	0.3
0.4	158	1.08	160	1.14	162	1.20	164	1.25	166	1.29	0.4
0.5	153	1.12	156	1.19	158	1.26	161	1.32	164	1.38	0.5
0.6	149	1.17	152	1.25	155	1.33	158	1.40	162	1.46	0.6
0.7	145	1.22	148	1.32	152	1.40	156	1.48	160	1.55	0.7
0.8	141	1.28	145	1.38	149	1.48	154	1.56	158	1.63	0.8
0.9	138	1.35	143	1.46	147	1.56	152	1.65	156	1.72	0.9
1.0	135	1.41	140	1.53	145	1.64	150	1.73	155	1.81	1.0
1.1	132	1.49	138	1.61	143	1.72	148	1.82	154	1.90	1.1
1.2	130	1.56	136	1.69	141	1.81	147	1.91	153	2.00	1.2
1.3	128	1.64	134	1.77	140	1.89	146	2.00	152	2.09	1.3
1.4	126	1.72	132	1.86	138	1.98	145	2.09	151	2.18	1.4
1.5	124	1.80	130	1.94	137	2.07	143	2.18	150	2.28	1.5
1.6	122	1.89	129	2.03	136	2.16	142	2.27	149	2.37	1.6
1.7	120	1.97	128	2.12	135	2.25	141	2.36	148	2.46	1.7
1.8	119	2.06	127	2.21	134	2.34	141	2.46	147	2.56	1.8
1.9	118	2.15	125	2.30	133	2.43	140	2.55	147	2.66	1.9
2.0	117	2.24	124	2.39	132	2.52	139	2.65	146	2.75	2.0
2.5	112	2.69	120	2.85	128	2.99	136	3.12	144	3.23	2.5
3.0	108	3.16	117	3.32	126	3.47	134	3.61	142	3.72	3.0
3.5	106	3.64	115	3.80	124	3.96	132	4.09	140	4.21	3.5
4.0	104	4.12	113	4.29	122	4.44	131	4.58	139	4.71	4.0
4.5	103	4.61	112	4.78	121	4.93	130	5.07	138	5.20	4.5
5.0	101	5.10	111	5.27	120	5.42	129	5.57	138	5.69	5.0
6.0	99	6.08	109	6.25	118	6.41	128	6.56	137	6.69	6.0
7.0	98	7.07	108	7.24	117	7.40	127	7.55	136	7.68	7.0
8.0	97	8.06	107	8.23	116	8.39	126	8.54	135	8.68	8.0
9.0	96	9.06	106	9.23	116	9.39	125	9.54	135	9.67	9.0
10.0	96	10.01	106	10.22	115	10.39	125	10.54	134	10.67	10.0

Apparent wind speed	140°		150°		160°		170°		180°		Apparent wind speed
0.0	180	1.00	180	1.00	180	1.00	180	1.00	180	1.00	0.0
0.1	177	1.08	177	1.09	178	1.09	179	1.10	180	1.10	0.1
0.2	174	1.16	175	1.18	177	1.19	178	1.20	180	1.20	0.2
0.3	171	1.24	173	1.27	175	1.29	178	1.30	180	1.30	0.3
0.4	169	1.33	172	1.36	174	1.38	177	1.40	180	1.40	0.4
0.5	167	1.42	170	1.45	173	1.48	177	1.50	180	1.50	0.5
0.6	165	1.51	169	1.55	173	1.58	176	1.60	180	1.60	0.6
0.7	164	1.60	168	1.64	172	1.68	176	1.69	180	1.70	0.7
0.8	162	1.69	167	1.74	171	1.77	176	1.79	180	1.80	0.8
0.9	161	1.79	166	1.84	171	1.87	175	1.89	180	1.90	0.9
1.0	160	1.88	165	1.93	170	1.97	175	1.99	180	2.00	1.0
1.1	159	1.97	164	2.03	170	2.07	175	2.09	180	2.10	1.1
1.2	158	2.07	164	2.13	169	2.17	175	2.19	180	2.20	1.2
1.3	157	2.16	163	2.22	169	2.27	174	2.29	180	2.30	1.3
1.4	157	2.26	162	2.32	168	2.36	174	2.39	180	2.40	1.4
1.5	156	2.36	162	2.42	168	2.46	174	2.49	180	2.50	1.5
1.6	155	2.45	161	2.52	168	2.56	174	2.59	180	2.60	1.6
1.7	155	2.55	161	2.61	167	2.66	174	2.69	180	2.70	1.7
1.8	154	2.65	161	2.71	167	2.76	174	2.79	180	2.80	1.8
1.9	154	2.74	160	2.81	167	2.86	173	2.89	180	2.90	1.9
2.0	153	2.84	160	2.91	167	2.96	173	2.99	180	3.00	2.0
2.5	151	3.33	158	3.40	166	3.46	173	3.49	180	3.50	2.5
3.0	150	3.82	157	3.90	165	3.95	172	3.99	180	4.00	3.0
3.5	149	4.31	157	4.39	164	4.45	172	4.49	180	4.50	3.5
4.0	148	4.81	156	4.89	164	4.95	172	4.99	180	5.00	4.0
4.5	147	5.31	155	5.39	164	5.45	172	5.49	180	5.50	4.5
5.0	146	5.80	155	5.89	163	5.95	172	5.99	180	6.00	5.0
6.0	145	6.80	154	6.88	163	6.95	171	6.99	180	7.00	6.0
7.0	145	7.79	154	7.88	162	7.95	171	7.99	180	8.00	7.0
8.0	144	8.79	153	8.88	162	8.95	171	8.99	180	9.00	8.0
9.0	144	9.79	153	9.88	162	9.95	171	9.99	180	10.00	9.0
10.0	143	10.78	153	10.88	162	10.95	171	10.98	180	11.00	10.0

TABLE 6

Conversion Tables for Thermometer Scales

F = Fahrenheit, C = Celsius (centigrade), K = Kelvin

F	C	K	F	C	K	C	F	K	K	F	C
-20	-28.9	244.3	+40	+4.4	277.6	-25	-13.0	248.2	250	-9.7	-23.2
19	28.3	244.8	41	5.0	278.2	24	11.2	249.2	251	7.9	22.2
18	27.8	245.4	42	5.6	278.7	23	9.4	250.2	252	6.1	21.2
17	27.2	245.9	43	6.1	279.3	22	7.6	251.2	253	4.3	20.2
16	26.7	246.5	44	6.7	279.8	21	5.8	252.2	254	2.5	19.2
-15	-26.1	247.0	+45	+7.2	280.4	-20	-4.0	253.2	255	-0.7	-18.2
14	25.6	247.6	46	7.8	280.9	19	2.2	254.2	256	+1.1	17.2
13	25.0	248.2	47	8.3	281.5	18	-0.4	255.2	257	2.9	16.2
12	24.4	248.7	48	8.9	282.0	17	+1.4	256.2	258	4.7	15.2
11	23.9	249.3	49	9.4	282.6	16	3.2	257.2	259	6.5	14.2
-10	-23.3	249.8	+50	+10.0	283.2	-15	+5.0	258.2	260	+8.3	-13.2
9	22.8	250.4	51	10.6	283.7	14	6.8	259.2	261	10.1	12.2
8	22.2	250.9	52	11.1	284.3	13	8.6	260.2	262	11.9	11.2
7	21.7	251.5	53	11.7	284.8	12	10.4	261.2	263	13.7	10.2
6	21.1	252.0	54	12.2	285.4	11	12.2	262.2	264	15.5	9.2
-5	-20.6	252.6	+55	+12.8	285.9	-10	+14.0	263.2	265	+17.3	-8.2
4	20.0	253.2	56	13.3	286.5	9	15.8	264.2	266	19.1	7.2
3	19.4	253.7	57	13.9	287.0	8	17.6	265.2	267	20.9	6.2
2	18.9	254.3	58	14.4	287.6	7	19.4	266.2	268	22.7	5.2
-1	18.3	254.8	59	15.0	288.2	6	21.2	267.2	269	24.5	4.2
0	-17.8	255.4	+60	+15.6	288.7	-5	+23.0	268.2	270	+26.3	-3.2
+1	17.2	255.9	61	16.1	289.3	4	24.8	269.2	271	28.1	2.2
2	16.7	256.5	62	16.7	289.8	3	26.6	270.2	272	29.9	1.2
3	16.1	257.0	63	17.2	290.4	2	28.4	271.2	273	31.7	-0.2
4	15.6	257.6	64	17.8	290.9	-1	30.2	272.2	274	33.5	+0.8
+5	-15.0	258.2	+65	+18.3	291.5	0	+32.0	273.2	275	+35.3	+1.8
6	14.4	258.7	66	18.9	292.0	+1	33.8	274.2	276	37.1	2.8
7	13.9	259.3	67	19.4	292.6	2	35.6	275.2	277	38.9	3.8
8	13.3	259.8	68	20.0	293.2	3	37.4	276.2	278	40.7	4.8
9	12.8	260.4	69	20.6	293.7	4	39.2	277.2	279	42.5	5.8
+10	-12.2	260.9	+70	+21.1	294.3	+5	+41.0	278.2	280	+44.3	+6.8
11	11.7	261.5	71	21.7	294.8	6	42.8	279.2	281	46.1	7.8
12	11.1	262.0	72	22.2	295.4	7	44.6	280.2	282	47.9	8.8
13	10.6	262.6	73	22.8	295.9	8	46.4	281.2	283	49.7	9.8
14	10.0	263.2	74	23.3	296.5	9	48.2	282.2	284	51.5	10.8
+15	-9.4	263.7	+75	+23.9	297.0	+10	+50.0	283.2	285	+53.3	+11.8
16	8.9	264.3	76	24.4	297.6	11	51.8	284.2	286	55.1	12.8
17	8.3	264.8	77	25.0	298.2	12	53.6	285.2	287	56.9	13.8
18	7.8	265.4	78	25.6	298.7	13	55.4	286.2	288	58.7	14.8
19	7.2	265.9	79	26.1	299.3	14	57.2	287.2	289	60.5	15.8
+20	-6.7	266.5	+80	+26.7	299.8	+15	+59.0	288.2	290	+62.3	+16.8
21	6.1	267.0	81	27.2	300.4	16	60.8	289.2	291	64.1	17.8
22	5.6	267.6	82	27.8	300.9	17	62.6	290.2	292	65.9	18.8
23	5.0	268.2	83	28.3	301.5	18	64.4	291.2	293	67.7	19.8
24	4.4	268.7	84	28.9	302.0	19	66.2	292.2	294	69.5	20.8
+25	-3.9	269.3	+85	+29.4	302.6	+20	+68.0	293.2	295	+71.3	+21.8
26	3.3	269.8	86	30.0	303.2	21	69.8	294.2	296	73.1	22.8
27	2.8	270.4	87	30.6	303.7	22	71.6	295.2	297	74.9	23.8
28	2.2	270.9	88	31.1	304.3	23	73.4	296.2	298	76.7	24.8
29	1.7	271.5	89	31.7	304.8	24	75.2	297.2	299	78.5	25.8
+30	-1.1	272.0	+90	+32.2	305.4	+25	+77.0	298.2	300	+80.3	+26.8
31	0.6	272.6	91	32.8	305.9	26	78.8	299.2	301	82.1	27.8
32	0.0	273.2	92	33.3	306.5	27	80.6	300.2	302	83.9	28.8
33	+0.6	273.7	93	33.9	307.0	28	82.4	301.2	303	85.7	29.8
34	1.1	274.3	94	34.4	307.6	29	84.2	302.2	304	87.5	30.8
+35	+1.7	274.8	+95	+35.0	308.2	+30	+86.0	303.2	305	+89.3	+31.8
36	2.2	275.4	96	35.6	308.7	31	87.8	304.2	306	91.1	32.8
37	2.8	275.9	97	36.1	309.3	32	89.6	305.2	307	92.9	33.8
38	3.3	276.5	98	36.7	309.8	33	91.4	306.2	308	94.7	34.8
39	3.9	277.0	99	37.2	310.4	34	93.2	307.2	309	96.5	35.8
+40	+4.4	277.6	+100	+37.8	310.9	+35	+95.0	308.2	310	+98.3	+36.8

TABLE 7

Relative Humidity

Dry-bulb temp. F	Difference between dry-bulb and wet-bulb temperatures														Dry-bulb temp. F
	1°	2°	3°	4°	5°	6°	7°	8°	9°	10°	11°	12°	13°	14°	
°	%	%	%	%	%	%	%	%	%	%	%	%	%	%	°
−20	7														−20
18	14														18
16	21														16
14	27														14
12	32														12
−10	37														−10
8	41	2													8
6	45	9													6
4	49	16													4
−2	52	22													−2
0	56	28													0
+2	59	33	7												+2
4	62	37	14												4
6	64	42	20												6
8	67	46	25	5											8
+10	69	50	30	11											+10
12	71	53	35	17											12
14	73	56	40	23	7										14
16	76	60	44	28	13										16
18	77	62	48	33	19	4									18
+20	79	65	51	37	24	10									+20
22	81	68	55	42	29	16	4								22
24	83	70	58	45	33	21	10								24
26	85	73	61	49	38	26	15	4							26
28	86	75	64	53	42	31	20	10							28
+30	88	77	66	56	45	35	25	15	6						+30
32	89	79	69	59	49	39	30	20	11	2					32
34	90	81	71	62	52	43	34	25	16	8					34
36	91	82	73	64	55	47	38	29	21	13	5				36
38	91	83	74	66	58	50	42	33	25	18	10	2			38
+40	92	84	76	68	60	52	45	37	30	22	15	7			+40
42	92	84	77	69	62	54	47	40	33	26	19	12	5		42
44	92	85	78	70	63	56	49	43	36	29	23	17	10	4	44
46	93	86	79	72	65	58	52	45	39	32	26	20	14	8	46
48	93	86	79	73	66	60	54	47	41	35	29	24	18	12	48
+50	93	87	80	74	68	61	55	49	44	38	32	27	21	16	+50
52	94	87	81	75	69	63	57	51	46	40	35	29	24	19	52
54	94	88	82	76	70	64	59	53	48	42	37	32	27	22	54
56	94	88	82	77	71	65	60	55	50	44	39	35	30	25	56
58	94	88	83	77	72	67	61	56	51	46	42	37	32	28	58
+60	94	89	83	78	73	68	63	58	53	48	43	39	34	30	+60
62	95	89	84	79	74	69	64	59	54	50	45	41	37	32	62
64	95	89	84	79	74	70	65	60	56	51	47	43	38	34	64
66	95	90	85	80	75	71	66	61	57	53	49	44	40	36	66
68	95	90	85	81	76	71	67	63	58	54	50	46	42	38	68
+70	95	90	86	81	77	72	68	64	59	55	51	48	44	40	+70
72	95	91	86	82	77	73	69	65	61	57	53	49	45	42	72
74	95	91	86	82	78	74	69	65	62	58	54	50	47	43	74
76	95	91	87	82	78	74	70	66	63	59	55	51	48	45	76
78	96	91	87	83	79	75	71	67	63	60	56	53	49	46	78
+80	96	91	87	83	79	75	72	68	64	61	57	54	50	47	+80
82	96	92	88	84	80	76	72	69	65	62	58	55	52	48	82
84	96	92	88	84	80	76	73	69	66	62	59	56	53	49	84
86	96	92	88	84	81	77	73	70	67	63	60	57	54	51	86
88	96	92	88	85	81	77	74	71	67	64	61	58	55	52	88
+90	96	92	89	85	81	78	74	71	68	65	61	58	55	52	+90
92	96	92	89	85	82	78	75	72	68	65	62	59	56	53	92
94	96	93	89	85	82	79	75	72	69	66	63	60	57	54	94
96	96	93	89	86	82	79	76	73	70	67	64	61	58	55	96
98	96	93	89	86	83	79	76	73	70	67	64	61	59	56	98
+100	96	93	90	86	83	80	77	74	71	68	65	62	59	57	+100

TABLE 7 (cont'd.)

Relative Humidity

Dry-bulb temp. F	Difference between dry-bulb and wet-bulb temperatures														Dry-bulb temp. F
	15°	16°	17°	18°	19°	20°	21°	22°	23°	24°	25°	26°	27°	28°	
°	%	%	%	%	%	%	%	%	%	%	%	%	%	%	°
+46	2														+46
48	7	1													48
+50	10	5													+50
52	14	9	4												52
54	17	12	7	3											54
56	20	16	11	7	2										56
58	23	19	14	10	6	2									58
+60	26	21	17	13	9	5	1								+60
62	28	24	20	16	12	8	4	1							62
64	30	26	22	19	15	11	8	4							64
66	32	29	25	21	17	14	10	7	4						66
68	34	31	27	23	20	16	13	10	7	3					68
+70	36	33	29	26	22	19	16	12	9	6	3				+70
72	38	34	31	28	24	21	18	15	12	9	6	3			72
74	40	36	33	30	26	23	20	17	14	11	8	6	3		74
76	41	38	35	31	28	25	22	19	16	14	11	8	5	3	76
78	43	39	36	33	30	27	24	21	18	16	13	10	8	5	78
+80	44	41	38	35	32	29	26	23	20	18	15	13	10	8	+80
82	45	42	39	36	33	30	28	25	22	20	17	15	12	10	82
84	46	43	40	38	35	32	29	27	24	21	19	17	14	12	84
86	48	45	42	39	36	33	31	28	26	23	21	18	16	14	86
88	49	46	43	40	37	35	32	30	27	25	22	20	18	16	88
+90	50	47	44	41	39	36	34	31	29	26	24	22	19	17	+90
92	51	48	45	42	40	37	35	32	30	28	25	23	21	19	92
94	51	49	46	44	41	39	36	34	31	29	27	25	23	20	94
96	52	50	47	45	42	40	37	35	33	30	28	26	24	22	96
98	53	51	48	45	43	41	38	36	34	32	29	27	25	23	98
+100	54	51	49	46	44	42	39	37	35	33	31	29	27	25	+100

Dry-bulb temp. F	Difference between dry-bulb and wet-bulb temperatures														Dry-bulb temp. F
	29°	30°	31°	32°	33°	34°	35°	36°	37°	38°	39°	40°	41°	42°	
°	%	%	%	%	%	%	%	%	%	%	%	%	%	%	°
+78	3														+78
+80	5	3													+80
82	7	5	3												82
84	10	7	5	3	1										84
86	11	9	7	5	3	1									86
88	13	11	9	7	5	3	1								88
+90	15	13	11	9	7	5	3	1							+90
92	17	15	13	11	9	7	5	3	1						92
94	18	16	14	12	11	9	7	5	3	2					94
96	20	18	16	14	12	10	9	7	5	4	2				96
98	21	19	17	16	14	12	10	9	7	5	4	2	1		98
+100	23	21	19	17	15	14	12	10	9	7	5	4	2	1	+100

TABLE 8
Dew Point

Dry-bulb temp. F	Difference between dry-bulb and wet-bulb temperatures														Dry-bulb temp. F
	1°	2°	3°	4°	5°	6°	7°	8°	9°	10°	11°	12°	13°	14°	
−20															−20
18	−52														18
16	45														16
14	39														14
12	34														12
−10	−29														−10
8	25	−75													8
6	22	50													6
4	18	39													4
−2	15	32													−2
0	−12	−26													0
+2	9	21	−49												+2
4	6	16	35												4
6	3	12	27												6
8	−1	9	20	−50											8
+10	+2	−5	−15	−34											+10
12	5	−2	10	24											12
14	7	+1	6	17											14
16	10	4	−2	11											16
18	12	7	+1	6											18
+20	+15	+10	+5	−2											+20
22	17	13	8	+2											22
24	20	16	11	6											24
26	22	18	14	10	+4	−4									26
28	24	21	17	13	8	+1	−8	−22							28
+30	+27	+24	+20	+16	+11	+6	−1	−12	−31						+30
32	29	26	23	19	15	10	+4	−4	16	−47					32
34	32	29	26	22	18	14	9	+2	−7	22					34
36	34	31	28	25	22	18	13	7	0	11	−30				36
38	36	33	31	28	25	21	17	12	+6	−2	14	−42			38
+40	+38	+35	+33	+30	+27	+24	+20	+16	+11	+4	−4	−18	−79		+40
42	40	38	35	33	30	27	23	19	15	10	+3	−7	23		42
44	42	40	37	35	32	29	26	23	19	14	9	+2	−9	−29	44
46	44	42	40	37	35	32	29	26	22	18	13	7	0	11	46
48	46	44	42	40	37	35	32	29	26	22	18	13	+6	−2	48
+50	+48	+46	+44	+42	+40	+37	+35	+32	+29	+25	+21	+17	+12	+5	+50
52	50	48	46	44	42	40	37	35	32	29	25	21	17	11	52
54	52	50	49	47	44	42	40	37	35	32	28	25	21	16	54
56	54	53	51	49	47	45	42	40	37	35	32	28	25	21	56
58	56	55	53	51	49	47	45	43	40	38	35	32	28	25	58
+60	+58	+57	+55	+53	+51	+49	+47	+45	+43	+40	+38	+35	+32	+28	+60
62	60	59	57	55	54	52	50	48	45	43	41	38	35	32	62
64	62	61	59	57	56	54	52	50	48	46	43	41	38	35	64
66	64	63	61	60	58	56	54	52	50	48	46	44	41	39	66
68	67	65	63	62	60	58	57	55	53	51	49	46	44	42	68
+70	+69	+67	+66	+64	+62	+61	+59	+57	+55	+53	+51	+49	+47	+45	+70
72	71	69	68	66	64	63	61	59	58	56	54	52	50	47	72
74	73	71	70	68	67	65	63	62	60	58	56	54	52	50	74
76	75	73	72	70	69	67	66	64	62	61	59	57	55	53	76
78	77	75	74	72	71	69	68	66	65	63	61	59	57	55	78
+80	+79	+77	+76	+74	+73	+72	+70	+68	+67	+65	+64	+62	+60	+58	+80
82	81	79	78	77	75	74	72	71	69	67	66	64	62	61	82
84	83	81	80	79	77	76	74	73	71	70	68	67	65	63	84
86	85	83	82	81	79	78	76	75	74	72	70	69	67	66	86
88	87	85	84	83	81	80	79	77	76	74	73	71	70	68	88
+90	+89	+87	+86	+85	+84	+82	+81	+79	+78	+76	+75	+73	+72	+70	+90
92	91	89	88	87	86	84	83	82	80	79	77	76	74	73	92
94	93	92	90	89	88	86	85	84	82	81	79	78	76	75	94
96	95	94	92	91	90	88	87	86	84	83	82	80	79	77	96
98	97	96	94	93	92	91	89	88	87	85	84	82	81	80	98
+100	+99	+98	+96	+95	+94	+93	+91	+90	+89	+87	+86	+85	+83	+82	+100

195

TABLE 8 (cont'd.)
Dew Point

Dry-bulb temp. F	Difference between dry-bulb and wet-bulb temperatures													Dry-bulb temp. F
	15°	16°	17°	18°	19°	20°	21°	22°	23°	24°	25°	26°	27°	28°	
+46	−36														+46
48	14	−45													48
+50	−3	−17	−78												+50
52	+4	−5	21												52
54	10	+3	−7	−25											54
56	16	10	+2	−8	−29										56
58	20	16	10	+2	−10	−34									58
+60	+25	+20	+15	+9	+1	−11	−39								+60
62	29	25	20	15	9	+1	−12	−45							62
64	32	29	25	20	15	9	0	−13	−52						64
66	36	33	29	25	21	15	+9	0	−14	−59					66
68	39	36	33	29	25	21	16	+9	0	−14	−68				68
+70	+42	+39	+36	+33	+30	+26	+21	+16	+9	0	−14	−76			+70
72	45	43	40	37	34	30	26	22	16	+10	+1	−14	−77		72
74	48	46	43	40	37	34	31	27	22	17	10	+1	−13	−70	74
76	51	48	46	44	41	38	35	31	27	23	17	11	+2	−12	76
78	53	51	49	47	44	41	38	35	32	28	23	18	11	+3	78
+80	+56	+54	+52	+50	+47	+45	+42	+39	+36	+32	+28	+24	+19	+12	+80
82	59	57	55	53	50	48	45	43	40	37	33	29	25	20	82
84	61	59	57	55	53	51	49	46	43	41	37	34	30	26	84
86	64	62	60	58	56	54	52	49	47	44	41	38	35	31	86
88	66	64	63	61	59	57	55	52	50	48	45	42	39	36	88
+90	+69	+67	+65	+63	+62	+60	+58	+55	+53	+51	+48	+46	+43	+40	+90
92	71	69	68	66	64	62	60	58	56	54	52	49	47	44	92
94	73	72	70	68	67	65	63	61	59	57	55	52	50	47	94
96	76	74	73	71	69	67	66	64	62	60	58	56	53	51	96
98	78	77	75	73	72	70	68	67	65	63	61	59	57	54	98
+100	+80	+79	+77	+76	+74	+73	+71	+69	+67	+66	+64	+62	+60	+57	+100

Dry-bulb temp. F	Difference between dry-bulb and wet-bulb temperatures													Dry-bulb temp. F
	29°	30°	31°	32°	33°	34°	35°	36°	37°	38°	39°	40°	41°	42°	
+76	−61														+76
78	−11	−53													78
+80	+4	−10	−45												+80
82	13	+5	−8	−39											82
84	20	14	+6	−6	−33										84
86	27	21	15	+7	−4	−28									86
88	32	27	22	16	+9	−2	−23								88
+90	+36	+33	+28	+24	+18	+10	0	−18							+90
92	41	37	34	30	25	19	+12	+2	−14						92
94	45	42	38	35	31	26	20	13	+4	−10					94
96	48	46	43	39	36	32	27	22	15	+6	−7	−43			96
98	52	49	47	44	40	37	33	28	23	17	+9	−4	−30		98
+100	+55	+53	+50	+47	+45	+41	+38	+34	+30	+25	+19	+11	0	−21	+100

APPENDIX B

SEA STATE

This appendix provides, by means of representative photographs illustrating the effects of the wind on the sea surface, a pictorial guide to mariners for estimating the wind speed at sea.

The photographs and associated text are taken from *State of Sea Photographs for the Beaufort Wind Scale*, Crown Copyright, Ottawa, 1975. The material is reprinted with minor changes through permission of the Atmospheric Environment Service, Department of the Environment, Canada.

State of Sea Photographs for the Beaufort Wind Scale

B1. Introduction.—This appendix presents the results of a project carried out on board the Canadian Ocean Weather Ships C.C.G.S. *St. Catharines* and C.C.G.S. *Stonetown*. The aim of the project was to collect photographs of the sea surface as it appears under the influence of the various ranges of wind speed defined by The Beaufort Scale of Wind Force (app. A). Word descriptions of the appearance of the sea for each Beaufort Force have been available for many years, but it was felt that photographs illustrating the conditions associated with each force would be of some assistance to ships' officers in estimating wind speed. Sea photographs taken by low-flying aircraft of the United States Navy were published by the Meteorological Branch under a circular memorandum dated December 20, 1957. However, these aerial photographs, while good, do not depict the aspect of the sea as viewed from the bridge of a ship. The apparent lack of good photographs of this nature prompted this project on the Ocean Weather Ships. A selection of the best photographs resulting from the project are presented.

B2. Estimating the wind at sea.—Observers on board ships at sea usually determine the speed of the wind by estimating its Beaufort Force, as merchant ships are not normally equipped with wind measuring instruments. Through experience, ships' officers have developed various methods of estimating this force. The effect of the wind on the observer himself, the ship's rigging, flags, etc., is used as a criterion; but, estimates based on these indications give the relative wind which must be corrected for the motion of the ship before an estimate of the true wind speed can be obtained.

The most common method involves the appearance of the sea surface. The state of the sea disturbance, i.e. the dimensions of the waves, the presence of white caps, foam or spray, depends principally on three factors:

1. *The wind speed.* The higher the speed of the wind, the greater is the sea disturbance.

2. *The duration of the wind.* At any point on the sea, the disturbance will increase the longer the wind blows at a given speed, until a maximum state of disturbance is reached.

3. *The fetch.* This is the length of the stretch of water over which the wind acts on the sea surface from the same direction. For a given wind speed and duration, the longer the fetch, the greater is the sea disturbance. If the fetch is short, say a few miles, the disturbance will be relatively small no matter how great the wind speed is or how long it has been blowing.

There are other factors which can modify the appearance of the sea surface caused by wind alone. These are strong currents, shallow water, swell, precipitation, ice, and wind shifts. Their affects will be described later.

A wind of a given Beaufort Force will, therefore, produce a characteristic appearance of the sea surface provided that it has been blowing for a sufficient length of time, and over a sufficiently long fetch. The effects of currents, shallow water, swell, precipitation, etc., should also be absent. The "Sea Criteria" associated with each Beaufort Force from force 0 to force 12 were agreed upon and drawn up by the International Meteorological Committee in 1939. These word descriptions of the state of the sea are known as the Sea Criterion of the Beaufort Scale of Wind Force.

The use of the sea criterion has the advantage that the speed of the ship need not be considered. In practice, the mariner observes the sea surface, noting the size of the waves, the white caps, spindrift, etc., and then finds the criterion (app. A) which best describes the sea surface as he saw it. This criterion is associated with a Beaufort number, for which a corresponding mean wind speed and range in knots are given. Since meteorological reports require that wind speeds be reported in knots, the mean speed for the Beaufort number may be reported, or an experienced observer may judge that the sea disturbance is such that a higher or lower speed within the range for the force is more accurate.

This method, while it appears simple, should be used with caution however. It should be borne in mind that the sea conditions described for each Beaufort Force (app. A) are "steady-state" conditions; i.e. the conditions which result when the wind has been blowing for a relatively long time, and over a great stretch of water. At any particular time at sea, though, the duration of the wind or the fetch, or both, may not have been great enough to produce these "steady-state" conditions. When a high wind springs up suddenly after previously calm or near calm conditions, it will require some hours, depending on the strength of the wind, to generate waves of maximum height. The height of the waves increases rapidly in the first few hours after the commencement of the blow, but increases at a much slower rate later on. Considering the effect of fetch, if the observer could start at the beginning of the fetch (say at a coastline when the wind is offshore) after the wind has been blowing for a long time, and proceed downwind, he would notice that the waves were quite small at the beginning, and increased in height rapidly over the first 50 miles or so of the fetch. Farther along he would notice that the rate of increase in height with distance would slow down, and after 500 miles or so from the beginning of the fetch there would be little or no increase in height.

To illustrate the duration of winds and the length of fetches required for various wind forces to build seas to 50 percent, 75 percent, and 90 percent of their theoretical maximum heights, table B2 is of interest.

The theoretical maximum wave heights represent the average heights of the highest third of the waves, as these waves are of the most practical significance.

It will be seen that winds of force 5 or less can build seas to 90 percent of their maximum height in less than 12 hours, provided the fetch is long enough. Higher winds require a much greater time—force 11 winds requiring 32 hours to build waves to 90 percent of their maximum height. The times given in table B2 represent those required to build waves starting from initially calm sea conditions. If waves are already present at the onset of the blow, the times would be somewhat less depending on the initial wave heights and their direction relative to the direction of the wind which has sprung up.

The first consideration when using the sea criterion to estimate wind speed, therefore, is to decide whether the wind has been blowing long enough from the same direction to produce a steady state sea condition. If not, then it is possible that the wind speed may be underestimated. For example, if a wind with an actual speed of force 9 has been blowing for only 7 hours, it may have generated a sea condition which would perhaps correspond to a steady state for force 6. If, in this case, the sea criterion was used blindly without considering the short duration, and force 6 was reported, then the wind would be underestimated by three Beaufort forces, or approximately 20 knots. This is an extreme example however, as it is very unlikely that even an inexperienced seaman could not distinguish a force 6 from a force 9 wind.

Beaufort force of wind.	Theoretical maximum wave height (ft) unlimited duration and fetch.	Duration jof winds, (hours), with unlimited fetch, to produce percent of maximum wave height indicated.			Fetch (nautical miles), with unlimited duration of blow, to produce percent of maximum wave height indicated.		
		50%	75%	90%	50%	75%	90%
3	2	1. 5	5	8	3	13	25
5	8	3. 5	8	12	10	30	60
7	20	5. 5	12	21	22	75	150
9	40	7	16	25	55	150	280
11	70	9	19	32	85	200	450

TABLE—B2

Experience has shown that the appearance of white caps, foam, spindrift, etc., reaches a steady state condition before the height of the waves attain their maximum value. It is a safe assumption that the appearance of the sea (as regards white caps, etc.) will reach a steady state in the time required to build the waves to 50—75 percent of their maximum height. Thus, from table B2, it is seen that a force 5 wind could require 8 hours at most to produce a characteristic appearance of the sea surface.

A second consideration when using the sea criterion is the length of the fetch over which the wind has been blowing to produce the present state of the sea. On the open sea, unless the mariner has a copy of the latest synoptic weather map available, he will not know the length of the fetch. It will be seen from table B2, though, that only relatively short fetches are required for the lower wind forces to generate their characteristic seas. On the open sea, the fetches associated with most storms and other weather systems are usually long enough so that even winds up to force 9 can build seas up to 90 percent or more of their maximum height, providing the wind blows from the same direction long enough.

When navigating close to a coast or in restricted waters, however, it may be necessary to make allowances for the shorter stretches of water over which the wind blows. For example, referring to table B2, if the ship is 22 miles from a coast and an offshore wind with an actual speed of force 7 is blowing, the waves at the ship will never attain more than 50 percent of their maximum height for this speed no matter how long the wind blows. Hence, if the sea criterion were used under these conditions without consideration of the short fetch, the wind speed would be underestimated. With an offshore wind, the sea criterion may be used with confidence if the distance to the coast is greater than the values given in the extreme right-hand column of table B2; again, provided that the wind has been blowing offshore for a sufficient length of time.

Other factors aside from the duration of the blow and the fetch affect the appearance of the sea surface, and should be considered if they are present.

B3. Tides and Currents.—A wind blowing against a tide or strong current causes a greater sea disturbance than normal, which may result in an overestimate of the wind speed. On the other hand, a wind blowing in the same direction as a tide or strong current causes less sea disturbance than normal, and may result in an underestimate of the wind speed.

B4. Shallow Water.—Waves running into shallow water increase in steepness, and hence, their tendency to break. With an onshore wind there will, therefore, be more white caps over the shallow waters than over the deeper water farther offshore. It is only over relatively deep water that the sea criterion can be used with confidence.

B5. Swell.—Swell is the name given to waves, generally of considerable length, which were raised in some distant area by winds blowing there, and which have moved into the vicinity of the ship; or to waves raised nearby and which continue to advance after the wind at the ship has abated or changed direction. The direction of swell waves is usually different from the direction of the wind and the sea waves. *Swell waves are not to be considered when estimating wind speed and direction. Only those waves raised by the wind blowing at the time are of any significance.* The wind-driven waves show a greater tendency to break when superimposed on the crests of swell, and hence more white caps may be formed than if the swell were absent. Under these conditions the use of the sea criterion may result in a slight overestimate of the wind speed.

B6. Precipitation.—Heavy rain has a damping or smoothing effect on the sea surface which must be mechanical in character. Since the sea surface will therefore appear less disturbed than would be the case without the rain, the wind speed may be underestimated unless the smoothing effect is taken into account.

B7. Ice.—Even small concentrations of ice floating on the sea surface will dampen waves considerably, and concentrations greater than about seven tenths average will eliminate waves altogether. Young sea ice, which in the early stages of formation has a thick soupy consistency, and later takes on a rubbery appearance, is very effective in dampening waves. Consequently, the sea criterion cannot be used with any degree of confidence when sea ice is present. In higher latitudes, the presence of an ice field some distance to windward of the ship may be suspected if, when the ship is not close to any coast, the wind is relatively strong but the seas abnormally underdeveloped. The edge of the ice field acts like a coastline, and the short fetch between the ice and the ship is not sufficient for the wind to develop the seas fully.

B8. Wind Shifts.—Following a rapid change in the direction of the wind, as occurs at the passage of a cold front, the new wind will flatten out to a great extent the waves which were present before the wind shift. This is so because the direction of the wind after the shift may differ by 90° or more from the direction of the waves, which does not change. Hence, the wind may oppose the progress of the waves and dampen them out quickly. At the same time the new wind begins to generate its own waves on top of this dissipating swell, and it is not long before the cross pattern of waves gives the sea a "choppy" or confused appearance. It is during the first few hours following the wind shift that the appearance of the sea surface may not provide a reliable indication of the wind speed. The wind is normally stronger than the sea would indicate, as old waves are being flattened out, and new waves are just beginning to be developed.

B9. Night Observations.—On a dark night, when it is impossible to see the sea clearly, the observer may estimate the apparent wind from its effect on the ship's rigging, flags, etc., or simply the "feel" of the wind. A guide to estimating the apparent wind is given in the Meteorological Branch publication *Manual of Marine Weather Observing* (MANMAR), to which the observer is referred. Tables for converting the apparent wind to true wind may also be found in MANMAR.

BEAUFORT WIND SCALE
WITH CORRESPONDING SEA STATE CODES

Beaufort number or force	Wind speed				World Meteorological Organization (1964)	Effects observed far from land	Effects observed near coast	Effects observed on land	Sea State	
	knots	mph	meters per second	km per hour					Term and height of waves, in meters	Code
0	under 1	under 1	0.0-0.2	under 1	Calm	Sea like mirror.	Calm.	Calm; smoke rises vertically.	Calm, glassy, 0	0
1	1-3	1-3	0.3-1.5	1-5	Light air	Ripples with appearance of scales; no foam crests.	Fishing smack just has steerage way.	Smoke drift indicates wind direction; vanes do not move.	Calm, rippled, 0-0.1	1
2	4-6	4-7	1.6-3.3	6-11	Light breeze	Small wavelets; crests of glassy appearance, not breaking.	Wind fills the sails of smacks which then travel at about 1-2 miles per hour.	Wind felt on face; leaves rustle; vanes begin to move.	Smooth, wavelets, 0.1-0.5	2
3	7-10	8-12	3.4-5.4	12-19	Gentle breeze	Large wavelets; crests begin to break; scattered whitecaps.	Smacks begin to careen and travel about 3-4 miles per hour.	Leaves, small twigs in constant motion; light flags extended.	Slight, 0.5-1.25	3
4	11-16	13-18	5.5-7.9	20-28	Moderate breeze	Small waves, becoming longer; numerous whitecaps.	Good working breeze, smacks carry all canvas with good list.	Dust, leaves, and loose paper raised up; small branches move.	Moderate, 1.25-2.5	4
5	17-21	19-24	8.0-10.7	29-38	Fresh breeze	Moderate waves, taking longer form; many whitecaps; some spray.	Smacks shorten sail.	Small trees in leaf begin to sway.	Rough, 2.5-4	5
6	22-27	25-31	10.8-13.8	39-49	Strong breeze	Larger waves forming; whitecaps everywhere; more spray.	Smacks have doubled reef in mainsail; care required when fishing.	Larger branches of trees in motion; whistling heard in wires.	Very rough, 4-6	6
7	28-33	32-38	13.9-17.1	50-61	Near gale	Sea heaps up; white foam from breaking waves begins to be blown in streaks.	Smacks remain in harbor and those at sea lie-to.	Whole trees in motion; resistance felt in walking against wind.		
8	34-40	39-46	17.2-20.7	62-74	Gale	Moderately high waves of greater length; edges of crests begin to break into spindrift; foam is blown in well-marked streaks.	All smacks make for harbor, if near.	Twigs and small branches broken off trees; progress generally impeded.	High, 6-9	7
9	41-47	47-54	20.8-24.4	75-88	Strong gale	High waves; sea begins to roll; dense streaks of foam; spray may reduce visibility.		Slight structural damage occurs; slate blown from roofs.		
10	48-55	55-63	24.5-28.4	89-102	Storm	Very high waves with overhanging crests; sea takes white appearance as foam is blown in very dense streaks; rolling is heavy and visibility reduced.		Seldom experienced on land; trees broken or uprooted; considerable structural damage occurs.	Very high, 9-14	8
11	56-63	64-72	28.5-32.6	103-117	Violent storm	Exceptionally high waves; sea covered with white foam patches; visibility still more reduced.		Very rarely experienced on land; usually accompanied by widespread damage.		
12	64 and over	73 and over	32.7 and over	118 and over	Hurricane	Air filled with foam; sea completely white with driving spray; visibility greatly reduced.			Phenomenal, over 14	9

Note: Since January 1, 1955, weather map symbols have been based upon wind speed in knots, at five-knot intervals, rather than upon Beaufort number.

The State of Sea Photographs

The photographs were taken by the Meteorological Branch personnel of the Canadian Ocean Weather Ships C.C.G.S. *St. Catharines* and C.C.G.S. *Stonetown* which occupy Ocean Weather Station "P" in the North Pacific Ocean at 50°N, 145°W, or approximately 1,000 miles west of Vancouver. The ships man the station for alternate periods of 6 weeks.

The photographs were taken between March, 1960 and May, 1961. In this period a total of 247 photographs were obtained. Zeiss Ikon (2¼×3¼) cameras and Kodak Verichrome Pan film were used.

Of the 247 pictures available, 2 pictures were chosen to illustrate conditions associated with each Beaufort force from force 1 to force 10. Only one picture was considered acceptable to illustrate force 0. No representative photographs were available to illustrate force 11 conditions, and no photographs were made of force 12 conditions.

In selecting the pictures for presentation here, it was considered that they should meet two requirements. Firstly, a picture illustrating the effects of a given wind force should conform as closely as possible to the Sea Criterion for that force. Secondly, the wind prior to the time of the picture should be relatively steady both in direction and at the given force over many hours to ensure that near steady-state sea conditions for that force at the time of the picture existed. A large percentage of the photographs available were rejected because the wind at the time of the picture had not been blowing long enough to produce a disturbance of near steady-state proportion.

The pictures which follow were judged to fulfill best the requirements stated above. Opposite each picture is the accompanying technical and other data appropriate to each. In addition to the wind at the time of the photograph, the wind at 3-hourly intervals over the previous 24 hours is also included. The height of the ships' anemometers above the sea surface was approximately 60 feet. The synoptic weather situation at the time of the picture is described briefly, and other comments are given when warranted.

BEAUFORT FORCE 0

Wind speed less than 1 knot

Sea Criterion: Sea like a mirror.

Date/Time of photograph: June 5, 1960, 2340 GMT.

Height of Camera above sea: 35 Feet.

Waves at time of picture

	Direction (°T)	Period (sec.)	Height (ft.)
Sea Waves	—	—	—
Swell	100°	5	2

Time (GMT)	Direction (°T)	Wind speed (kn)
2340	—	calm
2100	190	01
1800	080	03
1500	040	07
1200	040	09
0900	030	10
0600	030	11
0300	030	11
0000	040	10

Synoptic Situation

High pressure area centered at 49°N, 144°W, or approximately 50 nautical miles southeast of ship at time of picture. Center of high moving southeastward.

Remarks

Calm winds, and hence calm sea conditions, are rare at Station "P," occurring only about 1.5 percent of the time during the course of a year. Although the wind was essentially calm during the 3 hours prior to this photograph, it averaged 4 knots (force 2) over the previous 12 hours, and 7 knots (force 3) in the previous 24 hours. Air temperature 52°2F, sea temperature 45°7F.

Crown Copyright, Ottawa, 1975. Reprinted through permission of the Atmospheric Environment Service, Department of the Environment, Canada.

Crown Copyright, Ottawa, 1975. Reprinted through permission of the Atmospheric Environment Service, Department of the Environment, Canada.

BEAUFORT FORCE 0   SEA STATE 0

BEAUFORT FORCE 1

Wind Speed 1 to 3 knots, mean 2 knots

Sea Criterion: Ripples with the appearance of scales are formed, but without foam crests.

Date/Time of Photograph: May 22, 1960, 2000 GMT.

Height of camera above sea: 35 feet.

Waves at time of picture

	Direction (°T)	Period (sec.)	Height (ft.)
Sea Waves	—	—	—
Swell	290	10	3

Time (GMT)	Direction (°T)	Wind speed (kn)
2000	290	02
1800	—	calm
1500	220	04
1200	260	08
0900	280	02
0600	—	calm
0300	290	10
0000	290	12
2100	290	13

Synoptic situation

The station was near the center of a narrow ridge of high pressure which extended northward into the Gulf of Alaska from a high centered near 30°N, 147°W. A low pressure area, elongated in a north-south direction, lay just off the coast of Canada and the United States. Another low was centered in the eastern Bering Sea.

Remarks

Winds averaged less than 3 knots (force 1) over the previous 14 hours, and less than 6 knots (force 2) over the previous 24 hours. Air temperature 44°F, sea temperature 44°F.

Crown Copyright, Ottawa, 1975. Reprinted through permission of the Atmospheric Environment Service, Department of the Environment, Canada.

Crown Copyright, Ottawa, 1975. Reprinted through permission of the Atmospheric Environment Service, Department of the Environment, Canada.

BEAUFORT FORCE 1   SEA STATE 0

BEAUFORT FORCE 2

Wind speed 4 to 6 knots, mean 5 knots

Sea criterion: Small wavelets, still short but more pronounced—crests have a glassy appearance and do not break.

Date/Time of photograph: May 26, 1961, 1700 GMT.

Height of camera above sea: 45 ft.

Time (GMT)	Direction (°T)	Wind speed (kn)
1700	120	05
1500	100	03
1200	—	calm
0900	340	06
0600	010	07
0300	340	03
0000	320	04
2100	310	04
1800	310	03

Waves at time of picture

	Direction (°T)	Period (sec.)	Height (ft.)
Sea waves	120	—	—
Swell	050	6	1

Synoptic situation

A weak trough of low pressure, lying in a general northwest-southeast direction, was immediately west of the ship and passed by during the next hour. The centerline of the following ridge of high pressure was approximately 400 nautical miles west of the ship.

Remarks

This picture, taken from the C.C.G.S. *ST. CATHARINES*, shows her sister ship the C.C.G.S. *STONETOWN*, and was taken at 49.8N, 142.5W, or approximately 100 nautical miles east of the Station "P" position. The wind speed averaged close to 4 knots (force 2) during the previous 24 hours. Air temperature 46.5F; sea temperature 45.0F.

Crown Copyright, Ottawa, 1975. Reprinted through permission of the Atmospheric Environment Service, Department of the Environment, Canada.

Crown Copyright, Ottawa, 1975. Reprinted through permission of the Atmospheric Environment Service, Department of the Environment, Canada.

BEAUFORT FORCE 2   SEA STATE 1

BEAUFORT FORCE 3

Wind speed 7 to 10 knots, mean 9 knots

Sea criterion: Large wavelets. Crests begin to break. Foam of a glassy appearance. Perhaps scattered white caps.

Date/Time of photograph: Feb. 19, 1961, 2000 GMT.

Height of camera above sea: 45 ft.

Waves at time of picture

	Direction (°T)	Period (sec.)	Height (ft.)
Sea waves	—	—	—
Swell	180	7	8

Synoptic situation

Small low pressure area centered approximately 170 nautical miles west of station. The ship was approximately 40 nautical miles north of a warm front that extended eastward from the low center.

Remarks

Although the wind speed at the time of this picture was only 3 knots (force 1), it had decreased from 9 knots 2 hours previously, and had averaged almost 10 knots over the previous 24 hours. The direction had been constant from the south for 8 hours. Air temperature 44°2F; sea temperature 40°8F.

Time (GMT)	Direction (°T)	Wind speed (kn)
2000	190	03
1800	160	09
1500	180	10
1200	190	11
0900	260	08
0600	300	10
0300	320	10
0000	010	16
2100	090	10

Crown Copyright, Ottawa, 1975. Reprinted through permission of the Atmospheric Environment Service, Department of the Environment, Canada.

Crown Copyright, Ottawa, 1975. Reprinted through permission of the Atmospheric Environment Service, Department of the Environment, Canada.

BEAUFORT FORCE 3   **SEA STATE 2**

BEAUFORT FORCE 4

Wind speed 11 to 16 knots, mean 13 knots

Sea criterion: Small waves, becoming longer, fairly frequent white caps.

Date/Time of photograph: July 3, 1960, 2240 GMT.

Height of camera above sea: 35 ft.

Time (GMT)	Direction (°T)	Wind speed (kn)
2240	320	16
2100	320	16
1800	340	11
1500	270	05
1200	270	08
0900	140	08
0600	230	09
0300	260	09
0000	250	13

NOTE:—Between 0000 and 2240 GMT, the ship had traveled eastward from 49°3N, 133°7W to 48°8N, 128°2W, a distance of approximately 220 nautical miles.

Waves at time of picture

	Direction (°T)	Period (sec.)	Height (ft.)
Sea waves	310	5	3
Swell	—	—	—

Synoptic situation

High pressure area centered at 40°N, 137°W, with ridge line extending north-eastward to 100 nautical miles west of the ship, then northward to the northern section of the British Columbia coast. Low pressure area centered at 55°N, 150°W, in the Gulf of Alaska.

Remarks

This picture was made at 48°8N, 128°2W, or approximately 200 nautical miles west of Vancouver while the ship was returning from Station "P." The ship was moving into the area of northwesterly circulation east of the ridge of high pressure, and had passed the ridge line approximately 11 hours before the time of this picture. It is likely that force 4 winds had been acting on the area longer than would appear from the wind data, as the ship was moving eastward into the area. Air temperature 55°6F; sea temperature 52°7F.

Crown Copyright, Ottawa, 1975. Reprinted through permission of the Atmospheric Environment Service, Department of the Environment, Canada.

Crown Copyright, Ottawa, 1975. Reprinted through permission of the Atmospheric Environment Service, Department of the Environment, Canada.

BEAUFORT FORCE 4   SEA STATE 3

BEAUFORT FORCE 5

Wind speed 17 to 21 knots, mean 19 knots

Sea criterion: Moderate waves taking a more pronounced long form; many white caps are formed. (Chance of some spray.)

Waves at time of picture

	Direction (°T)	Period (sec.)	Height (ft.)
Sea waves	280	6	7
Swell	240	8	6

Date/Time of photograph: Apr. 7, 1961, 2315 GMT.

Height of camera above sea: 35 ft.

Time (GMT)	Direction (°T)	Wind speed (kn)
2315	280	23
2100	270	19
1800	240	20
1500	220	21
1200	220	24
0900	200	26
0600	190	25
0300	180	27
0000	180	30

Synoptic situation

High pressure area centered 550 nautical miles southeast of the station, which was in a northwesterly circulation 200 nautical miles northeast of the ridge line extending northwestward from the high. Low pressure area centered near 56°N, 136°W, in the Gulf of Alaska.

Remarks

The wind speed at the time of the picture was 23 knots (force 6), but had averaged less than 21 knots (force 5) over the previous 8 hours. In this period, the wind veered from southwest to west. Air temperature 41°0F; sea temperature 41°2F.

Crown Copyright, Ottawa, 1975. Reprinted through permission of the Atmospheric Environment Service, Department of the Environment, Canada.

Crown Copyright, Ottawa, 1975. Reprinted through permission of the Atmospheric Environment Service, Department of the Environment, Canada.

BEAUFORT FORCE 5   SEA STATE 4

BEAUFORT FORCE 6

Wind speed 22 to 27 knots, mean 24 knots

Sea criterion: Large waves begin to form; the white foam crests are more extensive everywhere. (Probably some spray.)

Date/Time of photograph: Feb. 10, 1961, 2115 GMT.

Height of camera above sea: 20 ft.

Time (GMT)	Direction (°T)	Wind speed (kn)
2115	280	25
1800	300	26
1500	300	14
1200	280	23
0900	280	26
0600	310	23
0300	300	26
0000	280	27

Waves at time of picture

	Direction (°T)	Period (sec.)	Height (ft.)
Sea waves	280	6	11
Swell	—	—	—

Synoptic situation

Station in a northwesterly flow south of a low pressure area centered at 56°N, 144°W, or 380 nautical miles north of the ship. Trough of low pressure extending southeastward from low to second low lying off west coast of the United States. Ridge of pressure 300 nautical miles west of the ship.

Remarks

Except for a period between 3 and 9 hours ago, the wind was fairly constant, in both speed and duration, over the previous 24 hours. The fetch of the wind at the time of the picture was at least 300 nautical miles. Air temperature 37.°4F; sea temperature 41.°0F.

Crown Copyright, Ottawa, 1975. Reprinted through permission of the Atmospheric Environment Service, Department of the Environment, Canada.

Crown Copyright, Ottawa, 1975. Reprinted through permission of the Atmospheric Environment Service, Department of the Environment, Canada.

BEAUFORT FORCE 6   SEA STATE 5

BEAUFORT FORCE 7

Wind speed 28 to 33 knots, mean 30 knots

Sea criterion: Sea heaps up and white foam from breaking waves begins to be blown in streaks along the direction of the wind.

Date/Time of photograph: Feb 28, 1961, 1900 GMT.

Height of camera above sea: 45 ft.

Time (GMT)	Direction (°T)	Wind speed (kn)
1900	300	30
1800	300	30
1500	260	35
1200	260	35
0900	270	30
0600	250	38
0300	220	39
0000	210	39
2100	180	30

Waves at time of picture

	Direction (°T)	Period (sec.)	Height (ft.)
Sea waves	300	6	13
Swell	250	9	10

Synoptic situation

A deep low pressure area centered at 60°N, 144°W dominated the northeastern Pacific east of 160°W and north of 45°N. Station in northwesterly circulation approximately 575 nautical miles south of the low center.

Remarks

Although the wind speed was force 7 at the time of the picture, it had decreased from low force 8 between 1 and 4 hours previously. The wind veered from south to northwest during the previous 24 hours as the low center passed west of the ship while moving northward into the Gulf of Alaska. The fetch of the wind at the time of the picture was approximately 350 nautical miles. Air temperature 38.4F; sea temperature 40.7F.

Crown Copyright, Ottawa, 1975. Reprinted through permission of the Atmospheric Environment Service, Department of the Environment, Canada.

Crown Copyright, Ottawa, 1975. Reprinted through permission of the Atmospheric Environment Service, Department of the Environment, Canada.

BEAUFORT FORCE 7   SEA STATE 6

BEAUFORT FORCE 8

Wind speed 34 to 40 knots, mean 37 knots

Sea criterion: Moderately high waves of greater length; edges of crests begin to break into spindrift. The foam is blown in well-marked streaks along the direction of the wind.

Date/Time of photograph: Jan. 15, 1961, 1955 GMT.

Height of camera above sea: 35 ft.

Time (GMT)	Direction (°T)	Wind speed (kn)
1955	250	45
1800	240	38
1500	240	38
1200	250	37
0900	250	35
0600	250	35
0300	250	43
0000	250	40
2100	240	30

Waves at time of picture

	Direction (°T)	Period (sec.)	Height (ft.)
Sea waves	260	7	18
Swell	—	—	—

Synoptic situation

Low pressure area centered at 58°N, 142°W, in the Gulf of Alaska. Station in southwesterly circulation 500 nautical miles south-southwest of the low center.

Remarks

Although the wind speed at the time of the picture was 45 knots (force 9), it had been force 8 for 12 hours previously, and had increased in speed less than 2 hours prior to the picture. The wind direction was remarkably constant at 240° to 250° over the previous 24 hours. Air temperature 39.0F.; sea temperature 41.2F.

Crown Copyright, Ottawa, 1975. Reprinted through permission of the Atmospheric Environment Service, Department of the Environment, Canada.

Crown Copyright, Ottawa, 1975. Reprinted through permission of the Atmospheric Environment Service, Department of the Environment, Canada.

BEAUFORT FORCE 8   SEA STATE 6

BEAUFORT FORCE 9

Wind speed 41 to 47 knots, mean 44 knots

Sea criterion: High waves. Dense streaks of foam along the direction of the wind. Crests of waves begin to topple, tumble, and roll over. Spray may affect visibility.

Date/Time of photograph: Jan. 17, 1961, 2130 GMT.

Height of camera above sea: 35 ft.

Time (GMT)	Direction (°T)	Wind speed (kn)
2130	120	50
2100	120	48
1800	120	46
1500	110	44
1200	110	30
0900	120	17
0600	140	08
0300	—	calm
0000	250	10

Waves at time of picture

	Direction (°T)	Period (sec.)	Height (ft.)
Sea waves	120	7	20
Swell	—	—	—

Synoptic situation

Station in strong southeasterly circulation, approximately 600 nautical miles east-northeast of a deep low, centered about 48°N, 160°W.

Remarks

Although the wind speed was 50 knots (force 10) at the time of the picture, the average speed over the previous 9½ hours was force 9. The wind had increased from calm conditions 18 hours ago. The appearance of the sea surface conforms well to the force 9 specification, but the reported wave heights are only 50 percent of the maximum theoretical height for force 9. It would be necessary for the wind to blow another 15 to 20 hours at force 9 before the waves approached their maximum height. Air temperature 42°.1F; sea temperature 41°.6F.

Crown Copyright, Ottawa, 1975. Reprinted through permission of the Atmospheric Environment Service, Department of the Environment, Canada.

Crown Copyright, Ottawa, 1975. Reprinted through permission of the Atmospheric Environment Service, Department of the Environment, Canada.

BEAUFORT FORCE 9   SEA STATE 6

BEAUFORT FORCE 10

Wind speed 48 to 55 knots, mean 52 knots

Sea criterion: Very high waves with long overhanging crests. The resulting foam, in great patches, is blown in dense white streaks along the direction of the wind. On the whole, the surface of the sea takes on a white appearance. The tumbling of the sea becomes heavy and shock-like. Visibility affected.

Date/Time of photograph: Mar. 14, 1961, 2330 GMT.

Height of camera above sea: 15 ft.

Time (GMT)	Direction (°T)	Wind speed (kn)
2330	340	55
2100	350	52
1800	000	55
1500	010	52
1200	010	44
0900	030	43
0600	030	29
0300	040	17
0000	050	12

Waves at time of picture

	Direction (°T)	Period (sec.)	Height (ft.)
Sea waves	340	9	22
Swell	—	—	—

Synoptic situation

Deep low pressure area centered at 49°N, 138°W, moving northeastward. Station in strong northerly circulation 250 nautical miles west-northwest of the low center.

Remarks

The wind had been blowing at force 10 for the past 8½ hours, the direction backing from 010° to 340°. Because of the relatively short duration of the force 10 winds, the wave heights have only attained roughly 50 percent of their maximum height for this force. The fetch of the wind at the time of the picture was estimated to lie between 150 and 200 nautical miles. Air temperature 37°6F.; sea temperature 40°0F.

Crown Copyright, Ottawa, 1975. Reprinted through permission of the Atmospheric Environment Service, Department of the Environment, Canada.

Crown Copyright, Ottawa, 1975. Reprinted through permission of the Atmospheric Environment Service, Department of the Environment, Canada.

BEAUFORT FORCE 10   **SEA STATE 7**

APPENDIX C

TIDE TABLES

Tide tables for the use of mariners have been published by the National Ocean Survey (formerly the Coast and Geodetic Survey) since 1853. For a number of years these tables appeared as appendixes to the annual reports of the Superintendent of the Survey, and consisted of more or less elaborated means for enabling the mariner to make his own prediction of tides as occasion arose.

The first tables to give predictions for each day were those for the year 1867. They gave the times and heights of high waters only and were published in two separate parts, one for the Atlantic coast and the other for the Pacific coast of the United States. Together they contained daily predictions for 19 stations and tidal differences for 124 stations. A few years later predictions for the low waters were also included, and for the year 1896 the tables were extended to include the entire maritime world, with full predictions for 70 ports and tidal differences for about 3,000 stations.

The tide tables are now issued in four volumes, as follows: *Europe and West Coast of Africa (including the Mediterranean Sea); East Coast of North and South America (including Greenland); West Coast of North and South America (including the Hawaiian Islands); Central and Western Pacific Ocean and Indian Ocean.* Together, they contain daily predictions for 196 reference ports and differences and other constants for about 6,000 stations. This edition of the *Tide Tables, East Coast of North and South America* contains full daily predictions for 48 reference ports and differences and other constants for about 2,000 stations in North and South America. It contains also a table for obtaining the approximate height of the tide at any time, a table of local mean time of sunrise and sunset for every 5th day of the year for different latitudes, a table for the reduction of local mean time to standard time, a table of moonrise and moonset for 8 places, and a table of the Greenwich mean time of the moon's phases, apogee, perigee, greatest north and south and zero declination, and the time of the solar equinoxes and solstices.

Up to and including the tide tables for the year 1884, all the tide predictions were computed by means of auxiliary tables and curves constructed from the results of tide observations at the different ports. From 1885 to 1911, inclusive, the predictions were generally made by means of the Ferrel tide-predicting machine. From 1912 to 1965, inclusive, they were made by means of the Coast and Geodetic Survey tide predicting machine No. 2. Since 1966, predictions have been made by electronic computer.

In accordance with cooperative arrangements for the exchange of tide predictions, the authorities given below have furnished the predictions for the following stations in the present issue:

Canadian Hydrographic Service. Harrington, Quebec, Halifax, St. John, Pictou, and Argentia.

Diretoria de Hidrografia e Navegacao, Brazil.—Recife, Rio de Janeiro, and Santos.

Servicio Hidrografico, Argentina.—Buenos Aires, Puerto Belgrano, Comodoro Rivadavia, and Punta Loyola.

In the preparation of these tables all available observations were used. In some cases, however, the observations were insufficient for obtaining final results, and as further information becomes available it will be included in subsequent editions. All persons using these tables are invited to send information or suggestions for increasing their usefulness to the Director, National Ocean Survey, Rockville, Md. 20852, U.S.A.

Reprinted from Tide Tables, east coast of North and South America. Available from National Oceanographic Survey.

TABLE 1.—DAILY TIDE PREDICTIONS

EXPLANATION OF TABLE

This table contains the predicted times and heights of the high and low waters for each day of the year at a number of places, which are designated as *reference stations*. By the application of tidal differences given in table 2 to the predictions for the reference stations, the mariner is enabled to obtain the approximate times and heights of the tide at many other places. High water is the maximum height reached by each rising tide and low water the minimum height reached by each falling tide. High and low waters can be selected from the predictions by comparison of consecutive heights. Because of diurnal inequality, however, at certain places there may be a difference of only a few tenths of a foot between one high water and low water of a day but a marked difference in height between the other high water and low water. It is essential, therefore, in using tide tables to carefully note the heights as well as the times of the tide.

Time.—The kind of time used for the predictions at each reference station is indicated by the time meridian at the bottom of each page.

Datum.—The datum from which the predicted heights are reckoned is the same as that used for the charts of the locality. The datum for the Atlantic coast of the United States is mean low water. For foreign coasts a datum approximating to mean low water springs, Indian spring low water, or the lowest possible low water is generally used. The depression of the datum below mean sea level for each of the reference stations of this volume is given on the preceding page.

Depth of water.—The nautical charts published by the United States and other maritime nations show the depth of water as referred to a low water datum corresponding to that from which the predicted tidal heights are reckoned. To find the actual depth of water at any time the height of the tide should be added to the charted depth. If the height of the tide is negative—that is, if there is a minus sign (−) before the tabular height—it should be subtracted from the charted depth. For any time between high and low water, the height of the tide may be estimated from the heights of the preceding and following tides, or table 3 may be used. On some foreign charts the depths are given in meters, and in such cases the heights of the tide may be reduced to meters by multiplying by 0.3 before being applied to the charted depths.

Variation in sea level.—Changes in winds and barometric conditions cause variations in sea level from day to day. In general, with onshore winds or a low barometer the heights of both the high and low waters will be higher than predicted, while with offshore winds or a high barometer they will be lower. There are also seasonal variations in sea level, but these variations have been included in the predictions for each station. At ocean stations the seasonal variation in sea level is usually less than half a foot.

At stations situated on tidal rivers the average seasonal variation in river level due to freshets and droughts may be considerably more than a foot. The predictions for these stations include an allowance for this seasonal variation representing average freshet and drought conditions. Unusual freshets or droughts, however, will cause the tides to be higher or lower, respectively, than predicted.

Number of tides.—There are usually two high and two low waters in a day. Tides follow the moon more closely than they do the sun, and the lunar or tidal day is about 50 minutes longer than the solar day. This causes the tide to occur later each day, and a tide that has occurred near the end of one calendar day will be followed by a corresponding tide that may skip the next day and occur in the early morning of the third day. Thus on certain days of each month only a single high or a single low water occurs. At some stations, during portions of each month, the tide becomes diurnal; that is, only one high and one low water occur during the period of a lunar day.

Relation of current to tide.—In using these tables of tide predictions it must be borne in mind that they give the times and heights of high and low waters and *not* the times of turning of the current or slack water. For stations on the outer coast there is usually but little difference between the time of high or low water and the beginning of ebb or flood current, but for places in narrow channels, landlocked harbors, or on tidal rivers, the time of slack water may differ by several hours from the time of high or low water stand. The relation of the times of high and low water to the turning of the current depends upon a number of factors, so that no simple or general rule can be given. For the predicted times of slack water reference should be made to the tidal current tables published by the National Ocean Survey in two separate volumes, one for the Atlantic coast of North America and the other for the Pacific coast of North America and Asia.

Typical tide curves.—The variations in the tide from day to day and from place to place are illustrated on the opposite page by the tide curves for representative ports along the Atlantic and Gulf coasts of the United States. It will be noted that the range of tide for stations along the Atlantic coast varies from place to place but that the type is uniformly semidiurnal with the principal variations following the changes in the moon's distance and phase. In the Gulf of Mexico, however, the range of tide is uniformly small but the type of tide differs considerably. At certain ports such as Pensacola there is usually but one high and one low water a day while at other ports such as Galveston the inequality is such that the tide is semidiurnal around the times the moon is on the Equator but becomes diurnal around the times of maximum north or south declination of the moon. In the Gulf of Mexico, consequently, the principal variations in the tide are due to the changing declination of the moon. Key West, at the entrance to the Gulf of Mexico, has a type of tide which is a mixture of semi-daily and daily types. Here the tide is semidiurnal but there is considerable inequality in the heights of high and low waters. By reference to the curves it will be seen that where the inequality is large there are times when there is but a few tenths of a foot difference between high water and low water.

TYPICAL TIDE CURVES FOR UNITED STATES PORTS

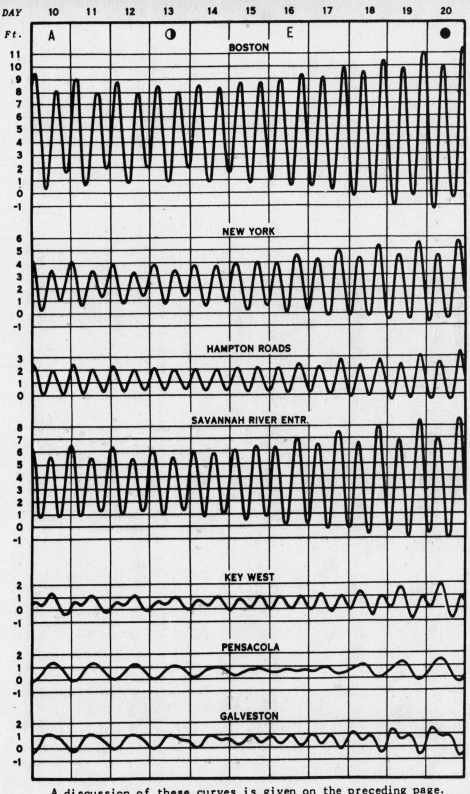

A discussion of these curves is given on the preceding page.

Lunar data:
A – moon in apogee
☾ – last quarter
E – moon on equator
● – new moon

EXTRACTS FROM TIDE TABLES

NEW YORK (THE BATTERY), N.Y., 1975

TIMES AND HEIGHTS OF HIGH AND LOW WATERS

JANUARY

DAY	TIME H.M.	HT. FT.	DAY	TIME H.M.	HT. FT.
1 W	0422	-0.8	16 TH	0431	0.0
	1043	5.1		1045	4.1
	1659	-1.1		1656	-0.2
	2321	4.6		2319	3.8
2 TH	0516	-0.6	17 F	0501	0.2
	1138	4.9		1120	3.9
	1749	-0.9		1722	0.0
				2357	3.7
3 F	0017	4.6	18 SA	0533	0.4
	0615	-0.4		1154	3.7
	1234	4.6		1746	0.2
	1847	-0.6			
4 SA	0111	4.6	19 SU	0029	3.7
	0724	-0.1		0613	0.6
	1329	4.3		1228	3.5
	1951	-0.4		1818	0.3
5 SU	0207	4.6	20 M	0108	3.8
	0835	0.0		0727	0.7
	1428	4.0		1311	3.4
	2053	-0.3		1917	0.4
6 M	0307	4.5	21 TU	0156	3.9
	0939	-0.1		0856	0.7
	1530	3.7		1404	3.3
	2153	-0.3		2050	0.4
7 TU	0407	4.5	22 W	0255	4.0
	1038	-0.2		1001	0.4
	1635	3.6		1515	3.2
	2247	-0.3		2157	0.2
8 W	0508	4.6	23 TH	0404	4.2
	1131	-0.3		1057	0.1
	1735	3.7		1637	3.4
	2339	-0.3		2256	0.0
9 TH	0603	4.7	24 F	0510	4.5
	1222	-0.4		1150	-0.2
	1828	3.8		1742	3.7
				2352	-0.3
10 F	0028	-0.4	25 SA	0609	4.9
	0649	4.8		1242	-0.6
	1310	-0.5		1838	4.0
	1915	3.9			
11 SA	0117	-0.4	26 SU	0048	-0.6
	0733	4.8		0701	5.2
	1354	-0.6		1330	-1.0
	1959	3.9		1929	4.4
12 SU	0202	-0.4	27 M	0140	-0.9
	0814	4.8		0749	5.4
	1436	-0.6		1419	-1.3
	2040	3.9		2019	4.7
13 M	0242	-0.4	28 TU	0231	-1.1
	0854	4.7		0840	5.5
	1516	-0.6		1505	-1.4
	2122	3.9		2110	5.0
14 TU	0321	-0.3	29 W	0321	-1.2
	0931	4.5		0931	5.4
	1552	-0.5		1550	-1.5
	2202	3.9		2203	5.0
15 W	0358	-0.1	30 TH	0410	-1.2
	1011	4.3		1024	5.2
	1625	-0.4		1635	-1.3
	2242	3.8		2258	5.0
			31 F	0459	-0.9
				1120	4.9
				1725	-1.0
				2354	4.9

FEBRUARY

DAY	TIME H.M.	HT. FT.	DAY	TIME H.M.	HT. FT.
1 SA	0554	-0.6	16 SU	0503	0.2
	1213	4.5		1111	3.8
	1818	-0.7		1704	0.1
				2338	4.0
2 SU	0047	4.8	17 M	0535	0.3
	0659	-0.2		1149	3.6
	1309	4.1		1733	0.2
	1919	-0.3			
3 M	0142	4.6	18 TU	0020	4.0
	0808	0.0		0621	0.5
	1405	3.8		1231	3.5
	2024	-0.1		1818	0.4
4 TU	0240	4.4	19 W	0108	4.1
	0915	0.1		0759	0.6
	1508	3.6		1329	3.4
	2129	0.0		1932	0.5
5 W	0341	4.3	20 TH	0211	4.1
	1015	0.0		0926	0.5
	1612	3.5		1439	3.3
	2225	0.0		2125	0.4
6 TH	0444	4.3	21 F	0327	4.2
	1110	-0.1		1026	0.2
	1715	3.5		1607	3.5
	2320	0.0		2233	0.1
7 F	0542	4.4	22 SA	0444	4.5
	1200	-0.2		1123	-0.2
	1811	3.7		1718	3.9
				2334	-0.3
8 SA	0009	-0.1	23 SU	0545	4.8
	0629	4.5		1214	-0.6
	1246	-0.3		1816	4.4
	1856	3.9			
9 SU	0057	-0.2	24 M	0030	-0.7
	0713	4.6		0640	5.2
	1329	-0.5		1306	-1.0
	1937	4.0		1909	4.8
10 M	0141	-0.3	25 TU	0123	-1.0
	0752	4.6		0731	5.4
	1410	-0.5		1354	-1.2
	2015	4.1		1959	5.2
11 TU	0222	-0.4	26 W	0215	-1.3
	0829	4.6		0820	5.4
	1449	-0.6		1441	-1.4
	2053	4.2		2048	5.4
12 W	0300	-0.4	27 TH	0305	-1.4
	0906	4.5		0912	5.3
	1523	-0.5		1526	-1.4
	2129	4.2		2139	5.4
13 TH	0335	-0.3	28 F	0352	-1.3
	0941	4.3		1003	5.1
	1555	-0.4		1612	-1.2
	2203	4.1		2232	5.3
14 F	0407	-0.2			
	1011	4.1			
	1622	-0.2			
	2235	4.1			
15 SA	0435	0.0			
	1043	3.9			
	1643	-0.1			
	2307	4.0			

MARCH

DAY	TIME H.M.	HT. FT.	DAY	TIME H.M.	HT. FT.
1 SA	0442	-1.0	16 SU	0416	-0.1
	1058	4.8		1013	4.0
	1658	-0.9		1613	0.0
	2325	5.1		2223	4.4
2 SU	0533	-0.6	17 M	0443	0.0
	1152	4.4		1043	3.8
	1749	-0.5		1637	0.2
				2259	4.4
3 M	0021	4.9	18 TU	0517	0.2
	0631	-0.2		1124	3.7
	1247	4.1		1708	0.3
	1847	0.0		2346	4.3
4 TU	0115	4.6	19 W	0602	0.4
	0739	0.1		1218	3.6
	1344	3.8		1752	0.5
	1956	0.3			
5 W	0212	4.3	20 TH	0042	4.3
	0848	0.3		0724	0.5
	1443	3.6		1318	3.6
	2103	0.4		1906	0.6
6 TH	0312	4.1	21 F	0146	4.3
	0949	0.3		0853	0.4
	1545	3.5		1429	3.6
	2204	0.4		2106	0.5
7 F	0415	4.1	22 SA	0300	4.3
	1041	0.2		0959	0.2
	1649	3.6		1548	3.9
	2258	0.3		2217	0.2
8 SA	0513	4.2	23 SU	0417	4.5
	1131	0.0		1056	-0.2
	1744	3.8		1658	4.3
	2347	0.1		2315	-0.2
9 SU	0603	4.3	24 M	0524	4.8
	1216	-0.1		1147	-0.5
	1830	4.1		1755	4.8
10 M	0033	0.0	25 TU	0011	-0.6
	0645	4.4		0620	5.0
	1259	-0.3		1238	-0.8
	1909	4.3		1848	5.2
11 TU	0117	-0.2	26 W	0107	-1.0
	0726	4.5		0712	5.2
	1341	-0.4		1328	-1.1
	1946	4.4		1938	5.6
12 W	0158	-0.3	27 TH	0157	-1.2
	0803	4.5		0802	5.3
	1418	-0.4		1416	-1.2
	2022	4.5		2025	5.7
13 TH	0237	-0.4	28 F	0247	-1.3
	0837	4.5		0852	5.2
	1453	-0.4		1502	-1.1
	2054	4.5		2114	5.7
14 F	0313	-0.4	29 SA	0334	-1.2
	0912	4.3		0942	4.9
	1525	-0.3		1547	-0.9
	2126	4.5		2206	5.5
15 SA	0346	-0.3	30 SU	0422	-0.9
	0941	4.2		1037	4.7
	1551	-0.1		1632	-0.6
	2152	4.5		2259	5.3
			31 M	0512	-0.6
				1132	4.3
				1721	-0.2
				2352	4.9

TIME MERIDIAN 75° W. 0000 IS MIDNIGHT. 1200 IS NOON.
HEIGHTS ARE RECKONED FROM THE DATUM OF SOUNDINGS ON CHARTS OF THE LOCALITY WHICH IS MEAN LOW WATER.

TABLE 2.—TIDAL DIFFERENCES AND OTHER CONSTANTS
EXPLANATION OF TABLE

The publication of full daily predictions is necessarily limited to a comparatively small number of stations. Tide predictions for many other places, however, can be obtained by applying certain differences to the predictions in table 1. The following pages give the places, called "subordinate stations," for which such predictions can be made and the differences and ratios to be used. These differences and ratios are to be applied to that station in table 1 which in table 2 is listed in bold face type above the differences for the subordinate station. The stations in this table are arranged in geographic order. The index at the end of the volume will assist in locating any station.

The data given in this table and their use are explained below.

Time difference.—To determine the time of high water or low water at any station listed in this table, the column headed "Differences, Time" gives the hours and minutes to be added to or subtracted from the time of high or low water at some reference station. A plus (+) sign indicates that the tide at the subordinate station is later than at the reference station and the difference should be added, a minus (−) sign that it is earlier and should be subtracted.

The results obtained by the application of the time differences will be in the kind of time indicated by the time meridian shown above the name of the subordinate station. Summer or daylight saving time is not used in the tide tables.

To obtain the tide at a subordinate station on any date apply the difference to the tide at the reference station for that same date, except that, in some cases, to obtain an a.m. tide it may be necessary to use the preceding day's p.m. tide at the reference station or to obtain a p.m. tide it may be necessary to use the following day's a.m. tide. For example, if a high water occurs at a reference station at 2200 on July 2 and the tide at the subordinate station occurs 3 hours later, then high water will occur at 0100 on July 3 at the subordinate station. For the second case, if a high water at a reference station occurs at 0200 on July 17 and the tide at the subordinate station occurs 5 hours earlier, the high water at the subordinate station will occur at 2100 on July 16.

Height differences.—The height of the tide, referred to the datum of charts, is obtained by means of the height differences or ratios. A plus (+) sign indicates that the difference should be added to the height at the reference station, and a minus (−) sign that it should be subtracted.

Ratio.—For some stations height differences would give unsatisfactory predictions. In such cases they have been omitted and one or two ratios are given. Where two ratios are given, one in the "height of high water" column and one in the "height of low water" column, the high waters and low waters at the reference station should be multiplied by these respective ratios. Where only one is given, the omitted ratio is either unreliable or unknown.

For certain stations there is given in parentheses a ratio to be applied to the heights of high and low waters at the reference station and a correction in feet to be applied to that product. As an example, in table 2, opposite Hare Bay, No. 245, the values in the height difference columns (*0.67+0.8) referred to Argentia. On a given morning the heights of high and low waters at Argentia are 6.7 and −0.3 feet, respectively. The predicted heights for Hare Bay are obtained as follows:

Argentia	6.7 ft. high water	−0.3 ft. low water
	×0.67 (ratio)	×0.67 (ratio)
	4.5	−0.2
	+0.8 (correction)	+0.8 (correction)
Hare Bay	5.3 ft. high water	+0.6 ft. low water

Range.—The *mean range* is the difference in height between mean high water and mean low water. The *spring range* is the average semidiurnal range occurring semimonthly as the result of the moon being new or full. It is larger than the mean range where the type of tide is either semidiurnal or mixed, and is of no practical significance where the type of tide is diurnal. Where the tide is chiefly of the diurnal type the table gives the *diurnal range*, which is the difference in height between mean higher high water and mean lower low water.

Caution.—For stations where the tide is chiefly diurnal the time differences and the height differences and ratios are intended primarily for predicting the higher high and lower low waters. When the lower high water and the higher low water at the reference station are nearly the same height the corresponding tides often cannot be obtained satisfactorily by means of the tidal differences.

Datum.—The datum of the predictions obtained through the height differences or ratios is also the datum of the largest scale chart for the locality. To obtain the depth at the time of high or low water, the predicted height should be added to the depth on the chart unless such height is negative (−), when it should be subtracted. To find the height at times between high and low water see table 3. On some foreign charts the depths are given in meters, and in such cases the heights of the tide can be reduced to meters by multiplying by 0.3 before being applied to the charted depths. The chart datum for the east coasts of the United States and for a part of the West Indies is *mean low water*. For the rest of the area covered by these tables the datums generally used are approximately *mean low water*, *mean low water springs*, *Indian spring low water*, or the *lowest possible low water*.

Mean Tide Level (Half Tide Level) is a plane midway between mean low water and mean high water. Tabular values are reckoned from chart datum.

Note.—Dashes are entered in the place of data which are unreliable, unknown, or given in another part of this book.

TABLE 2.—TIDAL DIFFERENCES AND OTHER CONSTANTS

No.	PLACE	POSITION		DIFFERENCES				RANGES		Mean Tide Level
				Time		Height				
		Lat.	Long.	High water	Low water	High water	Low water	Mean	Spring	
		° ′	° ′	h. m.	h. m.	feet	feet	feet	feet	feet
	NEW YORK and NEW JERSEY — Continued **Hudson River‡**	N.	W.		on NEW YORK, p.56					
	Time meridian, 75°W.									
1513	Jersey City, Pa. RR. Ferry, N. J	40 43	74 02	+0 07	+0 07	−0.1	0.0	4.4	5.3	2.2
1515	New York, Desbrosses Street	40 43	74 01	+0 10	+0 10	−0.1	0.0	4.4	5.3	2.2
1517	New York, Chelsea Docks	40 45	74 01	+0 17	+0 16	−0.2	0.0	4.3	5.2	2.1
1519	Hoboken, Castle Point, N. J	40 45	74 01	+0 17	+0 16	−0.2	0.0	4.3	5.2	2.1
1521	Weehawken, Days Point, N. J	40 46	74 01	+0 24	+0 23	−0.3	0.0	4.2	5.0	2.1
1523	New York, Union Stock Yards	40 47	74 00	+0 27	+0 26	−0.3	0.0	4.2	5.0	2.1
1525	New York, 130th Street	40 49	73 58	+0 37	+0 35	−0.5	0.0	4.0	4.8	2.0
1527	George Washington Bridge	40 51	73 57	+0 46	+0 43	−0.6	0.0	3.9	4.6	1.9
1529	Spuyten Duyvil, West of RR. bridge	40 53	73 56	+0 58	+0 53	−0.7	0.0	3.8	4.5	1.9
1531	Yonkers	40 56	73 54	+1 09	+1 10	−0.8	0.0	3.7	4.4	1.8
1533	Dobbs Ferry	41 01	73 53	+1 29	+1 40	−1.1	0.0	3.4	4.0	1.7
1535	Tarrytown	41 05	73 52	+1 45	+1 54	−1.3	0.0	3.2	3.7	1.6
1537	Ossining	41 10	73 52	+1 53	+2 14	−1.4	0.0	3.1	3.6	1.5
1539	Haverstraw	41 12	73 58	+1 59	+2 25	−1.6	0.0	2.9	3.4	1.4
1541	Peekskill	41 17	73 56	+2 24	+3 00	−1.3	+0.3	2.9	3.4	1.7
1543	West Point	41 24	73 57	+3 16	+3 37	−1.5	+0.3	2.7	3.1	1.6
1545	Newburgh	41 30	74 00	+3 42	+4 00	−1.5	+0.2	2.8	3.2	1.6
1547	New Hamburg	41 35	73 57	+4 00	+4 25	−1.5	+0.1	2.9	3.3	1.5
1549	Poughkeepsie	41 42	73 57	+4 30	+4 43	−1.3	+0.1	3.1	3.5	1.6
1551	Hyde Park	41 47	73 57	+4 56	+5 09	−1.3	0.0	3.2	3.6	1.6
1553	Kingston Point	41 56	73 58	+5 16	+5 31	−0.9	−0.1	3.7	4.2	1.7
1555	Tivoli	42 04	73 56	+5 46	+6 01	−0.8	−0.2	3.9	4.4	1.7
1557	Catskill	42 13	73 51	+6 37	+6 55	−0.7	−0.3	4.1	4.6	1.7
1559	Hudson	42 15	73 48	+6 54	+7 09	−0.9	−0.4	4.0	4.4	1.6
				on ALBANY, p.60						
1561	Coxsackie	42 21	73 48	−1 01	−1 38	−0.5	+0.2	3.9	4.3	2.1
1563	New Baltimore	42 27	73 47	−0 34	−0 56	−0.1	+0.4	4.1	4.5	2.4
1565	Castleton-on-Hudson	42 32	73 46	−0 17	−0 29	−0.2	+0.1	4.3	4.7	2.2
1567	ALBANY	42 39	73 45	Daily predictions				4.6	5.0	2.5
1569	Troy	42 44	73 42	+0 08	+0 10	+0.1	0.0	4.7	5.1	2.3
	The Kills and Newark Bay			on NEW YORK, p.56						
	Kill Van Kull									
1571	Constable Hook	40 39	74 05	−0 34	−0 21	0.0	0.0	4.5	5.4	2.2
1573	New Brighton	40 39	74 05	−0 12	−0 18	0.0	0.0	4.5	5.4	2.2
1575	Port Richmond	40 38	74 08	−0 03	+0 05	0.0	0.0	4.5	5.4	2.2
1577	Bergen Point	40 39	74 08	+0 03	+0 03	+0.1	0.0	4.6	5.5	2.3
1579	Shooters Island	40 39	74 10	+0 06	+0 18	+0.1	0.0	4.6	5.5	2.3
1581	Port Newark Terminal	40 41	74 08	−0 01	+0 18	+0.6	0.0	5.1	6.1	2.5
1583	Newark, Passaic River	40 44	74 10	+0 22	+0 52	+0.6	0.0	5.1	6.1	2.5
1585	Passaic, Gregory Ave. bridge	40 51	74 07	+0 49	+1 57	+0.6	0.0	5.1	6.1	2.5
	Hackensack River									
1586	Kearny Point	40 44	74 06	+0 09	+0 33	+0.5	0.0	5.0	6.0	2.5
1587	Secaucus	40 48	74 04	+1 13	+1 09	+0.6	0.0	5.1	6.1	2.6
1588	Little Ferry	40 51	74 02	+1 22	+1 14	+0.8	0.0	5.3	6.4	2.7
1589	Hackensack	40 53	74 02	+1 33	+1 58	+0.8	0.0	5.3	6.4	2.6
				on SANDY HOOK, p.64						
	Arthur Kill									
1591	Elizabethport	40 39	74 11	+0 25	+0 39	+0.3	0.0	4.9	5.9	2.4
1593	Chelsea	40 36	74 12	+0 24	+0 35	+0.4	0.0	5.0	6.0	2.5
1595	Carteret	40 35	74 13	+0 23	+0 31	+0.5	0.0	5.1	6.2	2.6
1597	Rossville	40 33	74 13	+0 17	+0 25	+0.7	0.0	5.3	6.4	2.6
1599	Tottenville	40 31	74 15	+0 13	+0 13	+0.7	0.0	5.3	6.4	2.6
1601	Perth Amboy	40 30	74 16	+0 13	+0 19	+0.6	0.0	5.2	6.3	2.6

‡Values for the Hudson River above the George Washington Bridge are based upon averages for the six months May to October, when the fresh-water discharge is a minimum.

TABLE 2.—TIDAL DIFFERENCES AND OTHER CONSTANTS

No.	PLACE	POSITION		DIFFERENCES				RANGES		Mean Tide Level
				Time		Height				
		Lat.	Long.	High water	Low water	High water	Low water	Mean	Spring	
		° ' N.	° ' W.	h. m.	h. m.	feet	feet	feet	feet	feet
	LABRADOR — Continued			on HALIFAX, p.20						
	Time meridian, 52°30'W.									
158	Hebron, Hebron Fjord	58 12	62 38	-0 49	-1 05	-1.2	-0.7	3.9	4.7	3.2
159	Nain	56 33	61 41	-0 32	-0 54	+0.5	-0.3	5.2	6.5	4.2
161	Hopedale Harbor	55 27	60 13	-0 46	-1 09	-0.2	-0.1	4.3	5.6	4.0
163	Webeck Harbor	54 54	58 02	-1 07	-1 38	-1.1	-0.6	3.9	5.0	3.3
	Hamilton Inlet and Lake Melville									
165	Indian Harbor	54 27	57 12	-0 37	-1 33	-0.8	-0.7	4.3	5.7	3.4
167	Ticoralak Island	54 17	58 12	-0 35	-0 55	-0.7	-0.3	4.0	4.9	3.7
169	Rigolet	54 11	58 25	+0 02	-0 17	-1.7	-0.8	3.5	4.5	2.8
170	Goose Bay	53 21	60 24	+4 22	+4 24	(*0.27	+0.5)	1.2	1.7	1.6
171	Cartwright Harbor	53 42	57 02	-0 03	-0 34	-1.1	-0.4	3.7	4.9	3.4
173	Curlew Harbor	53 45	56 33	-0 07	-0 38	-1.4	-0.7	3.7	4.9	3.1
175	Comfort Bight	53 09	55 46	-0 32	-1 03	-1.7	-0.8	3.5	4.6	2.9
177	Square Island Harbor	52 44	55 49	-0 34	-1 05	-1.8	-0.9	3.5	4.7	2.8
179	Port Marnham	52 23	55 44	-0 43	-1 14	-2.5	-0.8	2.7	3.6	2.5
180	Battle Harbor	52 16	55 36	-1 03	-1 30	-1.9	-0.1	2.6	3.8	3.1
	Strait of Bell Isle			on HARRINGTON, p.12						
181	Chateau Bay	52 00	55 50	-3 08	-3 19	*0.69	*0.81	2.4	3.1	2.5
183	Red Bay	51 43	56 25	-2 00	-1 55	*0.56	*0.56	2.1	2.6	2.0
185	Forteau Bay	51 27	56 53	-0 26	-0 17	*0.78	*0.81	2.9	3.7	2.8
	NEWFOUNDLAND, East Coast			on HALIFAX, p.20						
201	Pistolet Bay	51 30	55 44	-0 14	-0 28	*0.48	*0.32	2.4	3.1	1.8
203	Arriege Bay	51 10	56 00	-0 34	-0 34	-2.4	-1.3	3.3	4.3	2.3
205	Gouffre Harbor	50 42	56 10	-0 49	-1 01	-1.8	-0.9	3.5	4.7	2.8
207	Sops Island, White Bay	49 50	56 46	-0 49	-1 24	*0.48	*0.32	2.4	3.4	1.8
209	Exploits Harbor, Notre Dame Bay	49 32	55 04	-0 34	-1 09	-2.9	-1.1	2.6	3.5	2.1
211	Fogo Harbor	49 43	54 16	-0 34	-0 42	-2.4	-1.1	3.1	4.2	2.4
213	Valleyfield	49 10	53 37	-0 46	-1 13	*0.46	*0.37	2.2	2.9	1.8
215	Port Union	48 30	53 05	-0 53	-1 15	*0.51	*0.53	2.2	3.0	2.1
217	Randomhead Harbor, Trinity Bay	48 06	53 34	-0 53	-1 05	*0.49	*0.37	2.4	3.2	1.9
219	Harbour Grace, Conception Bay	47 41	53 12	-0 28	-0 46	*0.52	*0.37	2.6	3.5	2.0
221	St. John's	47 34	52 42	-0 34	-0 46	*0.54	*0.42	2.6	3.5	2.1
	NEWFOUNDLAND, South Coast			on ARGENTIA, p.4						
223	Trepassey Harbor	46 43	53 23	-0 19	-0 11	-1.0	-0.3	4.2	5.6	3.5
225	St. Mary Harbor, St. Mary Bay	46 55	53 35	-0 14	-0 06	-1.0	-0.3	4.2	5.6	3.5
	Placentia Bay									
227	ARGENTIA	47 18	53 59	Daily predictions				4.9	6.3	4.2
229	Woody Island	47 47	54 10	+0 09	+0 09	-0.3	-0.1	4.7	6.0	4.0
231	Mortier Bay	47 10	55 09	+0 15	+0 26	-0.8	-0.6	4.7	6.0	3.5
233	Great St. Lawrence Harbor	46 55	55 22	+0 28	+0 55	-0.5	+0.5	3.9	5.0	4.2
	Time meridian, 60°W.									
235	St. Pierre Hbr., St. Pierre I	46 47	56 10	-0 09	+0 13	-0.6	+0.4	3.9	5.0	4.1
	Time meridian, 52°30'W.									
	Fortune Bay									
237	Grand le Pierre Harbor	47 40	54 47	+1 09	+1 09	-0.8	+0.4	3.7	4.8	4.0
239	Belloram	47 32	55 25	+0 57	+0 57	(*0.67	+1.0)	3.3	4.3	3.8
241	Ship Cove, Despair Bay	47 52	55 50	+0 45	+0 53	-0.2	+0.2	4.5	5.5	4.2
243	Great Jervis Harbor, Despair Bay	47 39	56 11	+0 38	+1 05	-0.9	+0.3	3.7	4.8	3.9
245	Hare Bay	47 37	56 32	+0 41	+1 08	(*0.67	+0.8)	3.3	4.3	3.6
247	Grey (Little) River	47 34	57 07	+0 45	+1 12	(*0.63	+0.9)	3.1	4.0	3.5
249	Connoire Bay	47 40	57 54	+0 50	+0 50	(*0.59	+0.8)	2.9	3.8	3.3
251	La Poile Bay	47 40	58 24	+1 15	+1 15	(*0.63	+0.8)	3.1	4.0	3.4
				on HARRINGTON, p.12						
253	Port Aux Basques	47 35	59 09	-1 24	-1 28	*0.80	*0.75	3.1	4.0	2.8
255	Codroy Road	47 53	59 24	-1 22	-1 27	*0.74	*0.75	2.8	3.7	2.6

*Ratio. If ratio is accompanied by a correction multiply the heights of high and low waters at the reference station by the ratio and then apply the correction.

TABLE 2.—TIDAL DIFFERENCES AND OTHER CONSTANTS

No.	PLACE	POSITION		DIFFERENCES				RANGES		Mean Tide Level
				Time		Height				
		Lat.	Long.	High water	Low water	High water	Low water	Mean	Spring	
		° ' N.	° ' W.	h. m.	h. m.	feet	feet	feet	feet	feet
	NEWFOUNDLAND, West Coast *Time meridian, 52°30'W.*			on HARRINGTON, p.12						
257	St. Georges Harbor-----------------	48 27	58 30	-0 28	-0 38	*0.78	*0.88	2.8	3.5	2.8
259	Port-au-Port---------------------	48 33	58 45	+0 05	+0 10	-1.3	-1.0	3.5	4.5	2.4
261	Frenchman Cove, Bay of Islands------	49 04	58 10	+0 10	+0 10	-0.5	0.0	3.3	4.2	3.3
263	Norris Cove, Bonne Bay--------------	49 31	57 52	+0 10	+0 10	-0.7	-0.4	3.5	4.4	3.0
265	Portland Cove---------------------	50 11	57 36	+0 19	+0 19	-0.6	-0.4	3.6	4.6	3.0
267	Port Saunders---------------------	50 39	57 18	+0 07	+0 03	-0.3	-0.3	3.8	4.9	3.2
269	Castors Harbor, St. John Bay--------	50 55	56 59	+0 10	+0 10	*0.78	*0.75	3.0	4.1	2.7
271	Ste. Barbe Bay--------------------	51 12	56 46	0 00	0 00	*0.78	*0.56	3.3	4.4	2.6
	QUEBEC, Gulf of St. Lawrence *Time meridian, 60°W.*									
273	Bradore Bay----------------------	51 28	57 15	-0 35	-0 30	-0.6	-0.1	3.3	4.4	3.1
275	Mistanoque Harbor-----------------	51 16	58 12	-0 15	-0 15	-0.4	-0.1	3.5	4.6	3.3
277	HARRINGTON-----------------------	50 30	59 28	Daily predictions				3.8	4.9	3.5
279	Wapitagun Harbor------------------	50 12	60 01	+0 15	+0 15	-0.3	+0.1	3.4	4.4	3.4
281	Kegashka Bay---------------------	50 12	61 14	+0 40	+0 40	-0.9	-0.2	3.1	4.0	3.0
283	Natashquan Harbor-----------------	50 12	61 50	+1 00	+1 10	-0.8	-0.1	3.1	4.0	3.1
285	Betchewun Harbor------------------	50 14	63 11	+2 09	+2 13	-0.7	-0.4	3.5	4.6	3.0
287	Havre St. Pierre------------------	50 14	63 36	+2 23	+2 32	0.0	-0.1	3.9	4.8	3.5
301	Mingan Harbor--------------------	50 18	64 03	+2 35	+2 40	+0.9	0.0	4.7	5.8	3.9
	Anticosti Island									
303	Heath Point----------------------	49 05	61 42	+0 51	+0 52	(*0.61	+0.3)	2.3	3.0	2.4
305	Southwest Point-------------------	49 24	63 36	+3 21	+3 26	-0.3	0.0	3.5	4.4	3.4
307	Ellis Bay------------------------	49 48	64 22	+3 37	+3 38	+0.3	-0.5	4.6	5.7	3.4
309	Moisie Bay-----------------------	50 12	66 05	+3 43	+3 49	+2.3	+0.5	5.6	7.2	4.9
311	Seven Islands--------------------	50 13	66 24	+3 54	+3 58	+2.7	-0.1	6.6	8.6	4.8
313	Cawee Islands--------------------	49 50	67 00	+4 01	+4 07	+3.0	+0.6	6.2	8.0	5.3
	QUEBEC, St. Lawrence River *Time meridian, 75°W.*									
315	Ste. Anne des Monts---------------	49 08	66 29	+3 17	+3 19	+3.4	+0.6	6.6	8.6	5.5
317	Cap Chat-------------------------	49 06	66 45	+3 17	+3 21	+4.2	+1.0	7.0	9.0	6.1
319	Pointe des Monts------------------	49 20	67 22	+3 10	+3 16	+4.3	+0.8	7.3	9.6	6.1
321	Matane---------------------------	48 51	67 32	+3 18	+3 22	+4.7	+0.9	7.6	9.9	6.3
323	Little Metis---------------------	48 41	68 02	+3 24	+3 28	+5.4	+1.1	8.1	10.6	6.8
				on QUEBEC, p.16						
325	Bersimis River-------------------	48 53	68 39	-4 20	-5 08	-3.5	+1.7	8.5	11.2	7.3
327	Father Point---------------------	48 31	68 20	-4 22	-5 29	-3.1	+1.7	8.9	11.7	7.5
329	Bic Harbor-----------------------	48 22	68 44	-4 12	-5 14	-3.0	+1.7	9.0	11.8	7.5
331	Tadoussac, Saguenay River---------	48 08	69 43	-3 47	-4 54	-1.5	+1.1	11.1	14.0	8.0
333	Chicoutimi, Saguenay River--------	48 26	71 03	-3 28	-3 40	-1.1	+1.6	11.0	14.4	8.4
335	Brandypot Islands-----------------	47 52	69 41	-3 36	-4 40	-0.2	+2.5	11.0	14.5	9.3
337	Murray Bay-----------------------	47 39	70 08	-3 20	-4 22	+0.7	+2.6	11.8	15.3	9.8
339	Pointe aux Orignaux---------------	47 29	70 01	-2 47	-3 41	0.0	+2.5	11.2	14.7	9.4
341	Ile aux Coudres-------------------	47 26	70 19	-2 10	-3 21	+1.5	+2.3	12.9	15.8	10.1
343	L'Islet--------------------------	47 08	70 22	-1 17	-2 05	+0.3	+1.2	12.8	15.3	9.0
345	Beaujeu Channel-------------------	47 05	70 29	-1 10	-1 43	+0.9	+0.8	13.8	15.7	9.0
347	Grosse Isle----------------------	47 02	70 40	-0 57	-1 19	+1.6	+0.3	15.0	17.1	9.1
349	Berthier-------------------------	46 56	70 44	-0 47	-1 08	+1.6	+0.3	15.0	16.9	9.1
351	St. Laurent----------------------	46 52	71 00	-0 20	-0 30	+0.6	+0.5	13.8	15.6	8.7
353	QUEBEC---------------------------	46 49	71 11	Daily predictions				13.7	15.5	8.2
355	St. Nicholas---------------------	46 43	71 24	+0 35	+0 32	-0.4	-----	12.6	14.3	-----
357	St. Augustin Bar-----------------	46 43	71 28	+0 54	+0 53	-1.3	-----	11.8	13.3	-----
359	Ste. Croix-----------------------	46 37	71 45	+1 31	+2 00	-----	(‡)	11.8	13.3	-----
361	Pointe Platon--------------------	46 40	71 51	+1 43	+2 11	-----	(‡)	11.4	12.9	-----
363	Grondines------------------------	46 36	72 04	+2 14	+3 18	-----	(‡)	6.7	8.1	-----
365	Cap a la Roche--------------------	46 33	72 10	+2 37	+3 48	-----	(‡)	5.4	6.7	-----
367	Batiscan-------------------------	46 31	72 15	+3 32	+4 49	-----	(‡)	2.3	3.3	-----
369	Champlain------------------------	46 26	72 21	+4 08	+5 30	-----	(‡)	1.8	2.8	-----
371	Three Rivers---------------------	46 20	72 33	+4 45	+6 15	-----	(‡)	0.7	1.0	-----

*Ratio. If ratio is accompanied by a correction multiply the heights of high and low waters at the reference station by the ratio and then apply the correction.

‡Neap low water falls lower than spring low water.

TABLE 3.—HEIGHT OF TIDE AT ANY TIME

EXPLANATION OF TABLE

Most persons who desire to use this table will probably find the footnote sufficiently ample to follow without further explanations, but several examples are given here to further illustrate its use.

Example 1.—Required the height of the tide at 7^h 55^m at New York, N. Y., on a day when the predicted tides from the Tide Tables are as follows:

HIGH WATER			LOW WATER	
Time	Height		Time	Height
h. m.	ft.		h. m.	ft.
11 14	4. 2		5 22	0. 1
23 10	4. 1		17 41	0. 6

An inspection of the above times of tide shows that the given time is between the two morning tides.

The duration of rise is 11^h $14^m - 5^h$ $22^m = 5^h$ 52^m.
The time after low water for which the height is required is 7^h $55^m - 5^h$ $22^m = 2^h$ 33^m.
The range of tide is $4.2 - 0.1 = 4.1$ feet.

The duration of rise or fall in table 3 is given in heavy-faced type for each 20 minutes from 4^h 00^m to 10^h 40^m. The nearest tabular value to 5^h 52^m, the above duration of rise, is 6^h; and on the horizontal line of 6^h the nearest tabular time to 2^h 33^m after low water for which the height is required is 2^h 36^m. Following down the column in which this 2^h 36^m is found to its intersection with the line of the range 4.0 feet (which is the nearest tabular value to the above range of 4.1 feet) the correction is found to be 1.6 feet, which being reckoned from low water must be added, making $0.1 + 1.6 = 1.7$ feet, which is the required height above mean low water, the datum for New York.

Example 2.—Required the height of the tide at 10^h 45^m at Philadelphia, Pa., when the nearest predicted tides from the Tide Tables are as follows:

HIGH WATER		LOW WATER	
Time	Height	Time	Height
7^h 28^m	4.7 feet	14^h 33^m	-0.4 foot.

Here the duration of fall is 14^h $33^m - 7^h$ $28^m = 7^h$ 05^m.
The time after high water for which the height is required is 10^h $45^m - 7^h$ $28^m = 3^h$ 17^m.
The range of tide is $4.7 - (-0.4) = 5.1$ feet.

Entering table 3 at the duration of fall of 7^h, the nearest value on the horizontal line to 3^h 17^m is 3^h 16^m after high water. Following down this column to its intersection with a range of 5.0 feet which is the nearest tabular value to 5.1 feet, one obtains 2.2 feet, which being reckoned from high water must be subtracted from its height, making $4.7 - 2.2 = 2.5$ feet, which is the height required.

When the duration of rise and fall is greater than 10^h 40^m, enter the table with one-half the given duration and with one-half the time from the nearest high or low water; but if the duration of rise or fall is less than 4 hours, enter the table with double the given duration and with double the time from the nearest high or low water.

Similarly, when the range of tide is greater than 20 feet, enter the table with one-half the given range. The tabular correction should then be doubled before applying it to the given high or low water height. If the range of tide is greater than 40 feet, take one-third of the range and multiply the tabular correction by three.

If the height at any time is desired for a place listed in table 2, predictions of the high and low waters for the day in question should be obtained by the use of the differences given for the place in that table. With these predictions, the height for any intermediate time is obtained in the same manner as illustrated in the foregoing examples.

GRAPHICAL METHOD

If the height of the tide is required for a number of times on a certain day the full tide curve for the day may be obtained by the *one-quarter, one-tenth rule*. The procedure is as follows:

1. On cross section paper plot the high and low water points in the order of their occurrence for the day, measuring time horizontally and height vertically. These are the basic points for the curve.

2. Draw light straight lines connecting the points representing successive high and low waters.

3. Divide each of these straight lines into four equal parts. The halfway point of each line gives another point for the curve.

4. At the quarter point adjacent to high water draw a vertical line above the point and at the quarter point adjacent to low water draw a vertical line below the point, making the length of these lines equal to one-tenth of the range between the high and low waters used. The points marking the ends of these vertical lines give two additional intermediate points for the curve.

5. Draw a smooth curve through the points of high and low waters and the intermediate points, making the curve well rounded near high and low waters. This curve will closely approximate the actual tide curve, and heights for any time of the day may be readily scaled from it.

Caution. Both methods presented are based on the assumption that the rise and fall conform to simple cosine curves. Therefore the heights obtained will be approximate. The roughness of approximation will vary as the tide curve differs from a cosine curve.

TABLE 3.—HEIGHT OF TIDE AT ANY TIME

Time from the nearest high water or low water

Duration of rise or fall, see footnote

h. m.	h. m.	h. m.	h. m.	h. m.	h. m.	h. m.	h. m.	h. m.	h. m.	h. m.	h. m.	h. m.	h. m.	h. m.	h. m.
4 00	0 08	0 16	0 24	0 32	0 40	0 48	0 56	1 04	1 12	1 20	1 28	1 36	1 44	1 52	2 00
4 20	0 09	0 17	0 26	0 35	0 43	0 52	1 01	1 09	1 18	1 27	1 35	1 44	1 53	2 01	2 10
4 40	0 09	0 19	0 28	0 37	0 47	0 56	1 05	1 15	1 24	1 33	1 43	1 52	2 01	2 11	2 20
5 00	0 10	0 20	0 30	0 40	0 50	1 00	1 10	1 20	1 30	1 40	1 50	2 00	2 10	2 20	2 30
5 20	0 11	0 21	0 32	0 43	0 53	1 04	1 15	1 25	1 36	1 47	1 57	2 08	2 19	2 29	2 40
5 40	0 11	0 23	0 34	0 45	0 57	1 08	1 19	1 31	1 42	1 53	2 05	2 16	2 27	2 39	2 50
6 00	0 12	0 24	0 36	0 48	1 00	1 12	1 24	1 36	1 48	2 00	2 12	2 24	2 36	2 48	3 00
6 20	0 13	0 25	0 38	0 51	1 03	1 16	1 29	1 41	1 54	2 07	2 19	2 32	2 45	2 57	3 10
6 40	0 13	0 27	0 40	0 53	1 07	1 20	1 33	1 47	2 00	2 13	2 27	2 40	2 53	3 07	3 20
7 00	0 14	0 28	0 42	0 56	1 10	1 24	1 38	1 52	2 06	2 20	2 34	2 48	3 02	3 16	3 30
7 20	0 15	0 29	0 44	0 59	1 13	1 28	1 43	1 57	2 12	2 27	2 41	2 56	3 11	3 25	3 40
7 40	0 15	0 31	0 46	1 01	1 17	1 32	1 47	2 03	2 18	2 33	2 49	3 04	3 19	3 35	3 50
8 00	0 16	0 32	0 48	1 04	1 20	1 36	1 52	2 08	2 24	2 40	2 56	3 12	3 28	3 44	4 00
8 20	0 17	0 33	0 50	1 07	1 23	1 40	1 57	2 13	2 30	2 47	3 03	3 20	3 37	3 53	4 10
8 40	0 17	0 35	0 52	1 09	1 27	1 44	2 01	2 19	2 36	2 53	3 11	3 28	3 45	4 03	4 20
9 00	0 18	0 36	0 54	1 12	1 30	1 48	2 06	2 24	2 42	3 00	3 18	3 36	3 54	4 12	4 30
9 20	0 19	0 37	0 56	1 15	1 33	1 52	2 11	2 29	2 48	3 07	3 25	3 44	4 03	4 21	4 40
9 40	0 19	0 39	0 58	1 17	1 37	1 56	2 15	2 35	2 54	3 13	3 33	3 52	4 11	4 31	4 50
10 00	0 20	0 40	1 00	1 20	1 40	2 00	2 20	2 40	3 00	3 20	3 40	4 00	4 20	4 40	5 00
10 20	0 21	0 41	1 02	1 23	1 43	2 04	2 25	2 45	3 06	3 27	3 47	4 08	4 29	4 49	5 10
10 40	0 21	0 43	1 04	1 25	1 47	2 08	2 29	2 51	3 12	3 33	3 55	4 16	4 37	4 59	5 20

Correction to height

Range of tide, see footnote

Ft.	Ft.	Ft.	Ft.	Ft.	Ft.	Ft.	Ft.	Ft.	Ft.	Ft.	Ft.	Ft.	Ft.	Ft.	Ft.
0.5	0.0	0.0	0.0	0.0	0.0	0.0	0.1	0.1	0.1	0.1	0.1	0.2	0.2	0.2	0.2
1.0	0.0	0.0	0.0	0.0	0.1	0.1	0.1	0.2	0.2	0.2	0.3	0.3	0.4	0.4	0.5
1.5	0.0	0.0	0.0	0.1	0.1	0.1	0.2	0.2	0.3	0.4	0.4	0.5	0.6	0.7	0.8
2.0	0.0	0.0	0.0	0.1	0.1	0.2	0.3	0.3	0.4	0.5	0.6	0.7	0.8	0.9	1.0
2.5	0.0	0.0	0.1	0.1	0.2	0.2	0.3	0.4	0.5	0.6	0.7	0.9	1.0	1.1	1.2
3.0	0.0	0.0	0.1	0.1	0.2	0.3	0.4	0.5	0.6	0.8	0.9	1.0	1.2	1.3	1.5
3.5	0.0	0.0	0.1	0.2	0.2	0.3	0.4	0.6	0.7	0.9	1.0	1.2	1.4	1.6	1.8
4.0	0.0	0.0	0.1	0.2	0.3	0.4	0.5	0.7	0.8	1.0	1.2	1.4	1.6	1.8	2.0
4.5	0.0	0.0	0.1	0.2	0.3	0.4	0.6	0.7	0.9	1.1	1.3	1.6	1.8	2.0	2.2
5.0	0.0	0.1	0.1	0.2	0.3	0.5	0.6	0.8	1.0	1.2	1.5	1.7	2.0	2.2	2.5
5.5	0.0	0.1	0.1	0.2	0.4	0.5	0.7	0.9	1.1	1.4	1.6	1.9	2.2	2.5	2.8
6.0	0.0	0.1	0.1	0.3	0.4	0.6	0.8	1.0	1.2	1.5	1.8	2.1	2.4	2.7	3.0
6.5	0.0	0.1	0.2	0.3	0.4	0.6	0.8	1.1	1.3	1.6	1.9	2.2	2.6	2.9	3.2
7.0	0.0	0.1	0.2	0.3	0.5	0.7	0.9	1.2	1.4	1.8	2.1	2.4	2.8	3.1	3.5
7.5	0.0	0.1	0.2	0.3	0.5	0.7	1.0	1.2	1.5	1.9	2.2	2.6	3.0	3.4	3.8
8.0	0.0	0.1	0.2	0.4	0.5	0.8	1.0	1.3	1.6	2.0	2.4	2.8	3.2	3.6	4.0
8.5	0.0	0.1	0.2	0.4	0.6	0.8	1.1	1.4	1.8	2.1	2.5	2.9	3.4	3.8	4.2
9.0	0.0	0.1	0.2	0.4	0.6	0.9	1.2	1.5	1.9	2.2	2.7	3.1	3.6	4.0	4.5
9.5	0.0	0.1	0.2	0.4	0.6	0.9	1.2	1.6	2.0	2.4	2.8	3.3	3.8	4.3	4.8
10.0	0.0	0.1	0.2	0.4	0.7	1.0	1.3	1.7	2.1	2.5	3.0	3.5	4.0	4.5	5.0
10.5	0.0	0.1	0.3	0.5	0.7	1.0	1.3	1.7	2.2	2.6	3.1	3.6	4.2	4.7	5.2
11.0	0.0	0.1	0.3	0.5	0.7	1.1	1.4	1.8	2.3	2.8	3.3	3.8	4.4	4.9	5.5
11.5	0.0	0.1	0.3	0.5	0.8	1.1	1.5	1.9	2.4	2.9	3.4	4.0	4.6	5.1	5.8
12.0	0.0	0.1	0.3	0.5	0.8	1.1	1.5	2.0	2.5	3.0	3.6	4.1	4.8	5.4	6.0
12.5	0.0	0.1	0.3	0.5	0.8	1.2	1.6	2.1	2.6	3.1	3.7	4.3	5.0	5.6	6.2
13.0	0.0	0.1	0.3	0.6	0.9	1.2	1.7	2.2	2.7	3.2	3.9	4.5	5.1	5.8	6.5
13.5	0.0	0.1	0.3	0.6	0.9	1.3	1.7	2.2	2.8	3.4	4.0	4.7	5.3	6.0	6.5
14.0	0.0	0.2	0.3	0.6	0.9	1.3	1.8	2.3	2.9	3.5	4.2	4.8	5.5	6.3	7.0
14.5	0.0	0.2	0.4	0.6	1.0	1.4	1.9	2.4	3.0	3.6	4.3	5.0	5.7	6.5	7.2
15.0	0.0	0.2	0.4	0.6	1.0	1.4	1.9	2.5	3.1	3.8	4.4	5.2	5.9	6.7	7.5
15.5	0.0	0.2	0.4	0.7	1.0	1.5	2.0	2.6	3.2	3.9	4.6	5.4	6.1	6.9	7.8
16.0	0.0	0.2	0.4	0.7	1.1	1.5	2.1	2.6	3.3	4.0	4.7	5.5	6.3	7.2	8.0
16.5	0.0	0.2	0.4	0.7	1.1	1.6	2.1	2.7	3.4	4.1	4.9	5.7	6.5	7.4	8.2
17.0	0.0	0.2	0.4	0.7	1.1	1.6	2.2	2.8	3.5	4.2	5.0	5.9	6.7	7.6	8.5
17.5	0.0	0.2	0.4	0.8	1.2	1.7	2.2	2.9	3.6	4.4	5.2	6.0	6.9	7.8	8.8
18.0	0.0	0.2	0.4	0.8	1.2	1.7	2.3	3.0	3.7	4.5	5.3	6.2	7.1	8.1	9.0
18.5	0.1	0.2	0.5	0.8	1.2	1.8	2.4	3.1	3.8	4.6	5.5	6.4	7.3	8.3	9.2
19.0	0.1	0.2	0.5	0.8	1.3	1.8	2.4	3.1	3.9	4.8	5.6	6.6	7.5	8.6	9.5
19.5	0.1	0.2	0.5	0.8	1.3	1.9	2.5	3.2	4.0	4.9	5.8	6.7	7.7	8.7	9.8
20.0	0.1	0.2	0.5	0.9	1.3	1.9	2.6	3.3	4.1	5.0	5.9	6.9	7.9	9.0	10.0

Obtain from the predictions the high water and low water, one of which is before and the other after the time for which the height is required. The difference between the times of occurrence of these tides is the duration of rise or fall, and the difference between their heights is the range of tide for the above table. Find the difference between the nearest high or low water and the time for which the height is required.

Enter the table with the duration of rise or fall, printed in heavy-faced type, which most nearly agrees with the actual value, and on that horizontal line find the time from the nearest high or low water which agrees most nearly with the corresponding actual difference. The correction sought is in the column directly below, on the line with the range of tide.

When the nearest tide is high water, subtract the correction.

When the nearest tide is low water, add the correction.

TABLE 4.—CURRENT DIFFERENCES AND OTHER CONSTANTS

EXPLANATION OF TABLE

The principal purpose of this table is to furnish data which will enable one to determine the approximate time of slack water and the time and velocity of maximum current at numerous stations on the Atlantic coast of North America. These stations are arranged in geographic order, beginning with the Bay of Fundy. For each station there are given the latitude and longitude to the nearest minute, the differences in time of slack water and maximum current and the flood and ebb velocity ratios with respect to some station for which daily predictions are given in the first part of this publication; and the direction and average velocity of maximum flood and ebb.

Reference station.—In this table it is convenient to designate as reference stations those stations for which daily predictions are given in table 1. For the same reason all other stations in this table are designated as subordinate stations. Referring the time of the current at subordinate stations to the predicted times of the current at some reference station makes it possible to determine, in advance, the approximate times of the current at all the stations listed in this table.

Latitude and longitude.—For convenience in locating on a chart any of the stations listed in this table, the latitude and longitude of each station are given to the nearest minute. Since a minute of latitude is about a mile, it is obvious that the latitude and longitude of a station to the nearest minute do not indicate the exact position of the station. This is to be borne in mind, especially in the case of a narrow stream, where the nearest minute of latitude and longitude may locate a station inland. In every such case it is to be remembered that, unless the description locates the station elsewhere, reference is made to the current in the center of the channel. In some instances the charts do not present a convenient name for locating a station. In these cases the position is described by a true bearing from a prominent place on a chart of the locality.

Time difference.—To determine the time of slack water or maximum current at any station listed in this table, there are given, in the columns headed "Time differences," the times that are to be added to or subtracted from the times of slack water or maximum current at the reference station listed directly above the differences for the subordinate station desired. A plus (+) sign indicates that the time of current at a station is later than at the reference station, and a minus (−) sign that it is earlier. Allowance has been made for any difference in the kind of time used at the reference and the subordinate station, and the results obtained by the application of the time differences will be in the kind of time ordinarily used at the station as indicated in the table. To obtain a particular phase of the current at a subordinate station the time difference should be applied to the corresponding phase at the reference station. Thus the time of slack before flood at a subordinate station is obtained by applying the time difference to the predicted time of slack before flood at the reference station, etc. The direction of the flood or ebb current thus obtained is given by the column "Flood Direction" or "Ebb Direction" under "Maximum Currents." It is important to note that although a given phase of the current at the subordinate station is always obtained from the corresponding phase at the reference station the direction of the current at the two places may differ considerably. The direction of the current at the subordinate station should consequently not be taken the same as that at the reference station but should always be obtained from the direction given for the subordinate station in table 2.

Velocity ratio.—The approximate velocity of the current at the time of maximum at a subordinate station may be obtained by multiplying the predicted velocity for the corresponding maximum current at the reference station by the appropriate flood or ebb velocity ratio.

At maximum current.—Near the coast and in inland tidal waters the current from slack water increases in velocity for a period of about 3 hours when the maximum velocity or the strength of the current is reached. The velocity then decreases for another period of about 3 hours when slack water is again reached and the current begins a similar cycle in the opposite direction. The current that flows toward the coast or up a stream is known as the flood current, while the one that sets from the coast or down a stream is known as the ebb current. In the columns headed "Maximum Currents" there are given the flood direction, average velocity, ebb direction and average velocity at strength of current. The flood and ebb directions are given in degrees, true, reading clockwise from 000° at north to 359° and are the directions toward which the current flows. The average velocities given represent the mean velocities of all strengths of flood and ebb currents.

Owing to the variations in the velocities at strength of current from day to day, due to astronomical causes, accurate predictions of the maximum currents for any particular day can be best obtained by reference to a station where daily predictions are given; the velocities at the subordinate station being obtained by the velocity ratios.

Example.—Suppose it is desired to find the time of slack water and the time and velocity of maximum current at Hull Gut, Massachusetts. From table 2 we obtain the following information for Hull Gut: Reference station, Boston Harbor; slack water time difference -0^h35^m; maximum flood and ebb tide difference -0^h50^m; flood velocity ratio 1.2; ebb velocity ratio 1.8; flood direction 155° true; ebb direction 350° true. This means that the times of slack water are 35 minutes earlier and maximum flood and ebb 50 minutes earlier at Hull Gut than at Boston Harbor (Deer Island Light); that the velocity at strength of flood is 1.2 times and velocity at strength of ebb is 1.8 times the corresponding velocities at Boston Harbor. On a given morning the predicted currents at Boston Harbor are as given in the table below. The predictions for Hull Gut are obtained as follows:

	Slack water	Maximum ebb		Slack water	Maximum flood	
	h. m.	h. m.	kn.	h. m.	h. m.	kn.
Boston Harbor	0132	0413	1. 5	0807	1119	1. 3
	−0035	−0050	x1. 8 ratio	−0035	−0050	x1. 2 ratio
Hull Gut	0057	0323	2. 7	0732	1029	1. 6

It is important to note that at Hull Gut the directions of the flood and ebb currents are about 155° and 350° respectively whereas at Boston Harbor the directions of corresponding phases are 260° and 085°.

Note.—Dashes are entered in the place of data which are unreliable, unknown, or given in another part of this book.

TABLE 4.—CURRENT DIFFERENCES AND OTHER CONSTANTS

No.	PLACE	POSITION		TIME DIF-FERENCES		VELOCITY RATIOS		MAXIMUM CURRENTS			
								Flood		Ebb	
		Lat.	Long.	Slack water	Maximum current	Maximum flood	Maximum ebb	Direction (true)	Average velocity	Direction (true)	Average velocity
		° ′	° ′	h. m.	h. m.			deg.	knots	deg.	knots
	MASSACHUSETTS COAST—Continued	N.	W.	on BOSTON HARBOR, p.16							
	Time meridian, 75°W.										
390	Merrimack River entrance--------------	42 49	70 49	+0 40	(*)	1.3	1.1	285	2.2	105	1.4
395	Newburyport, Merrimack River----------	42 49	70 52	+1 10	+0 40	0.9	1.1	290	1.5	100	1.4
400	Plum Island Sound entrance------------	42 42	70 47	+0 15	-0 10	0.9	1.2	315	1.6	185	1.5
405	Annisquam Harbor Light----------------	42 40	70 41	+0 20	-0 05	0.8	0.8	200	1.0	015	1.3
410	Gloucester Harbor entrance------------	42 35	70 40	*Current too weak and variable to be predicted.*							
415	Blynman Canal ent., Gloucester Hbr-----	42 37	70 40	-0 40	-0 45	1.8	2.5	310	3.0	130	3.3
420	Marblehead Channel--------------------	42 30	70 49	+0 40	+0 40	0.3	0.2	285	0.4	105	0.4
425	Nahant, off East Point----------------	42 25	70 54	-0 20	-0 20	0.5	0.5	235	0.8	085	0.7
430	Lynn Harbor entrance------------------	42 25	70 57	-0 05	-0 05	0.3	0.3	325	0.5	170	0.5
435	Winthrop Beach, 1.2 miles east of-----	42 23	70 57	-0 05	-0 30	0.2	0.2	195	0.4	095	0.2
	BOSTON HARBOR APPROACHES										
440	Stellwagen Bank-----------------------	42 24	70 24	*Current too weak and variable to be predicted.*							
445	Boston Lightship, 3 miles SSE. of-----	42 20	70 45	*Current too weak and variable to be predicted.*							
450	North Channel, off Great Faun---------	42 21	70 56	-0 05	-0 25	0.7	1.1	200	1.2	025	1.4
455	Hypocrite Channel---------------------	42 21	70 54	-0 30	-0 30	0.7	0.7	255	1.2	070	1.0
460	Nantasket Roads entrance--------------	42 19	70 53	0 00	-0 20	1.0	1.0	260	1.4	085	1.5
465	Black Rock Channel--------------------	42 19	70 55	-0 15	-0 15	0.7	0.9	220	1.2	035	1.2
	BOSTON HARBOR										
470	The Narrows, off Georges Island-------	42 19	70 56	(¹)	(¹)	0.5	1.0	280	0.9	110	1.3
475	The Narrows, off Gallups Island-------	42 20	70 56	+0 10	-0 05	0.4	0.2	125	0.6	315	0.2
480	Between Gallups I. and Georges I------	42 19	70 56	(²)	(²)	0.4	0.6	265	0.6	040	0.8
485	Nubble Channel------------------------	42 20	70 57	+0 10	+0 10	0.5	0.6	195	0.8	015	0.8
490	BOSTON HARBOR (Deer Island Light)-----	42 20	70 57	Daily predictions				260	1.7	085	1.3
495	Point Shirley, 0.5 mile SW. of--------	42 21	70 59	*Current too weak and variable to be predicted.*							
500	Main Ship Chan., off Ft. Independence-	42 21	71 01	-0 05	-0 25	0.4	0.5	300	0.6	110	0.6
505	Fort Point Channel entrance-----------	42 21	71 03	+0 10	-0 05	0.1	0.3	210	0.2	025	0.4
510	Between Boston and East Boston--------	42 22	71 03	+0 10	0 00	0.2	0.5	340	0.4	160	0.6
515	Charles River entrance----------------	42 22	71 04	*Current too weak and variable to be predicted.*							
520	Between Charlestown and East Boston---	42 22	71 03	+0 10	+0 05	0.2	0.5	025	0.4	200	0.6
525	Chelsea River entrance----------------	42 23	71 02	+0 05	+0 05	0.3	0.5	100	0.4	290	0.6
530	Mystic River entrance-----------------	42 23	71 03	+0 15	0 00	0.2	0.4	270	0.4	090	0.5
535	Between Spectacle I. and Castle I------	42 20	71 00	-0 20	-0 20	0.4	0.5	250	0.6	070	0.6
540	Between Spectacle I. and Thompson I---	42 19	71 00	-2 25	-2 50	0.2	0.4	310	0.3	120	0.5
545	Dorchester Bay, off Thimble Island----	42 18	71 02	0 00	-0 20	0.5	0.7	230	0.8	065	0.9
550	Neponset River (railroad bridge)------	42 17	71 02	0 00	-0 30	0.4	0.5	230	0.6	050	0.7
555	Between Long I. and Spectacle I--------	42 19	70 58	+0 05	-0 10	0.2	0.4	185	0.4	025	0.5
560	Between Thompson I. and Moon Head-----	42 18	71 00	-0 35	-1 20	0.2	0.4	260	0.4	085	0.5
565	Between Moon Head and Long Island-----	42 19	70 59	*Current weak and rotary, turning counterclockwise.*							
570	Between Long I. and Rainsford I-------	42 19	70 58	-0 05	-0 10	0.5	0.8	225	0.8	045	1.0
575	Nantasket Roads, off Georges Island---	42 19	70 55	-0 05	-0 35	0.8	1.5	250	1.3	040	1.9
580	Nantasket Roads, off Rainsford I------	42 18	70 57	+0 20	+0 15	0.7	0.8	240	1.1	060	1.0
585	Squantum, 0.3 mile SE. of-------------	42 17	71 00	*Current too weak and variable to be predicted.*							0.2
590	Nut Island, 0.8 mile west of----------	42 17	70 58	-1 45	-2 25	0.2	0.2	275	0.3	055	0.2
595	Hull Gut-----------------------------	42 18	70 55	-0 35	-0 50	1.2	1.8	155	2.0	350	2.4
600	Between Peddocks I. and Nut I---------	42 17	70 57	+0 25	+0 05	0.9	1.1	125	1.6	315	1.4
605	Between Peddocks I. and Sheep I-------	42 17	70 56	+2 00	+0 55	0.3	0.3	040	0.5	220	0.4
610	Between Raccoon I. and Grape I--------	42 16	70 56	+0 10	-0 05	0.3	0.3	185	0.5	030	0.4

*For flood, +0ʰ 35ᵐ. For ebb, -0ʰ 50ᵐ.
¹For flood and slack before ebb, -1ʰ 20ᵐ. For ebb and slack before flood, -0ʰ 25ᵐ.
²For flood and slack before ebb, -2ʰ 00ᵐ. For ebb and slack before flood, -0ʰ 20ᵐ.

TABLE 4.—CURRENT DIFFERENCES AND OTHER CONSTANTS

| No. | PLACE | POSITION | | TIME DIFFERENCES | | VELOCITY RATIOS | | MAXIMUM CURRENTS | | | |
| | | | | | | | | Flood | | Ebb | |
		Lat.	Long.	Slack water	Maximum current	Maximum flood	Maximum ebb	Direction (true)	Average velocity	Direction (true)	Average velocity
		° ′ N.	° ′ W.	h. m.	h. m.			deg.	knots	deg.	knots
	BOSTON HARBOR—Continued *Time meridian, 75°W.*				on BOSTON HARBOR, p.16						
620	Weymouth Fore River, Quincy Point-----	42 15	70 58	-0 1C	-0 20	0.4	0.7	225	0.6	045	0.9
625	Weymouth Back River, bridge-----------	42 15	70 56	0 00	0 00	0.7	0.9	165	1.2	345	1.2
630	Hull, Channel SE. of------------------	42 18	70 54	-1 15	-1 15	0.2	0.2	060	0.2	215	0.3
635	Strawberry Hill, 0.5 mile west of-----	42 17	70 54	Current too weak and variable to be predicted.							
640	Off Bumkin Island---------------------	42 16	70 54	+0 05	+0 05	0.5	0.8	130	0.9	325	1.1
645	Worlds End, North of------------------	42 16	70 53	0 00	0 00	0.6	0.9	085	1.0	275	1.2
650	Off Crow Point-----------------------	42 16	70 54	+0 15	f-1 05	0.5	0.8	150	0.9	335	1.0
	CAPE COD BAY										
655	Race Point, 7 miles north of----------	42 11	70 16	-0 30	-0 30	0.9	1.2	290	1.5	----	1.5
660	Race Point, 1 mile northwest of-------	42 05	70 15	-0 35	-0 35	0.6	0.7	225	1.0	060	0.9
665	Provincetown Harbor-------------------	42 03	70 10	-0 25	-0 25	0.3	0.3	315	0.6	135	0.4
670	Wellfleet Harbor----------------------	41 54	70 03	-0 20	-0 20	0.4	0.4	020	0.7	200	0.5
675	Barnstable Harbor--------------------	41 44	70 16	-0 10	+0 15	0.7	1.1	190	1.2	005	1.4
680	Sandwich Harbor-----------------------	41 46	70 29	Current too weak and variable to be predicted.							
	Cape Cod Canal (see page 139)---------	-----	-----	-----	-----	---	---	---	---	---	---
685	Sagamore Beach------------------------	41 48	70 31	Current too weak and variable to be predicted.							
690	Ellisville Harbor, 1 mile east of-----	41 51	70 30	-0 15	-0 15	0.2	0.2	200	0.3	020	0.3
695	Manomet Point-------------------------	41 56	70 32	-0 25	-0 25	0.7	0.7	155	1.1	010	0.9
700	Gurnet Point, 1 mile east of----------	42 00	70 35	-0 35	-0 35	0.8	0.8	250	1.4	----	1.0
705	Plymouth Harbor-----------------------	41 58	70 39	-0 25	-0 25	0.3	0.3	245	0.5	010	0.4
710	Farnham Rock, 1 mile east of----------	42 06	70 35	-0 50	-0 50	0.7	0.7	180	1.1	010	0.9
	MASSACHUSETTS COAST—Continued				on POLLOCK RIP CHANNEL, p.28						
715	Nauset Beach Light, 5 miles NE. of----	41 56	69 54		See table 5.						
720	Georges Bank and vicinity-------------	-----	-----		See table 5.						
725	Davis Bank---------------------------	-----	-----		See table 5.						
730	Nantucket Shoals---------------------	40 37	69 37		See table 5.						
735	Old Man Shoal, Nantucket Shoals-------	41 14	69 59	+1 20	+1 10	0.9	0.9	080	1.9	225	1.6
740	Miacomet Pond, 3.0 miles SSE. of-----	41 11	70 06	+2 20	+2 10	0.6	0.8	080	1.3	280	1.4
745	Tuckernuck I., 4.2 miles SSW. of------	41 14	70 17	(1)	+3 35	0.3	0.6	090	0.5	280	1.0
750	Martha's Vineyard, 1.4 miles S. of----	41 20	70 40	(2)	-2 50	0.1	0.1	230	0.3	095	0.3
	NANTUCKET SOUND ENTRANCE										
755	Pollock Rip Lightship-----------------	41 36	69 51		See table 5.						
760	Pollock Rip Channel, east end--------	41 34	69 55	-0 20	-0 40	1.0	1.1	055	2.0	210	1.8
765	POLLOCK RIP CHANNEL (Butler Hole)-----	41 33	69 59	Daily predictions				035	2.0	225	1.8
770	Great Round Shoal Channel-------------	-----	-----		See table 5.						
	NANTUCKET SOUND										
775	Monomoy Pt., channel 0.2 mile W. of---	41 33	70 01	+0 10	+0 10	0.8	1.2	170	1.7	345	2.0
780	Chatham Roads-------------------------	41 39	70 02	Current too weak and variable to be predicted.							
785	Stage Harbor, west of Morris Island---	41 39	69 58	+2 45	+3 00	0.3	0.6	335	0.5	145	1.0
790	Dennis Port, 2.2 miles south of-------	41 37	70 07	+0 55	+1 00	0.2	0.2	075	0.3	270	0.3
795	Monomoy Point, 6 miles west of-------	41 34	70 09	+1 15	+1 35	0.2	0.3	090	0.5	275	0.5
800	Handkerchief Lighted Whistle Buoy "H"	41 29	70 04	+1 00	+1 05	0.6	0.8	080	1.3	250	1.3
805	Halfmoon Shoal, 1.9 miles NE. of------	41 29	70 12	+1 35	+1 45	0.4	0.3	110	0.8	265	0.6
810	Halfmoon Shoal, 3.5 miles east of-----	41 28	70 09	+1 10	+1 15	0.5	0.6	090	1.1	295	1.0
815	Great Point, 0.5 mile west of--------	41 24	70 04	+0 50	+1 05	0.6	0.7	030	1.1	195	1.2
820	Great Point, 3 miles west of---------	41 23	70 07	+1 05	+1 15	0.4	0.5	065	0.8	250	0.8
825	Tuckernuck Shoal, off east end--------	41 24	70 10	+1 15	+1 20	0.5	0.5	115	0.9	285	0.9

[1]Flood begins, +4h 10m; ebb begins, +2h 15m.
[2]Times of slack are indefinite.
f= flood only; for ebb, use +1h 10m.

APPENDIX D

TIDAL CURRENT TABLES
INTRODUCTION

Current tables for the use of mariners have been published by the National Ocean Survey (formerly the Coast and Geodetic Survey) since 1890. Tables for the Atlantic coast first appeared as a part of the tide tables and consisted of brief directions for obtaining the times of the current for a few locations from the times of high and low waters. Daily predictions of slack water for five stations were given for the year 1916, and by 1923 the tables had so expanded that they were then issued as a spearate publication entitled *Current Tables, Atlantic Coast*. A companion volume, *Current Tables, Pacific Coast*, was also issued that year. In 1930 the predictions for the Atlantic coast were extended to include the times and velocities of maximum current.

In the preparation of these tables, all available observations were used. In some cases, however, the observations were insufficient for obtaining final results, and as further information becomes available it will be included in subsequent editions. All persons using these tables are invited to send information or suggestions for increasing their usefulness to the Director, National Ocean Survey, Rockville, Md. 20852, U.S.A. The data for lightship stations are based on observations obtained through the cooperation of the U.S. Coast Guard. By cooperative arrangements, full predictions for Bay of Fundy Entrance (Grand Manan Channel) were furnished by the Canadian Hydrographic Service.

Daily predicted times of slack water and predicted times and velocities of maximum current (flood and ebb) are presented in table 1 for a number of reference stations. Similar predictions for many other locations may be obtained by applying the correction factors listed in table 2 to the predictions of the appropriate reference station. The velocity of a current at times between slack water and maximum current may be approximated by the use of table 3. The duration of weak current near the time of slack water may be computed by the use of table 4.

Reprinted from Tidal Current Tables. Available from National Oceanographic Survey.

241

TABLE 1.—DAILY CURRENT PREDICTIONS

EXPLANATION OF TABLE

This table gives the predicted times of slack water and the predicted times and velocities of maximum current—flood and ebb—for each day of the year at a number of stations on the Atlantic coast of North America. The times are given in hours and minutes and the velocities in knots.

Time.—The kind of time used for the predictions at each reference station is indicated by the time meridian at the bottom of each page.

Slack water and maximum current.—The columns headed "Slack water" contain the predicted times at which there is no current; or, in other words, the times at which the current has stopped setting in a given direction and is about to begin to set in the opposite direction. Offshore, where the current is rotary, slack water denotes the time of minimum current. Beginning with the slack water before flood the current increases in velocity until the strength or maximum velocity of the flood current is reached; it then decreases until the following slack water or slack before ebb. The ebb current now begins, increases to a maximum velocity, and then decreases to the next slack. The predicted times and velocities of maximum current are given in the columns headed "Maximum Current." Flood velocities are marked with an "F," the ebb velocities with an "E." An entry in the "Slack Water" column will be *slack, flood begins* if the maximum current which follows it is marked "F." Otherwise the entry will be *slack, ebb begins*.

Directions of set.—As the terms flood and ebb do not in all cases clearly indicate the direction of the current, the approximate directions toward which the currents flow are given at the top of each page to distinguish the two streams.

Number of slacks and strengths.—There are usually four slacks and four maximums each day. When a vacancy occurs in any day, the slack or maximum that seems to be missing will be found to occur soon after midnight as the first slack or maximum of the following day. At some stations where the diurnal inequality is large, there may be on certain days a continuous flood or ebb current with varying velocity throughout half the day giving only two slacks and two maximums on that particular day.

Current and tide.—It is important to notice that the predicted slacks and strengths given in this table refer to the horizontal motion of the water and not to the vertical rise and fall of the tide. The relation of current to tide is not constant, but varies from place to place, and the time of slack water does not generally coincide with the time of high or low water, nor does the time of maximum velocity of the current usually coincide with the time of most rapid change in the vertical height of the tide. At stations located on a tidal river or bay the time of slack water may differ from 1 to 3 hours from the time of high or low water. The times of high and low waters are given in the tide tables published by the National Ocean Survey.

Variations from predictions.—In using this table it should be borne in mind that actual times of slack or maximum occasionally differ from the predicted times by as much as half an hour and in rare instances the difference may be as much as an hour. Comparisons of predicted with observed times of slack water indicate that more than 90 percent of the slack waters occurred within half an hour of the predicted times. To make sure, therefore, of getting the full advantage of a favorable current or slack water, the navigator should reach the entrance or strait at least half an hour before the predicted time of the desired condition of current. Currents are frequently disturbed by wind or variations in river discharge. On days when the current is affected by such disturbing influences the times and velocities will differ from those given in the table, but local knowledge will enable one to make proper allowance for these effects.

Typical current curves.—The variations in the tidal current from day to day and from place to place are illustrated on the opposite page by the current curves for representative ports along the Atlantic and Gulf Coasts of the United States. Flood current is represented by the solid line curve above the zero velocity (slack water) line and the ebb current by the broken line curve below the slack water line. The curves show clearly that the currents along the Atlantic coast are semi-daily (two floods and two ebbs in a day) in character with their principal variations following changes in the moon's distance and phase. In the Gulf of Mexico, however, the currents are daily in character. As the dominant factor is the change in the moon's declination the currents in the Gulf tend to become semi-daily when the moon is near the equator. By reference to the curves it will be noted that with this daily type of current there are times when the current may be erratic (marked with an asterisk), or one flood or ebb current of the day may be quite weak. Therefore in using the predictions of the current it is essential to carefully note the velocities as well as the times.

TYPICAL CURRENT CURVES FOR REFERENCE STATIONS
(Flood: Solid line. Ebb: Broken Line.)

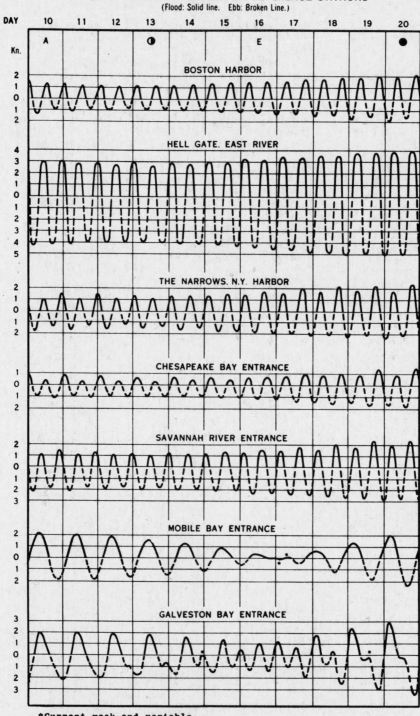

*Current weak and variable.
A discussion of these curves is given on the preceding page.

Lunar data: A — moon in apogee
               ◑ — last quarter
               E — moon on equator
               ● — new moon

EXTRACTS FROM TIDAL CURRENT TABLES

THE NARROWS, NEW YORK HARBOR, N.Y., 1975

F-FLOOD, DIR. 340° TRUE   E-EBB, DIR. 160° TRUE

JANUARY

DAY	SLACK WATER TIME H.M.	MAX CURRENT TIME H.M.	VEL. KNOTS	DAY	SLACK WATER TIME H.M.	MAX CURRENT TIME H.M.	VEL. KNOTS
1 W		0254	2.4E	16 TH	0002	0310	1.8E
	0617	0857	2.2F		0641	0907	1.6F
	1211	1526	2.5E		1217	1532	2.0E
	1901	2132	2.0F		1917	2139	1.5F
2 TH	0037	0345	2.3E	17 F	0046	0351	1.7E
	0716	0952	2.1F		0730	0954	1.5F
	1301	1614	2.4E		1257	1611	1.9E
	1954	2227	2.0F		2000	2224	1.5F
3 F	0132	0440	2.2E	18 SA	0131	0437	1.6E
	0819	1049	1.9F		0823	1043	1.4F
	1352	1708	2.3E		1339	1654	1.8E
	2050	2321	2.0F		2045	2309	1.5F
4 SA	0230	0542	2.1E	19 SU	0218	0530	1.6E
	0924	1147	1.7F		0920	1130	1.3F
	1445	1807	2.1E		1424	1745	1.7E
	2146				2132	2358	1.5F
5 SU		0018	1.9F	20 M	0310	0631	1.5E
	0331	0648	2.0E		1017	1220	1.2F
	1027	1244	1.5F		1514	1842	1.6E
	1543	1909	2.0E		2220		
	2242						
6 M		0119	1.9F	21 TU		0047	1.6F
	0434	0751	2.0E		0407	0728	1.6E
	1129	1352	1.4F		1114	1313	1.1F
	1643	2007	2.0E		1610	1940	1.6E
	2337				2310		
7 TU		0232	1.9F	22 W		0140	1.6F
	0537	0850	2.0E		0505	0823	1.7E
	1230	1515	1.3F		1211	1410	1.1F
	1744	2102	2.0E		1710	2033	1.7E
8 W	0033	0343	1.9F	23 TH	0001	0237	1.7F
	0635	0946	2.0E		0602	0917	1.9E
	1328	1623	1.4F		1305	1509	1.2F
	1841	2154	1.9E		1808	2124	1.8E
9 TH	0127	0442	2.0F	24 F	0055	0337	1.9F
	0727	1037	2.1E		0655	1007	2.0E
	1423	1714	1.4F		1357	1612	1.4F
	1932	2242	1.9E		1903	2215	2.0E
10 F	0219	0527	2.0F	25 SA	0148	0434	2.1F
	0814	1126	2.1E		0745	1059	2.2E
	1511	1757	1.5F		1445	1703	1.6F
	2020	2336	1.9E		1955	2309	2.1E
11 SA	0307	0609	2.0F	26 SU	0240	0523	2.3F
	0857	1215	2.1E		0834	1150	2.4E
	1556	1836	1.5F		1531	1751	1.8F
	2106				2046		
12 SU		0023	1.9E	27 M		0003	2.3E
	0351	0638	2.0F		0330	0611	2.4F
	0938	1300	2.1E		0922	1241	2.5E
	1637	1906	1.5F		1615	1838	2.0F
	2150				2138		
13 M		0108	1.9E	28 TU		0057	2.4E
	0434	0709	1.9F		0420	0658	2.5F
	1018	1340	2.2E		1011	1329	2.6E
	1717	1937	1.5F		1659	1925	2.2F
	2235				2231		
14 TU		0151	1.9E	29 W		0148	2.5E
	0515	0743	1.8F		0510	0745	2.4F
	1058	1418	2.1E		1100	1416	2.7E
	1756	2010	1.5F		1745	2015	2.2F
	2318				2324		
15 W		0231	1.9E	30 TH		0237	2.6E
	0557	0822	1.7F		0602	0838	2.3F
	1137	1455	2.1E		1149	1503	2.6E
	1836	2051	1.5F		1833	2106	2.2F
				31 F	0017	0327	2.5E
					0659	0932	2.1F
					1238	1551	2.5E
					1925	2201	2.1F

FEBRUARY

DAY	SLACK WATER TIME H.M.	MAX CURRENT TIME H.M.	VEL. KNOTS	DAY	SLACK WATER TIME H.M.	MAX CURRENT TIME H.M.	VEL. KNOTS
1 SA	0111	0420	2.3E	16 SU	0058	0404	1.8E
	0759	1027	1.9F		0751	1011	1.4F
	1328	1643	2.3E		1307	1615	1.8E
	2021	2258	2.1F		1958	2236	1.6F
2 SU	0207	0517	2.1E	17 M	0144	0451	1.7E
	0902	1124	1.7F		0846	1100	1.3F
	1420	1738	2.1E		1350	1656	1.7E
	2118	2355	2.0F		2046	2325	1.6F
3 M	0305	0621	2.0E	18 TU	0233	0547	1.6E
	1005	1222	1.5F		0944	1151	1.2F
	1516	1839	1.9E		1439	1800	1.6E
	2216				2138		
4 TU		0056	1.8F	19 W		0015	1.6F
	0408	0727	1.9E		0329	0650	1.6E
	1107	1331	1.3F		1041	1242	1.1F
	1617	1943	1.8E		1535	1903	1.6E
	2314				2234		
5 W		0207	1.7F	20 TH		0108	1.6F
	0511	0830	1.9E		0429	0752	1.7E
	1208	1455	1.2F		1138	1337	1.2F
	1720	2039	1.8E		1638	2004	1.7E
					2331		
6 TH	0011	0322	1.7F	21 F		0205	1.7F
	0611	0925	1.9E		0529	0848	1.9E
	1306	1602	1.3F		1233	1439	1.2F
	1820	2133	1.8E		1741	2100	1.9E
7 F	0107	0422	1.8F	22 SA	0030	0308	1.8F
	0704	1016	1.9E		0627	0941	2.1E
	1359	1655	1.4F		1325	1544	1.5F
	1913	2226	1.8E		1840	2154	2.1E
8 SA	0200	0511	1.8F	23 SU	0127	0408	2.0F
	0751	1104	2.0E		0720	1032	2.2E
	1447	1741	1.5F		1415	1639	1.7F
	2001	2313	1.8E		1935	2248	2.2E
9 SU	0248	0552	1.9F	24 M	0222	0504	2.2F
	0833	1149	2.0E		0810	1123	2.4E
	1530	1818	1.5F		1502	1730	2.0F
	2045				2027	2343	2.4E
10 M		0000	1.9E	25 TU	0315	0552	2.4F
	0333	0625	1.9F		0859	1214	2.5E
	0912	1232	2.1E		1547	1818	2.2F
	1609	1845	1.6F		2119		
	2128						
11 TU		0044	1.9E	26 W		0037	2.6E
	0414	0650	1.8F		0405	0641	2.4F
	0951	1312	2.1E		0947	1304	2.6E
	1647	1911	1.6F		1632	1903	2.4F
	2209				2211		
12 W		0127	2.0E	27 TH		0129	2.7E
	0455	0718	1.8F		0455	0726	2.4F
	1029	1350	2.1E		1036	1353	2.7E
	1723	1942	1.7F		1718	1952	2.4F
	2251				2303		
13 TH		0207	2.0E	28 F		0218	2.7E
	0534	0757	1.7F		0547	0816	2.2F
	1108	1426	2.1E		1125	1439	2.6E
	1800	2019	1.7F		1805	2041	2.3F
	2332				2356		
14 F		0245	2.0E				
	0616	0838	1.6F				
	1147	1501	2.0E				
	1836	2102	1.7F				
15 SA	0015	0323	1.9E				
	0700	0923	1.5F				
	1226	1536	1.9E				
	1915	2149	1.6F				

TABLE 2.—CURRENT DIFFERENCES AND OTHER CONSTANTS

No.	PLACE	POSITION		TIME DIFFERENCES		VELOCITY RATIOS		MAXIMUM CURRENTS			
								Flood		Ebb	
		Lat.	Long.	Slack water	Maximum current	Maximum flood	Maximum ebb	Direction (true)	Average velocity	Direction (true)	Average velocity
		° ′	° ′	h. m.	h. m.			deg.	knots	deg.	knots
	LONG ISLAND, South Coast—Continued	**N.**	**W.**	on THE NARROWS, p.52							
				Time meridian, 75°W.							
2250	Shinnecock Inlet———————————	40 51	72 29	−0 20	−0 40	1.5	1.2	350	2.5	170	2.3
2255	Fire I. Inlet, 0.5 mi. S. of Oak Beach	40 38	73 18	+0 15	0 00	1.4	1.2	80	2.4	245	2.4
2260	Jones Inlet———————————————	40 35	73 34	−1 00	−0 55	1.8	1.3	35	3.1	215	2.6
2265	Long Beach, inside, between bridges———	40 36	73 40	−0 10	+0 10	0.3	0.3	75	0.5	275	0.6
2270	East Rockaway Inlet—————————	40 35	73 45	−1 25	−1 35	1.3	1.2	40	2.2	225	2.3
2275	Ambrose Light——————————————	40 27	73 49	See table 5.							
2280	Sandy Hook App. Lighted Horn Buoy 2A——	40 27	73 55	See table 5.							
	JAMAICA BAY										
2285	Rockaway Inlet————————————	40 34	73 56	−1 45	−2 15	1.1	1.3	85	1.8	245	2.7
2290	Barren Island, east of————————	40 35	73 53	−2 00	−2 25	0.7	0.9	5	1.2	190	1.7
2295	Canarsie (midchannel, off Pier)———	40 38	73 53	−1 35	−1 50	0.3	0.3	45	0.5	220	0.7
2300	Beach Channel (bridge)———————	40 35	73 49	−1 20	−1 20	1.1	1.0	60	1.9	225	2.0
2305	Grass Hassock Channel———————	40 37	73 47	−1 10	−1 00	0.6	0.5	50	1.0	230	1.0
	NEW YORK HARBOR ENTRANCE										
2310	Ambrose Channel entrance——————	40 30	73 58	−1 10	−1 05	1.0	1.2	310	1.7	110	2.3
2315	Ambrose Channel, SE. of West Bank Lt—	40 32	74 01	(¹)	−0 25	0.8	0.9	310	1.3	170	1.8
2320	Coney Island Lt., 1.6 miles SSW. of—	40 33	74 01	−0 10	(²)	0.5	0.8	330	0.8	145	1.5
2325	Ambrose Channel, north end——————	40 34	74 02	+0 05	+0 15	0.8	0.9	330	1.3	175	1.9
2330	Coney Island, 0.2 mile west of————	40 35	74 01	−0 55	−0 55	0.9	1.0	330	1.5	170	2.0
2335	Ft. Lafayette, channel east of————	40 36	74 02	(³)	(³)	0.6	0.5	345	1.1	195	0.9
2340	THE NARROWS, midchannel——————	40 37	74 03	Daily predictions				340	1.7	160	2.0
	NEW YORK HARBOR, Upper Bay										
2345	Tompkinsville————————————	40 38	74 04	−0 10	+0 20	0.9	1.0	5	1.6	170	2.0
2350	Bay Ridge Channel—————————	40 39	74 02	−0 35	−0 45	0.6	0.6	40	1.0	220	1.1
2355	Red Hook Channel——————————	40 40	74 01	−0 35	−0 35	0.6	0.4	355	1.0	170	0.7
2360	Robbins Reef Light, east of—————	40 39	74 03	+0 10	+0 20	0.8	0.8	15	1.3	205	1.6
2365	Red Hook, 1 mile west of——————	40 41	74 02	+0 45	+1 00	0.8	1.2	25	1.3	205	2.3
2370	Statue of Liberty, east of—————	40 42	74 02	+0 55	+1 00	0.8	1.0	30	1.4	205	1.9
	HUDSON RIVER, Midchannel[4]										
2375	The Battery, northwest of—————	40 43	74 02	+1 30	+1 35	0.9	1.2	15	1.5	195	2.3
2380	Desbrosses Street—————————	40 43	74 01	+1 35	+1 40	0.9	1.2	10	1.5	————	2.3
2385	Chelsea Docks———————————	40 45	74 01	+1 30	+1 40	1.0	1.0	20	1.7	185	2.0
2390	Forty-second Street—————————	40 46	74 00	+1 35	+1 45	1.0	1.2	30	1.7	————	2.3
2395	Ninety-sixth Street—————————	40 48	73 59	+1 40	+1 50	1.0	1.2	30	1.7	————	2.3
2400	Grants Tomb, 123d Street——————	40 49	73 58	+1 45	+1 55	0.9	1.2	25	1.6	————	2.3
2405	George Washington Bridge——————	40 51	73 57	+1 45	+2 00	0.9	1.1	20	1.6	200	2.2
2410	Spuyten Duyvil———————————	40 53	73 56	+2 00	+2 10	0.9	1.1	20	1.6	————	2.1
2415	Riverdale—————————————	40 54	73 55	+2 05	+2 20	0.8	1.0	15	1.4	200	2.0
2420	Dobbs Ferry—————————————	41 01	73 53	+2 25	+2 40	0.8	0.9	10	1.3	————	1.7
2425	Tarrytown——————————————	41 05	73 53	+2 40	+2 55	0.6	0.8	0	1.1	————	1.5
2430	Ossining——————————————	41 10	73 54	+2 55	+3 10	0.5	0.7	320	0.9	————	1.3
2435	Haverstraw—————————————	41 12	73 57	+3 05	+3 15	0.5	0.7	335	0.8	————	1.3
2440	Peekskill——————————————	41 17	73 57	+3 20	+3 35	0.5	0.6	0	0.8	————	1.2
2445	Bear Mountain Bridge———————	41 19	73 59	+3 25	+3 40	0.5	0.6	0	0.8	————	1.1
2450	Highland Falls———————————	41 22	73 58	+3 35	+3 50	0.6	0.6	5	1.0	185	1.2
2455	West Point, off Duck Island—————	41 24	73 57	+3 40	+3 55	0.5	0.6	10	1.0	————	1.1

[1]Current is rotary, turning clockwise. Minimum current of 0.9 knot sets SW. about time of "Slack, flood begins" at The Narrows. Minimum current of 0.5 knot sets NE. about 1 hour before "Slack, ebb begins" at The Narrows.

[2]Maximum flood, −0ʰ 50ᵐ; maximum ebb, +0ʰ 55ᵐ.

[3]Flood begins, −2ʰ 15ᵐ; maximum flood, −0ʰ 05ᵐ; ebb begins, +0ʰ 05ᵐ; maximum ebb, −1ʰ 50ᵐ.

[4]The values for the Hudson River are for the summer months, when the fresh-water discharge is a minimum.

TABLE 3.— VELOCITY OF CURRENT AT ANY TIME

EXPLANATION

Though the predictions in this publication give only the slacks and maximum currents, the velocity of the current at any intermediate time can be obtained approximately by the use of this table. Directions for its use are given below the table.

Before using the table for a place listed in table 2, the predictions for the day in question should first be obtained by means of the differences and ratios given in table 2. The examples below follow the numbered steps in the directions.

Example 1.—Find the velocity of the current in The Race at 6:00 on a day when the predictions which immediately precede and follow 6:00 are as follows:

(1)	Slack Water	Maximum (Flood)	
	Time	*Time*	*Velocity*
	4:18	7:36	3.2 knots

Directions under the table indicate table A is to be used for this station.

(2) Interval between slack and maximum flood is $7:36 - 4:18 = 3^h18^m$. Column heading nearest to 3^h18^m is 3^h20^m.

(3) Interval between slack and time desired is $6:00 - 4:18 = 1^h42^m$. Line labeled 1^h40^m is nearest to 1^h42^m.

(4) Factor in column 3^h20^m and on line 1^h40^m is 0.7. The above flood velocity of 3.2 knots multiplied by 0.7 gives a flood velocity of 2.24 knots (or 2.2 knots, since one decimal is sufficient) for the time desired.

Example 2.—Find the velocity of the current in the Harlem River at Broadway Bridge at 16:30 on a day when the predictions (obtained using the difference and ratio in table 2) which immediately precede and follow 16:30 are as follows:

(1)	Maximum (Ebb)		Slack Water
	Time	*Velocity*	*Time*
	13:49	2.5 knots	17:25

Directions under the table indicate table B is to be used, since this station in table 2 is referred to Hell Gate

(2) Interval between slack and maximum ebb is $17:25 - 13:49 = 3^h36^m$. Hence, use column headed 3^h40^m.

(3) Interval between slack and time desired is $17:25 - 16:30 = 0^h55^m$. Hence, use line labeled 1^h00^m.

(4) Factor in column 3^h40^m and on line 1^h00^m is 0.5. The above ebb velocity of 2.5 knots multiplied by 0.5 gives an ebb velocity of 1.2 knots for the desired time.

When the interval between slack and maximum current is greater than 5^h40^m, enter the table with one-half the interval between slack and maximum current and one-half the interval between slack and the desired time and use the factor thus found.

TABLE 3.—VELOCITY OF CURRENT AT ANY TIME

TABLE A

Interval between slack and desired time	Interval between slack and maximum current													
h. m.	h. m. 1 20	h. m. 1 40	h. m. 2 00	h. m. 2 20	h. m. 2 40	h. m. 3 00	h. m. 3 20	h. m. 3 40	h. m. 4 00	h. m. 4 20	h. m. 4 40	h. m. 5 00	h. m. 5 20	h. m. 5 40
	f.	f.	f.	f.	f.	f.	f.	f.	f.	f.	f.	f.	f.	f.
0 20	0.4	0.3	0.3	0.2	0.2	0.2	0.2	0.1	0.1	0.1	0.1	0.1	0.1	0.1
0 40	0.7	0.6	0.5	0.4	0.4	0.3	0.3	0.3	0.3	0.2	0.2	0.2	0.2	0.2
1 00	0.9	0.8	0.7	0.6	0.6	0.5	0.5	0.4	0.4	0.4	0.3	0.3	0.3	0.3
1 20	1.0	1.0	0.9	0.8	0.7	0.6	0.6	0.5	0.5	0.5	0.4	0.4	0.4	0.4
1 40	----	1.0	1.0	0.9	0.8	0.8	0.7	0.7	0.6	0.6	0.5	0.5	0.5	0.4
2 00	----	----	1.0	1.0	0.9	0.9	0.8	0.8	0.7	0.7	0.6	0.6	0.6	0.5
2 20	----	----	----	1.0	1.0	0.9	0.9	0.8	0.8	0.7	0.7	0.7	0.6	0.6
2 40	----	----	----	----	1.0	1.0	1.0	0.9	0.9	0.8	0.8	0.7	0.7	0.7
3 00	----	----	----	----	----	1.0	1.0	1.0	0.9	0.9	0.8	0.8	0.8	0.7
3 20	----	----	----	----	----	----	1.0	1.0	1.0	0.9	0.9	0.9	0.8	0.8
3 40	----	----	----	----	----	----	----	1.0	1.0	1.0	0.9	0.9	0.9	0.9
4 00	----	----	----	----	----	----	----	----	1.0	1.0	1.0	1.0	0.9	0.9
4 20	----	----	----	----	----	----	----	----	----	1.0	1.0	1.0	1.0	0.9
4 40	----	----	----	----	----	----	----	----	----	----	1.0	1.0	1.0	1.0
5 00	----	----	----	----	----	----	----	----	----	----	----	1.0	1.0	1.0
5 20	----	----	----	----	----	----	----	----	----	----	----	----	1.0	1.0
5 40	----	----	----	----	----	----	----	----	----	----	----	----	----	1.0

TABLE B

Interval between slack and desired time	Interval between slack and maximum current													
h. m.	h. m. 1 20	h. m. 1 40	h. m. 2 00	h. m. 2 20	h. m. 2 40	h. m. 3 00	h. m. 3 20	h. m. 3 40	h. m. 4 00	h. m. 4 20	h. m. 4 40	h. m. 5 00	h. m. 5 20	h. m. 5 40
	f.	f.	f.	f.	f.	f.	f.	f.	f.	f.	f.	f.	f.	f.
0 20	0.5	0.4	0.4	0.3	0.3	0.3	0.3	0.3	0.2	0.2	0.2	0.2	0.2	0.2
0 40	0.8	0.7	0.6	0.5	0.5	0.5	0.4	0.4	0.4	0.4	0.3	0.3	0.3	0.3
1 00	0.9	0.8	0.8	0.7	0.7	0.6	0.6	0.5	0.5	0.5	0.4	0.4	0.4	0.4
1 20	1.0	1.0	0.9	0.8	0.8	0.7	0.7	0.6	0.6	0.6	0.5	0.5	0.5	0.5
1 40	----	1.0	1.0	0.9	0.9	0.8	0.8	0.7	0.7	0.7	0.6	0.6	0.6	0.6
2 00	----	----	1.0	1.0	0.9	0.9	0.9	0.8	0.8	0.7	0.7	0.7	0.7	0.6
2 20	----	----	----	1.0	1.0	1.0	0.9	0.9	0.8	0.8	0.8	0.7	0.7	0.7
2 40	----	----	----	----	1.0	1.0	1.0	0.9	0.9	0.9	0.8	0.8	0.8	0.7
3 00	----	----	----	----	----	1.0	1.0	1.0	0.9	0.9	0.9	0.9	0.8	0.8
3 20	----	----	----	----	----	----	1.0	1.0	1.0	1.0	0.9	0.9	0.9	0.8
3 40	----	----	----	----	----	----	----	1.0	1.0	1.0	1.0	0.9	0.9	0.9
4 00	----	----	----	----	----	----	----	----	1.0	1.0	1.0	1.0	0.9	0.9
4 20	----	----	----	----	----	----	----	----	----	1.0	1.0	1.0	1.0	0.9
4 40	----	----	----	----	----	----	----	----	----	----	1.0	1.0	1.0	1.0
5 00	----	----	----	----	----	----	----	----	----	----	----	1.0	1.0	1.0
5 20	----	----	----	----	----	----	----	----	----	----	----	----	1.0	1.0
5 40	----	----	----	----	----	----	----	----	----	----	----	----	----	1.0

Use Table A for all places except those listed below for Table B.
Use Table B for Cape Cod Canal, Hell Gate, Chesapeake and Delaware Canal and all stations in Table 2 which are referred to them.

1. From predictions find the time of slack water and the time and velocity of maximum current (flood or ebb), one of which is immediately before and the other after the time for which the velocity is desired.

2. Find the interval of time between the above slack and maximum current, and enter the top of Table A or B with the interval which most nearly agrees with this value.

3. Find the interval of time between the above slack and the time desired, and enter the side of Table A or B with the interval which most nearly agrees with this value.

4. Find, in the table, the factor corresponding to the above two intervals and multiply the maximum velocity by this factor. The result will be the approximate velocity at the time desired.

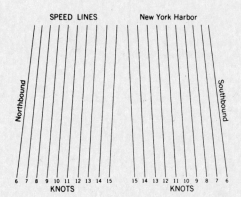

SPEED LINES   New York Harbor

Northbound

Southbound

6 7 8 9 10 11 12 13 14 15   15 14 13 12 11 10 9 8 7 6

KNOTS   KNOTS

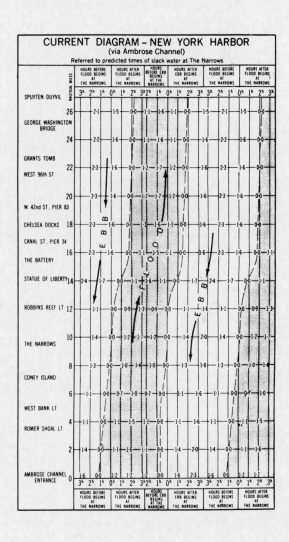

CURRENT DIAGRAM – NEW YORK HARBOR
(via Ambrose Channel)
Referred to predicted times of slack water at The Narrows

A NAVIGATION CLASSIC

Bowditch's Coastal Navigation

Navigation is both an art and a science — a process requiring a well-informed person to interpret the information offered by ship, sea, and sky. When Nathaniel Bowditch produced the first **AMERICAN PRACTICAL NAVIGATOR** in 1802, he intended it to be a compendium of navigational material that was within the reach of every mariner. The yachtsman's best guide to piloting.

In BOWDITCH'S COASTAL NAVIGATION, editor James W. Morrison follows the same tradition. He has selected the fourteen chapters, five appendices, and nine sets of tables most directly concerned with navigating coastal waters. Among the topics covered are:

Basic Definitions	The Sailings
Chart Projections	Piloting
Visual and Audible Aids	Use of Sextant in Piloting
The Nautical Chart	Tide and Current Predictions
Instruments for Piloting	Sailing Directions and Light Lists
Compass Error	Tide Tables
Dead Reckoning	Radio Direction Finding

BOWDITCH'S COASTAL NAVIGATION presents the essential elements of coastal navigation from the exhaustive (and massive) **AMERICAN PRACTICAL NAVIGATOR** making them accessible, well-organized, and handy — in a word, usable.

The book contains a complete Arco Nautical Chart for practical navigation. Made of the miracle plastic Kimdura, the chart is impervious to water. It can be erased or wiped clean after plotting a course on the chart with a soft lead pencil.

7⅞" x 10¼" paperbound

384 pages $9.95

Little Ship Meteorology

M.J. RANTZEN

The book is written so as to give the small-boat man a basic understanding of weather lore; the information in it has been reduced to essentials, and the texts is illustrated with diagrams. This edition has been revised to take into account, as far as possible, recent changes, broadcast weather reporting, charts, wave-lengths, etc.

5¼" x 8"/164 pages/Illustrated/
B & J ISBN 0 214 20006 x
ADON 8042 cloth $9.95

ARCO PUBLISHING, INC.

Little Ship Astro-Navigation
Fourth edition

M.J. RANTZEN

The best short manual on the subject is designed to give amateurs a full and practical grasp of the latest methods of position-finding by observations of the sun, moon, stars and planets, combining simplicity with professional standards of accuracy.

5¼" x 8"/164 pages/Illustrated/
B & J ISBN 0 214 20373 5
ADON 8039 cloth $10.95

Send payment (plus $1.00 for postage and shipping) to:
219 Park Ave. South, New York, N.Y. 10003